U.S. ARMY WEAPONS SYSTEMS 2014–2015

DEPARTMENT OF THE ARMY

Skyhorse Publishing

All inquiries should be addressed to Skyhorse Publishing, 307 West 36th Street, 11th Floor, New York, NY 10018.

Skyhorse Publishing books may be purchased in bulk at special discounts for sales promotion, corporate gifts, fund-raising, or educational purposes. Special editions can also be created to specifications. For details, contact the Special Sales Department, Skyhorse Publishing, 307 West 36th Street, 11th Floor, New York, NY 10018 or info@skyhorsepublishing.com.

Skyhorse® and Skyhorse Publishing® are registered trademarks of Skyhorse Publishing, Inc.®, a Delaware corporation.

Visit our website at www.skyhorsepublishing.com.

10 9 8 7 6 5 4 3 2 1

Library of Congress Cataloging-in-Publication Data is available on file.

ISBN: 978-1-62914-401-6
Ebook ISBN: 978-1-63220-162-1

Printed in China

Dear Reader:

Thank you for your interest in this annual publication, which details the Army's major weapon system programs and illustrates our ongoing efforts to empower, unburden and protect our Soldiers. The Army's Acquisition community is charged with the solemn responsibility of maintaining our Soldiers' unprecedented edge against current and future threats.

With program descriptions, status and specifications, projected activities, and names and locations of large and small contractors, this book will provide you with a better understanding of our efforts to provide Soldiers with the best, most advanced and sustainable equipment possible. To this end, we are mindful of the public trust imposed by the use of taxpayer resources. We continuously seek to improve our business practices to meet the needs of our Soldiers on an efficient and timely basis.

In providing our Soldiers with world-class capabilities, Army acquisition's most important asset is our people. Our skilled and dedicated professionals, working in Program Executive Offices and program offices throughout the nation, execute diverse responsibilities to enable the disciplined management of an extensive acquisition portfolio with programs that range from Soldiers systems to precision fires and from air and mission defense to ground combat systems. The responsibility of safeguarding future Army capabilities is a significant honor for the acquisition community and is one that we do not take lightly.

Heidi Shyu
Assistant Secretary of the Army
(Acquisition, Logistics, and Technology)
and Army Acquisition Executive

Table of Contents

How to Use this Book...VI

Introduction 1

Weapon Systems 7

120M Motor Grader 8

2.75 Inch Rocket Systems (Hydra-70)...10

Abrams Tank Upgrade 12

Advanced Field Artillery Tactical Data System (AFATDS).............14

Advanced Threat Infrared Countermeasures (ATIRCM) and Common
Missile Warning System (CMWS) Programs and Pre-MDAP Common Infrared
Countermeasure (CIRCM) 16

Air Soldier System (Air SS) 18

Air Warrior (AW)..20

Air/Missile Defense Planning and Control System (AMDPCS)......22

Airborne and Maritime/Fixed Station Joint Tactical Radio System (AMF JTRS) 24

Airborne Reconnaissance Low (ARL) 26

Anti-Personnel Mine Clearing System, Remote Control M160 28

AN/TPQ-53 (formerly known as the Enhanced AN/TPQ-36) 30

Armored Multi-Purpose Vehicle (AMPV).....................................32

Army Integrated Air and Missile Defense (AIAMD) 34

Army Key Management System (AKMS) 36

Artillery Ammunition...38

Assault Breacher Vehicle (ABV)..40

Assembled Chemical Weapons Alternatives (ACWA) 42

Aviation Combined Arms Tactical Trainer (AVCATT)...................44

Biometric Enabling Capability (BEC)..46

Black Hawk/UH/HH-60 48

Bradley Fighting Vehicle Systems Upgrade 50

Calibration Sets Equipment (CALSETS)......................................52

Capability Set 13 (CS 13) 54

CH-47F Chinook 56

Chemical Biological Medical Systems (CBMS)–Prophylaxis........58

Chemical Biological Medical Systems (CBMS)–Diagnostics........60

Chemical Biological Medical Systems (CBMS)–Therapeutics......62

Chemical Biological Protective Shelter (CBPS) M8E1 64

Chemical, Biological, Radiological, Nuclear Dismounted Reconnaissance
Sets, Kits, and Outfits (CBRN DR SKO) 66

Clip-on Sniper Night Sight (CoSNS), AN/PVS-30 68

Close Combat Tactical Trainer (CCTT)...70

Combat Service Support Communications (CSS Comms) 72

Command Post Systems and Integration (CPS&I) Standardized Integrated
Command Post Systems (SICPS) 74

Common Hardware Systems (CHS)..76

Common Remotely Operated Weapon Station (CROWS)...............78

Computer Hardware, Enterprise Software and Solutions (CHESS)...80

Counter Defilade Target Engagement (CDTE)–XM25 82

Countermine...84

Counter-Rocket, Artillery, Mortar (C-RAM) / Indirect Fire Protection Capability (IFPC)...86

Cryptographic Systems..88

Defense Enterprise Wideband SATCOM System (DEWSS)............90

Distributed Common Ground System–Army (DCGS-A)..................92

Distributed Learning System (DLS) 94

Dry Support Bridge (DSB)...96

Enhanced Medium Altitude Reconnaissance and Surveillance System (EMARSS).........98

Enterprise Email (EE) 100

Excalibur (M982)...102

Family of Medium Tactical Vehicles (FMTV) 104

Fixed Wing..106

Force Protection Systems 108

Force Provider (FP) 110

Force XXI Battle Command Brigade and Below (FBCB2) 112

Forward Area Air Defense Command and Control (FAAD C2) 114

General Fund Enterprise Business Systems (GFEBS) 116

Global Combat Support System–Army (GCSS-Army) 118

Global Command and Control System–Army (GCCS-A) 120

Ground Combat Vehicle (GCV) 122

Guardrail Common Sensor (GR/CS) 124

Guided Multiple Launch Rocket System (GMLRS) DPICM/Unitary/
Alternative Warhead (Tactical Rockets) 126

Harbormaster Command and Control Center (HCCC) 128

Heavy Expanded Mobility Tactical Truck (HEMTT)/HEMTT Extended
Service Program (ESP) 130

Heavy Loader 132

HELLFIRE Family of Missiles 134

Helmet Mounted Night Vision Devices (HMNVD) 136

High Mobility Artillery Rocket System (HIMARS) M142 138

High Mobility Engineer Excavator (HMEE) I and III 140

High Mobility Multipurpose Wheeled Vehicle (HMMWV) Recapitalization
(RECAP) Program 142

Improved Environmental Control Unit (IECU) 144

Improved Ribbon Bridge (IRB) 146

Improved Target Acquisition System (ITAS) 148

Improvised Explosive Device Defeat/Protect Force (IEDD/PF) 150

Installation Information Infrastructure Modernization Program (I3MP) 152

Instrumentable Multiple Integrated Laser Engagement System (I-MILES) 154

Integrated Family of Test Equipment (IFTE) 156

Integrated Personnel and Pay System–Army (IPPS-A) 158

Interceptor Body Armor (IBA) 160

Javelin 162

Joint Air-to-Ground Missile (JAGM) 164

Joint Battle Command–Platform (JBC-P) 166

Joint Biological Point Detection System (JBPDS) 168

Joint Biological Tactical Detection System (JBTDS) 170

Joint Chem/Bio Coverall for Combat Vehicle Crewman (JC3) 172

Joint Chemical Agent Detector (JCAD) M4A1 174

Joint Effects Model (JEM) 176

Joint Effects Targeting System (JETS) Target Location Designation System (TLDS) ... 178

Joint Land Attack Cruise Missile Defense Elevated Netted Sensor System (JLENS) 180

Joint Land Component Constructive Training Capability (JLCCTC) 182

Joint Light Tactical Vehicle (JLTV) 184

Joint Personnel Identification Version 2 (JPIv2) 186

Joint Precision Airdrop System (JPADS) 188

Joint Service Aircrew Mask–Rotary Wing (JSAM RW) (MPU-5) 190

Joint Service General Purpose Mask (JSGPM) M-50/M-51 192

Joint Service Transportable Small Scale Decontaminating
Apparatus (JSTSS DA) M26 194

Joint Tactical Ground Station (JTAGS) 196

Joint Tactical Radio System Ground Mobile Radios (JTRS GMR) 198

Joint Tactical Radio System Handheld, Manpack, Small Form Fit (JTRS HMS) 200

Joint Tactical Radio System Network Enterprise Domain (JTRS NED) 202

Joint Warning and Reporting Network (JWARN) 204

Kiowa Warrior 206

Korea Transformation, Yongsan Relocation Plan, Land Partnership
Plan (KT/YRP/LPP) 208

Lakota/UH-72A 210

Light Capability Rough Terrain Forklift (LCRTF) 212

Lightweight 155mm Howitzer System (LW155) 214

Lightweight Counter Mortar Radar (LCMR) 216

Lightweight Laser Designator Rangefinder (LLDR) AN/PED-1 & AN/PED-1A 218

Table of Contents

Line Haul Tractor 220

Load Handling System Compatible Water Tank Rack (Hippo) 222

Longbow Apache (AH-64D) (LBA) 224

M109 Family of Vehicles (FOV) (Paladin/FAASV, PIM SPH/CAT) 226

M1200 Armored Knight 228

Medical Communications for Combat Casualty Care (MC4) 230

Medical Simulation Training Center (MSTC) 232

Medium Caliber Ammunition (MCA) 234

Medium Extended Air Defense System (MEADS) 236

Meteorological Measuring Set-Profiler (MMS-P)/Computer Meteorological
Data-Profiler (CMD-P) 238

Mine Protection Vehicle Family (MPVF), Area Mine Clearing System
(AMCS), Interrogation Arm 240

Mine Resistant Ambush Protected Vehicles (MRAP), Army 242

Modular Fuel System (MFS) 244

Mortar Systems 246

Movement Tracking System (MTS) 248

MQ-1C Gray Eagle Unmanned Aircraft System (UAS) 250

Multiple Launch Rocket System (MLRS) M270A1 252

NAVSTAR Global Positioning System (GPS) 254

Nett Warrior (NW) 256

Night Vision Thermal Systems–Thermal Weapon Sight (TWS) 258

Non-Intrusive Inspection Systems (NIIS) 260

Nuclear Biological Chemical Reconnaissance Vehicle (NBCRV)
–Stryker Sensor Suites 262

One Semi-Automated Force (OneSAF) 264

Palletized Load System (PLS) and PLS Extended Service Program (ESP) 266

PATRIOT Advanced Capability–3 (PAC-3) 268

Precision Guidance Kit (PGK) 270

Prophet 272

Rocket, Artillery, Mortar (RAM) Warn 274

Rough Terrain Container Handler (RTCH) 276

RQ-11B Raven Small Unmanned Aircraft System (SUAS) 278

RQ-7B Shadow Tactical Unmanned Aircraft System (TUAS) 280

Secure Mobile Anti-Jam Reliable Tactical Terminal (SMART-T) 282

Sentinel 284

Single Channel Ground and Airborne Radio System (SINCGARS) 286

Small Arms–Crew Served Weapons 288

Small Arms–Individual Weapons 290

Small Arms–Precision Weapons 292

Small Caliber Ammunition 294

Stryker Family of Vehicles 296

Sustainment System Mission Command (SSMC) 298

T-9 Medium Dozer 300

Tactical Electric Power (TEP) 302

Tactical Mission Command (TMC)/Maneuver Control System (MCS) 304

Tank Ammunition 306

Test Equipment Modernization (TEMOD) 308

Transportation Coordinators'–Automated Information for Movement
System II (TC-AIMS II) 310

Tube-Launched, Optically-Tracked, Wire-Guided (TOW) Missiles 312

Unified Command Suite (UCS) 314

Unit Water Pod System (Camel II) 316

Warfighter Information Network–Tactical (WIN-T) Increment 1 318

Warfighter Information Network–Tactical (WIN-T) Increment 2 320

Warfighter Information Network–Tactical (WIN-T) Increment 3 322

XM1216 & XM1216 E1 Small Unmanned Ground System (SUGV) 324

XM7 Spider 326

Science and Technology 328

Army S&T Mission 328

Science and Technology Tenets 329

Resourcing S&T 330

Army S&T in Action 331

S&T Portfolios – Defining the Army's Capabilities of Tomorrow 331

Soldier S&T Portfolio 332

    Major Efforts 333

Ground S&T Portfolio 334

    Major Efforts 334

Air Portfolio 336

    Major Efforts 336

Command, Control, Communications & Intelligence (C3I) Portfolio 337

    Major Efforts 338

Innovation Enablers Portfolio 339

Basic Research Portfolio 340

    Major Efforts 340

Technology Transition – A Key Metric of Performance 342

Summary 343

Appendices 344

Army Combat Organizations 345

Glossary of Terms 346

Systems by Contractors 350

Points of Contact 363

How to Use this Book

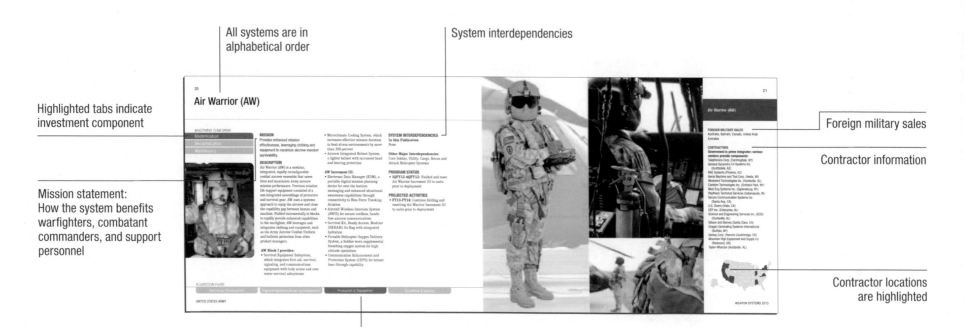

All systems are in alphabetical order

System interdependencies

Highlighted tabs indicate investment component

Mission statement: How the system benefits warfighters, combatant commanders, and support personnel

Foreign military sales

Contractor information

Contractor locations are highlighted

Highlighted tabs indicate acquisition phase

WHAT ARE SYSTEM INTERDEPENDENCIES?

The purpose of the **System Interdependencies** section is to identify which other weapon systems or components (if any) the main system works in concert with or relies upon for its operation. We categorize the interdependencies in two ways: 1) under the heading "In this Publication," which is a listing of systems in this 2014 edition and 2) "Other Major Interdependencies," which is a listing of systems that are not included in this publication.

WHAT ARE INVESTMENT COMPONENTS?

Modernization programs develop and/or procure new systems with improved warfighting capabilities.

Recapitalization programs rebuild or provide selected upgrades to currently fielded systems to ensure operational readiness and a zero-time, zero-mile system.

Maintenance programs include the repair or replacement of end items, parts, assemblies, and subassemblies that wear out or break.

WHAT ARE ACQUISITION PHASES?

Technology Development refers to the development of a materiel solution to an identified, validated need. During this phase, the Mission Needs Statement is approved, technology issues are considered, and possible alternatives are identified. This phase includes:
- Concept exploration
- Decision review
- Component advanced development

Engineering & Manufacturing Development is the phase in which a system is developed, program risk is reduced, operational supportability and design feasibility are ensured, and feasibility and affordability are demonstrated. This is also the phase in which system integration,

interoperability, and utility are demonstrated. It includes:
- System integration
- System demonstration
- Interim progress review

Production & Deployment achieves an operational capability that satisfies mission needs. Components of this phase are:
- Low-rate initial production
- Full-rate production decision review
- Full-rate production and deployment
- Military equipment valuation

Operations & Support ensures that operational support performance requirements and sustainment of systems are met in the most cost-effective manner. Support varies but generally includes:
- Supply
- Maintenance
- Transportation
- Sustaining engineering
- Data management
- Configuration management
- Human factors engineering

- Personnel
- Manpower
- Training
- Habitability
- Survivability
- Safety and occupational health
- Information technology supportability
- Environmental management functions
- Anti-tamper provisions
- Interoperability
- Disposal/demilitarization

Because the Army is spiraling technology to the troops as soon as it is feasible, some programs and systems may be in all four phases at the same time. Mature programs are often only in one phase, such as operations and support, while newer systems are only in concept and technology development.

For additional information and definitions of these categories and terms, please see the Glossary.

DESIGN / DEVELOP / DELIVER / DOMINATE
TODAY AND TOMORROW

UNITED STATES ARMY

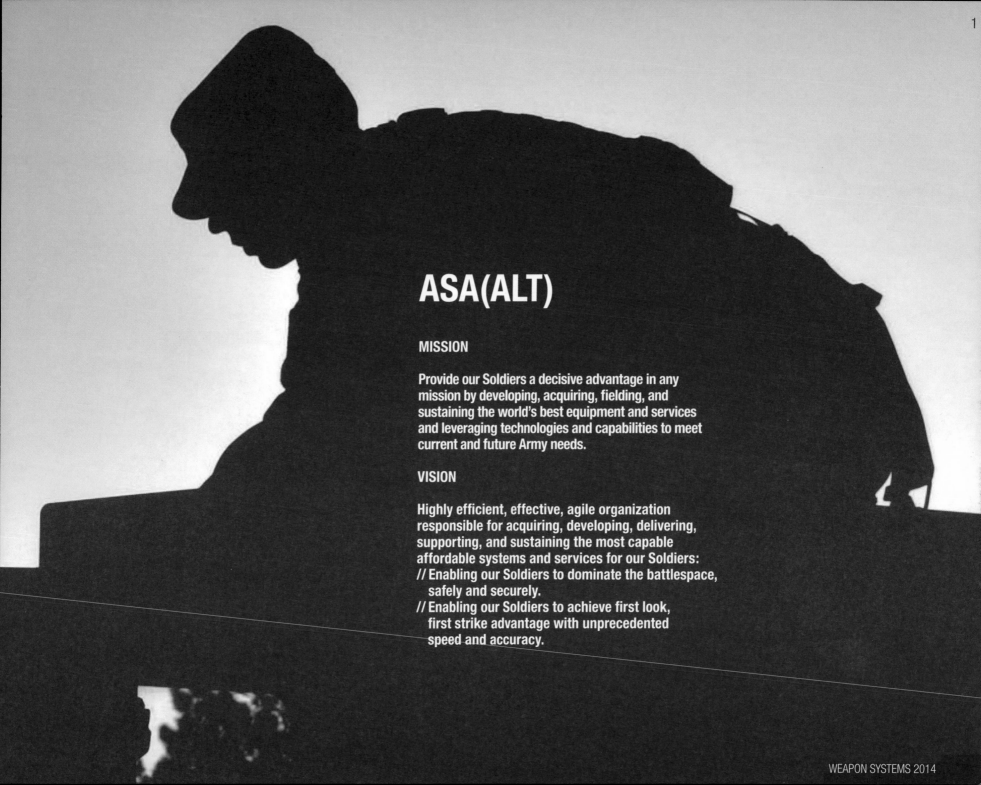

ASA(ALT)

MISSION

Provide our Soldiers a decisive advantage in any mission by developing, acquiring, fielding, and sustaining the world's best equipment and services and leveraging technologies and capabilities to meet current and future Army needs.

VISION

Highly efficient, effective, agile organization responsible for acquiring, developing, delivering, supporting, and sustaining the most capable affordable systems and services for our Soldiers:
// Enabling our Soldiers to dominate the battlespace, safely and securely.
// Enabling our Soldiers to achieve first look, first strike advantage with unprecedented speed and accuracy.

STRATEGIC CONTEXT

The U.S. Army is involved in combat operations around the world against adaptive enemies able to take advantage of the ever-increasing pace of technological change. Concurrently, we are facing an increasingly constrained fiscal environment. In this challenging environment, our goal in the Acquisition, Logistics, and Technology community is to do everything we can to provide the best equipment and services to our Soldiers.

Our Soldiers require comprehensive capabilities that allow them to communicate, engage, and disengage. Our troops must continue to operate with confidence in their equipment, operational capabilities, communication technology, enhanced situational awareness, and force protection. We must provide our Soldiers a decisive advantage in every fight so they return safely from every operation and engagement.

Modernizing the Army enables us to counter rapidly emerging threats that change the nature of battlefield operations. Through lessons learned, the Army will develop and field new capabilities or sustain, improve, or divest current systems based on operational value, capabilities shortfalls, and available resources.

After 10 years of combat, today's Army is significantly more capable than the Army of 2001. As we draw down from Iraq and Afghanistan, we must remain flexible, adaptable, and agile enough to respond and meet the needs of the combatant commanders. Our objective is to equip and maintain an Army with the latest most advanced weaponry to win and return home quickly.

The right foundation for success is based on sound planning – we cannot succeed unless requirements are matched with stable and well-planned resources under sound program management. This necessary collaboration does not end when programs are launched – and we have learned that it DOES take this collaboration to even get them launched, in the case of the Ground Combat Vehicle (GCV) – it must continue throughout the acquisition lifecycle. We have also reviewed our ongoing programs to mitigate risk by embracing competition, adopting sensible acquisition strategies that reflect more realistic assessments of what a program will cost, and address technological maturity.

The Assistant Secretary of the Army for Acquisition, Logistics, and Technology (ASA(ALT)) is deeply invested in developing, delivering, and sustaining the best weapons technology available. With the Soldier as the key focus, ASA(ALT) seeks to equip Soldiers with the best in cutting-edge technology and effectively manage over 600+ programs that are vital to success in combat.

ASA(ALT)'s focus is closely aligned with the Army Modernization Strategy, which outlines a series of key goals—such as the continued development of new technologies engineered to provide Soldiers with the decisive edge in battle. These technologies in development span a range of new capability to include sensors, Unmanned Aircraft Systems (UAS), missiles and missile guidance systems, emerging combat platforms such as the GCV, and key technologies such as the Army's maturing network, designed to connect Soldiers, sensors, and multiple nodes to one another in real-time to improve operational effectiveness across the full spectrum of combat operations.

The Army's equipping strategy is designed to counter changing threats and addresses the emergence of hybrid threats—the dynamic combination of conventional, irregular, terrorist, and criminal capabilities. The Army seeks to train, develop, and equip Soldiers who are able to stay in front of an adaptive, fast-changing adversary. By emphasizing the best design, delivery, and sustainment of Army equipment, ASA(ALT) will remain focused on harnessing scientific innovations in order to identify and develop the most promising new technologies.

We are focused upon preserving investment in our Science and Technology (S&T) efforts, to identify, leverage and deliver critical innovations which will better equip, empower and enable our force for the future. We strive to develop and sustain a near, mid and far-term S&T investment strategy so that we can spiral in new capabilities and technologies as they emerge and also identify disruptive or paradigm changing next-generation systems and solutions.

The Army is also implementing a more "Agile" acquisition and modernization process by conducting Network Evaluation Exercises (NIEs). The NIEs place emerging technologies in the hands of Soldiers in a combat-realistic environment in order to harness their feedback, keep pace with the speed of technological change, and in some cases, blend commercial-off-the-shelf technologies with existing programs of record. The heart of the NIE exercises is using the best available technologies to move information, voice, video, data, and images faster, further, and more efficiently across the force, and developing systems within a Common Operating Environment (COE), meaning they are built on software foundations that enable the maximum amount of interoperability. The Army's network will make it possible for Soldiers in a vehicle on-the-move to view and share real-time feeds from a nearby robot, ground sensor, or UAS—instantaneously providing them combat-relevant information and enabling them to share that information with other units on-the-move, dismounted Soldiers, and higher echelons of the force.

TRANSFORMING ARMY ACQUISITION AND BUSINESS PRACTICES

The Army remains focused on finding ways to continually examine and improve the acquisition process while increasing efficiency and serving as a full partner in the Department of Defense's Better Buying Power Initiatives.

A major challenge to acquisition continues to be the need to properly prioritize, streamline, and collaborate on requirements at the front end of the process in order to emphasize technological maturity, affordability, and productivity. The revised Request for Proposal for the GCV is an excellent demonstration of how we approached reform in this area; requirements were properly "tiered" and industry was given "trade space" designed to encourage innovation.

Also, we have learned of the importance of streamlining and at times challenging requirements in order to emphasize technological maturity and lower costs wherever possible. For instance, in our Joint Light Tactical Vehicle program, the Army worked with industry participants to "trade-off" certain requirements in order to lower costs and meet scheduling goals. Through this process, the Army was able to substantially lower the unit price of the vehicle while simultaneously ensuring the platform will succeed in delivering important next-generation capabilities.

The goal of our acquisition initiatives is to work with our industry and academic partners to more efficiently develop and deliver capabilities needed by the Soldier. A key aspect of this is an effort to identify and address inefficiencies discovered in the acquisition process.

A system-of-systems approach is vital to these ongoing efforts to transform business practices. The Army will continue to look at developing, managing, and acquiring technologies in the most efficient way possible, an approach which includes the need to understand the interdependencies among systems. We place an emphasis upon maturing the capability to synchronize programs and integrate schedules, deliveries, and other developments across the acquisition process.

As a result of these and other practices, the acquisition community remains acutely aware of its need to further the transformation of its business efforts. These initiatives help the Army transform as an institution and ensure that the best value possible is provided to the taxpayer and the Soldier—who is at the very center of these efforts.

COMMUNICATING AND COLLABORATING WITH INDUSTRY

ASA(ALT) will continue to foster, develop, and enhance its relationships with vital industry partners as a way to ensure the best possible development of new and emerging systems. With this as an organizing principle, ASA(ALT) has an industry outreach engagement program squarely focused on furthering partnerships with industry and facilitating constructive dialogue designed to achieve the best results for Soldiers in combat. Recognizing the importance of revitalizing industry engagement, the Army continues to nurture this outreach program, fostering and preserving strong relationships between the Army and its key industry partners.

Often there are circumstances where procurement sensitivities and ongoing competition may preclude the occasion to dialogue with industry. There are, nonetheless, ample opportunities for positive, proactive, and constructive engagement with industry partners. While placing a premium upon the importance of properly defining the parameters for discussion with industry partners, ASA(ALT) seeks to foster an environment of open dialogue.

The rationale behind this approach is based on the effort to minimize misunderstandings and "eleventh hour" reactions. This industry program is designed to anticipate future developments, recognize and communicate industry trends, and identify the evolution of key technologies that will support and protect our Soldiers in combat.

PATH FORWARD

We will provide whatever it takes to achieve the Nation's objectives in the current fight. At the same time, we will develop a shared vision to build tomorrow's Army—designing and preparing units, developing Soldiers, and growing leaders to win in an increasingly competitive learning environment. We will continue to maintain battlefield dominance but remain versatile and adaptable to any task our Nation may call upon us to perform. Continuous modernization is key to transforming Army capabilities and maintaining a technological advantage over our adversaries across the full spectrum of operations. ASA(ALT) looks forward to collaboration with all stakeholders to achieve the Army's broad modernization goals while supporting a cost-conscious culture.

The systems listed in this book are not isolated, individual products. Rather, they are part of an integrated Army system-of-systems investment designed to equip the Army of the future to successfully face any challenges. Each system and capability is important. Our goal is to develop and field a versatile and affordable mix of equipment that will enable Soldiers to succeed and maintain our decisive advantage over any enemy we face.

WEAPON SYSTEMS

LISTED IN ALPHABETICAL ORDER

120M Motor Grader

INVESTMENT COMPONENT

Modernization

Recapitalization

Maintenance

MISSION

Provides coarse and fine grading, low- and high-bank sloping, flat- and V-ditching, and scarifying materials to support military road, bridge, and airfield construction.

DESCRIPTION

The 120M Motor Grader (MG) is a six-wheeled, commercial, construction grader for use by engineering units. It features all-wheel drive, articulated frame steer, pneumatic tires, electro-hydraulic joystick control operation, and an automatic power shift transmission with eight forward and six reverse speeds. The MG is equipped with a completely enclosed rollover protective structure/falling object protective structure (ROPS/FOPS) cab, with air conditioning, heating, one primary door, and one secondary exit. It includes these commercial options: radial tires, turn signals, emergency flashers, a secondary steering system, and a blade leveling system. The MG with ROPS/FOPS removed is designed to be capable of roll-on/roll-off transportation by C-130J, C-5, and C-17 military aircraft.

SYSTEM INTERDEPENDENCIES

None

PROGRAM STATUS

- **Current:** In production and fielding

PROJECTED ACTIVITIES

- **FY13:** Airdrop testing
- **FY13-FY15:** 480 systems projected to be fielded

ACQUISITION PHASE

Technology Development | Engineering & Manufacturing Development | Production & Deployment | Operations & Support

UNITED STATES ARMY

120M Motor Grader

FOREIGN MILITARY SALES
Afghanistan, Iraq

CONTRACTORS
Caterpillar Inc. (Peoria, IL)
BAE Systems (Cincinnati, OH)

2.75 Inch Rocket Systems (Hydra-70)

INVESTMENT COMPONENT

Modernization

Recapitalization

Maintenance

MISSION

Provides air-to-ground suppression, smoke screening, illumination, and direct and indirect fires to defeat area materiel and personnel targets at close and extended ranges.

DESCRIPTION

The Hydra-70 Rocket System of 2.75-inch air-launched rockets is employed by tri-service and special operating forces on both fixed- and rotary-wing aircraft and is inherently immune to countermeasures. This highly modular rocket family incorporates several different mission-oriented warheads for the Hydra-70 variant, including high-explosive, anti-personnel, multipurpose submunition, red phosphorus smoke, flechette, training, visible-light illumination flare, and infrared illumination flare.

Diameter: 2.75 inches
Weight: 23–27 pounds (depending on warhead)
Length: 55–70 inches (depending on warhead)
Range: 300–8,000 meters
Velocity: 700+ meters per second
Area suppression: No precision

SYSTEM INTERDEPENDENCIES

None

PROGRAM STATUS

- **Current:** Producing annual replenishment for training, theater combat expenditures, and war reserve requirements

PROJECTED ACTIVITIES

- **FY13:** Continue Hydra-70 production and safety, reliability, and producibility program activities

2.75 Inch Rocket Systems (Hydra-70)

FOREIGN MILITARY SALES
Colombia, Egypt, Japan, Jordan, Kuwait, Netherlands, Saudi Arabia, Singapore, Thailand, Taiwan, Tunisia, United Arab Emirates, United Kingdom

CONTRACTORS
Prime System:
General Dynamics (Burlington, VT)
Grain:
BAE Systems (Radford, VA)
Warhead Fuzes:
Action Manufacturing (Philadelphia, PA)
Shipping Container (Fastpack):
CONCO (Louisville, KY)
Fin & Nozzle:
General Dynamics Ordnance and Tactical Systems (Anniston, AL)

Abrams Tank Upgrade

INVESTMENT COMPONENT

Modernization

Recapitalization

Maintenance

MISSION

Provides mobility, firepower, and shock effect with lethality, survivability, and fightability necessary to close with and destroy enemy forces on the integrated battlefield.

DESCRIPTION

The Abrams Tank Upgrade includes two powerful variants, the M1A1 SA (Situational Awareness) and the M1A2 SEP (System Enhancement Program) Version 2. The 1,500-horsepower AGT turbine engine, the 120mm main gun, and special armor make the Abrams tank particularly lethal against heavy armor forces. Both variants feature the new Block I 2nd generation forward-looking infrared (FLIR) technology and ballistic solution upgrades for the M829A3 kinetic and the M1028 canister rounds.

M1A1 SA: Improvements include the Gunner's Primary Sight (GPS), with the new Block I 2nd generation FLIR technology and the Stabilized Commander's Weapon Station (SCWS). Modifications include Blue Force Tracking (BFT), a digital command and control system that gives commanders information about their location relative to friendly forces; and the Power Train Improvement and Integration Optimization Program (Total InteGrated Engine Revitalization (TIGER) engine and improved transmission), which provides more reliability and durability. Survivability improvements include frontal armor and turret side armor upgrades.

M1A2 SEP v2: Lethality improvements include the Common Remotely Operated Weapon Station (CROWS). The M1A2 SEP v2 has enhanced microprocessors, color flat-panel displays, better Soldier-machine interface, and a new open operating system designed to run the Common Operating Environment (COE) software. Both the GPS and the Commander's Independent Thermal Viewer (CITV) tank include Block I FLIR technology. The M1A2 SEP has improved frontal and side armor for enhanced crew survivability. The vehicle is also equipped with battery-based auxiliary power, TIGER, and an upgraded transmission for improved automotive reliability and durability.

SYSTEM INTERDEPENDENCIES

Other Major Interdependencies

Army Battle Command System (ABCS), Blue Force Tracker (BFT), FM Voice-Advanced SINCGARS Improvement Program (ASIP) Radio, System of Systems Common Operating Environment (SOSCOE)

PROGRAM STATUS

- **Current:** 3rd Armored Cavalry Regiment; 4th Infantry Division; 1st Cavalry Division; 1st Armored Division; Army Prepositioned Stocks 4 and 5; 1st Brigade, 2nd Infantry Division; 1st Armored Division; 3rd Infantry Division; 2nd Brigade, 1st Infantry Division, and the Training and Doctrine Command, Fort Benning GA are equipped with the Abrams M1A2 SEP v2.
- **Current:** Abrams production of the M1A2 SEP v2 tank continues for both the Active Army and the Army National Guard (ARNG) through June 2014.

PROJECTED ACTIVITIES

- **FY13:** Development of the next version of the Abrams tank begins. The upgrade will focus on increasing the electrical power margin, improving survivability with improved armor protection and advanced counter-IED protection, integrating the new Army network, electronic component improvements, a new auxiliary power unit, and an ammunition data link.
- **FY13-FY14:** Continue M1A2 SEP v2 production with final delivery in June 2014.
- **FY13-FY15:** 116th Heavy Brigade Combat Team (HBCT), ID ARNG will be fielded with the Abrams M1A2 SEP v2 tank; M1A1 SA fielding continues to the 30th HBCT, NC ARNG; 81st HBCT, WA ARNG; 155th HBCT, MS ARNG; 1-34 HBCT, MN ARNG; 278th HBCT, TN ARNG, 55-28th HBCT, PA ARNG; 1-118th Combined Arms Army Battalion (CAB), SC ARNG; 1-145th CAB, OH ARNG, 2-137th CAB, KS ARNG: 11th ACR; and Regional Training

FOREIGN MILITARY SALES
M1A1:
Australia (59), Egypt (1,130), Iraq (140)
M1A2:
Kuwait (218)
M1A2/M1A2S:
Saudi Arabia (329)

CONTRACTORS
Prime:
General Dynamics Land Systems
    (Sterling Heights, MI)
Engine:
Honeywell (Phoenix, AZ)
Transmission:
Allison Transmission (Indianapolis, IN)
Anniston Army Depot (Anniston, AL)

Combat weight (tons): M1A1 - 68.59; M1A2 SEP v1 - 68.57;
M1A2 SEP v2 - 69.29
Speed: 42 mph, 30 mph cross-country
Main gun/rounds (basic load): M1 - 105mm/55 rounds
M1A1 - 120mm/40 rounds; M1A2 - 120mm/42 rounds
Machine guns: .50 caliber 900 rounds, 7.62mm 11,400 rounds

Advanced Field Artillery Tactical Data System (AFATDS)

INVESTMENT COMPONENT

Modernization

Recapitalization

Maintenance

MISSION
Provides the Army, Navy, and Marine Corps with automated fire support command, control, and communications.

DESCRIPTION
The Advanced Field Artillery Tactical Data System (AFATDS) functions as the land component's automated Fire Support Command and Control (FSC2) system. The AFATDS software is used in the Fires Warfighting Function (WFF) to plan, execute, and deliver lethal and non-lethal effects within the overall Mission Command and Control enterprise. AFATDS interoperates and integrates with over 80 different battlefield systems, to include Navy and Air Force command and control weapon systems and the German, French, Turkish, and Italian fire support systems. AFATDS fuses the essential situational awareness (SA) data, intelligence information and targeting

data, in near real time, in order to effectively manage target selection and target engagement in accordance with Mission Command guidance and priorities.

AFATDS pairs targets to weapons to provide optimum use of fire support assets and timely execution of fire missions. AFATDS automates the planning, coordinating, and controlling of all fire support assets (field artillery, mortars, close air support, naval gunfire, attack helicopters, offensive electronic warfare, fire support meteorological systems, forward observers, and fire support radars). The system is in use in operations in Afghanistan.

SYSTEM INTERDEPENDENCIES
In this Publication
Battle Command Sustainment Support System (BCS3), Distributed Common Ground System–Army (DCGS-A) Global Command and Control System–Army (GCCS-A), Tactical Battle Command (TBC)/Maneuver Control System (MCS), Joint Battle Command–Platform

(JBC-P), Lightweight 155mm Howitzer System (LW155), Multiple Launch Rocket System (MLRS) M270A1

Other Major Interdependencies
Lightweight Forward Entry Device (LFED), Pocket-Sized Forward Entry Device (PFED), Joint Automated Deep Operations Coordination System (JADOCS), Theater Battle Management Core System (TBMCS), Naval Fires Control System (NFCS), Command and Control Personal Computer / Joint Tactical COP Workstation (C2PC / JTCW)

PROGRAM STATUS
- **3QFY12:** Developed web-based capabilities to allow remote access to AFATDS fires and functions to align with emerging Common Operating Environment (COE) guidance. This capability will be fielded in AFATDS version 6.8 (1QFY13).
- **3QFY12:** Awarded the AFATDS Version 6.8.X software contract
- **3QFY12:** Network Integration Evaluation (NIE) 12.2
- **4QFY12:** AGILE Fire VI: Integrated Tactical Airspace (ITA) Project. Demonstrated improved airspace

management between U.S. Air Force Airspace Command and Control nodes and Army applications utilizing newly productized Net-Centric Airspace Operations services
- **Continued** fielding of AFATDS

PROJECTED ACTIVITY
- **1QFY13:** AFATDS Increment 2 Materiel Development Decision
- **1QFY13:** NIE 13.1
- **2QFY13:** Full Materiel Release of AFATDS 6.8.0 (BC 13)
- **2QFY13:** Completion of hardware transition to Windows Operating System
- **3QFY13:** AFATDS Increment 2 Milestone B Decision Review
- **3QFY13:** NIE 13.2
- **1QFY14:** NIE 14.1
- **3QFY14:** NIE 14.2

Advanced Field Artillery Tactical Data System (AFATDS)

FOREIGN MILITARY SALES
Australia, Bahrain, Egypt, Jordan, Portugal, Taiwan, and Turkey

CONTRACTORS
Raytheon (Fort Wayne, IN)
General Dynamics (Taunton, MA)
Northrop Grumman (Aberdeen, MD)
Computer Sciences Corp. (CSC)
 (Eatontown, NJ; Belcamp, MD)
CACI (Eatontown, NJ)

Advanced Threat Infrared Countermeasures (ATIRCM) and Common Missile Warning System (CMWS) Programs and Pre-MDAP Common Infrared Countermeasure (CIRCM)

Modernization

Recapitalization

Maintenance

MISSION

Detects missile launches/flights, protects aircraft from infrared (IR) guided missiles, and provides threat awareness and IR countermeasures using an airborne self-protection system.

DESCRIPTION

Advanced Threat Infrared Countermeasures (ATIRCM) and Common Missile Warning System (CMWS) Programs and Pre-MDAP Common Infrared Countermeasure (CIRCM) integrates defensive infrared countermeasures capabilities into existing, current-generation aircraft to engage and defeat IR-guided missile threats.

The U.S. Army operational requirements concept for IR countermeasure systems is the Suite of Integrated Infrared Countermeasures (SIIRCM). It mandates an integrated warning and countermeasure system to enhance aircraft survivability against infrared guided threat missile systems. The ATIRCM and CMWS subprograms form the core element of the SIIRCM concept.

An April 2009 Acquisition Decision Memorandum from the Defense Acquisition Executive directed the Army to establish a subprogram for the Next Generation ATIRCM (now called CIRCM). ATIRCM and CMWS have a modular configuration consisting of an integrated ultraviolet missile warning system, an Infrared Laser Jammer, and Improved Countermeasure Dispensers (ICMDs). This configuration can vary with aircraft and type.

CMWS can function as a stand-alone system with the capability to detect missiles and provide audible and visual warnings to pilots. When installed with the Advanced IRCM Munitions and ICMDs, it activates expendables to decoy/defeat infrared-guided missiles.

ATIRCM adds the Directed Energy Laser Countermeasure Technology to CMWS and is a key for Future Force Army aircraft. ATIRCM is the Army's latest Aircraft Survivability Equipment (ASE) initiative to protect crews and aircraft from advanced threat Man Portable Air Defense Systems (MANPADS) until CIRCM is fielded.

The CIRCM subprogram is being developed to replace the ATIRCM system. CIRCM is expected to be lighter in weight, more reliable, and achieve more affordable life-cycle costs. CIRCM is being designed to operate with the Army CMWS missile warning system and the evolving Navy Joint and Allied Threat Awareness System to provide protection for rotary wing, tilt-rotor and small fixed wing aircraft across DoD.

SYSTEM INTERDEPENDENCIES
In this Publication
None

Other Major Interdependencies
AH-64D, C-12R/T/U, C-23, C-26, Constant Hawk-A, Constant Hawk-I, DHC-7, HH-60L, HH-60M, MH-47E/G, MH-60K/L/M, RC-12/C-12, RC-12K/N/P/Q, UC-35

PROGRAM STATUS
- **Current:** All aircraft deployed to Operation Enduring Freedom equipped with CMWS prior to deployment; OH-58D, Kiowa Warrior is latest platform to integrate CMWS
- **Current:** In process, next-generation Electronic Control Unit (ECU) and Missile Warning Algorithms for all aircraft
- **1QFY13:** CIRCM system approved for Milestone A
- **2QFY13:** Two contracts awarded for the Technology Development Phase effort

PROJECTED ACTIVITIES
ATIRCM
- **4QFY13:** Continue fielding to CH-47D/F; Complete all A-Kit installation

CMWS
- **4QFY13:** Start fielding Next Generation ECU

CIRCM
- **2QFY14:** Pre-EMD Review
- **1QFY15:** Milestone B planned

Technology Development	Engineering & Manufacturing Development	Production & Deployment	Operations & Support

FOREIGN MILITARY SALES
United Kingdom

CONTRACTORS
ATIRCM and CMWS (Prime):
BAE Systems (Nashua, NH)
Logistics Support:
DATA Inc. (Huntsville, AL)
Software Configuration Management Support:
Science Applications International Corp. (SAIC)
   (Huntsville, AL)
CMWS-GTRI E2E Data Analysis/SIL Development:
Georgia Tech Applied Research Corp. (Atlanta, GA)
OH-58D Product Documentation Update:
Bell Helicopter Textron (Fort Worth, TX)
Test Support Data Analysis:
MacAulay-Brown Inc. (Dayton, OH)
UH-60A/L P31 Upgrade:
Rockwell Collins (Cedar Rapids, IA)
Engineering/Tech Production Support:
Computer Sciences Corp. (CSC) (Huntsville, AL)
Open Architecture Translator System (OATS):
David H. Pollock Consultants (Eatontown, NJ)
CIRCM (Primes):
BAE Systems (Nashua, NH)
Northrop Grumman (Rolling Meadows, IL)
Logistics Support:
DATA Inc. (Huntsville, AL)
Software Configuration Management Support:
SAIC (Huntsville, AL)
Engineering/Tech Production Support:
Computer Sciences Corp. (CSC) (Huntsville, AL)
OATS:
David H. Pollock Consultants (Eatontown, NJ)

Air Soldier System (Air SS)

INVESTMENT COMPONENT

Modernization

Recapitalization

Maintenance

MISSION

Integrates aircrew protective clothing, electronics, and survival equipment to reduce body-worn bulk and weight by 25% and provides a helmet-mounted display with symbology to improve flight safety in conditions of poor visibility.

DESCRIPTION

The Air Soldier System (Air SS) is flight crew clothing and individual equipment that improves mission effectiveness and duration, and provides the aviation Soldier with the ability to survive, evade hostile forces, and safely return to friendly forces. It is a mission and survival gear carriage system with an integrated capability to sustain the Soldier in water immersion, extreme heat or cold, and protects the Soldier from flash fire, crash impact, and chemical and ballistic threats. Air SS reduces the bulk and weight borne by the aviation Soldier by eliminating layers of protective clothing without compromising protection, and combining functions of current standalone hardware components, batteries, and displays with a new body-mounted electronics and power management system for the physiological support, communications, and life support equipment worn by the aviation Soldier. Air SS provides Army pilots with a new day/night flight helmet-mounted display with enhanced symbology and head tracking to prevent aircraft mishaps in degraded visual environments such as darkness, fog, or blowing sand, dust, or snow. The Air SS reduces combat load and improves situational awareness and safety.

SYSTEM INTERDEPENDENCIES

In this Publication

None

Other Major Interdependencies

Core Soldier, Utility, Cargo, Recon and Attack Helicopter Systems

PROGRAM STATUS

- **1QFY12:** Milestone B, entry into the Engineering and Manufacturing Development phase

PROJECTED ACTIVITIES

- **3QFY14:** Milestone C, entry into the Production and Deployment phase

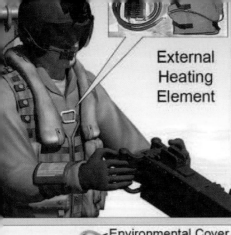

External Heating Element

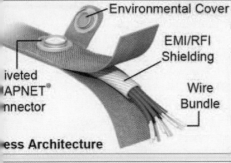

Environmental Cover

EMI/RFI Shielding

iveted APNET nnector

Wire Bundle

ess Architecture

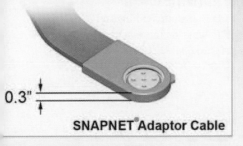

0.3"

SNAPNET Adaptor Cable

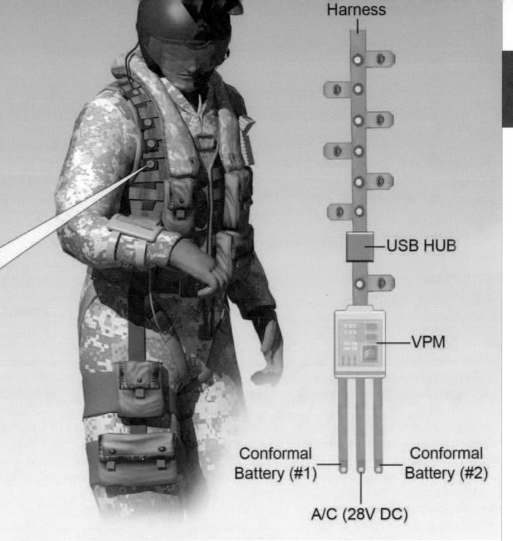

Harness

USB HUB

VPM

Conformal Battery (#1)

Conformal Battery (#2)

A/C (28V DC)

Personal Display Module (PDM)

Radio Interface Card Module (RICM)

Soldier Computer Module (SCM)

Conformal Battery

4"

6"

Air Warrior (AW)

Modernization

Recapitalization

Maintenance

MISSION
Provides enhanced mission effectiveness, leveraging clothing and equipment to maximize aircrew member survivability.

DESCRIPTION
Air Warrior (AW) is a modular, integrated, rapidly reconfigurable combat aircrew ensemble that saves lives and maximizes Army aircrew mission performance. Previous aviation life support equipment consisted of a non-integrated assemblage of protective and survival gear. AW uses a systems approach to equip the aircrew and close the capability gap between human and machine. Fielded incrementally in blocks to rapidly provide enhanced capabilities to the warfighter, AW leverages and integrates clothing and equipment, such as the Army Aircrew Combat Uniform and ballistic protection from other product managers.

AW Block I provides:
- Survival Equipment Subsystem, which integrates first aid, survival, signaling, and communications equipment with body armor and over-water survival subsystems
- Microclimate Cooling System, which increases effective mission duration in heat-stress environments by more than 350 percent
- Aircrew Integrated Helmet System, a lighter helmet with increased head and hearing protection

AW Increment III:
- Electronic Data Manager (EDM), a portable digital-mission planning device for over-the-horizon messaging and enhanced situational awareness capabilities through connectivity to Blue Force Tracking, Aviation
- Aircraft Wireless Intercom System (AWIS) for secure cordless, hands-free aircrew communications
- Survival Kit, Ready Access, Modular (SKRAM) Go-Bag with integrated hydration
- Portable Helicopter Oxygen Delivery System, a Soldier-worn supplemental breathing oxygen system for high-altitude operations
- Communication Enhancement and Protection System (CEPS) for helmet hear-through capability

SYSTEM INTERDEPENDENCIES
In this Publication
None

Other Major Interdependencies
Core Soldier, Utility, Cargo, Recon and Attack Helicopter Systems

PROGRAM STATUS
- **1QFY12-4QFY12:** Fielded and reset Air Warrior Increment III to units prior to deployment

PROJECTED ACTIVITIES
- **FY13-FY14:** Continue fielding and resetting Air Warrior Increment III to units prior to deployment

Air Warrior (AW)

FOREIGN MILITARY SALES
Australia, Bahrain, Canada, United Arab Emirates

CONTRACTORS
Government is prime integrator; various vendors provide components:
Telephonics Corp. (Farmingdale, NY)
General Dynamics C4 Systems Inc.
  (Scottsdale, AZ)
BAE Systems (Phoenix, AZ)
Aerial Machine and Tool Corp. (Vesta, VA)
Westwind Technologies Inc. (Huntsville, AL)
Carleton Technologies Inc. (Orchard Park, NY)
Med-Eng Systems Inc. (Ogdensburg, NY)
Raytheon Technical Services (Indianapolis, IN)
Secure Communication Systems Inc.
  (Santa Ana, CA)
U.S. Divers (Vista, CA)
CEP Inc. (Enterprise, AL)
Science and Engineering Services Inc. (SESI)
  (Huntsville, AL)
Gibson and Barnes (Santa Clara, CA)
Oxygen Generating Systems International
  (Buffalo, NY)
Gentex Corp. (Rancho Cucamonga, CA)
Mountain High Equipment and Supply Co.
  (Redmond, OR)
Taylor-Wharton (Huntsville, AL)

Air/Missile Defense Planning and Control System (AMDPCS)

MISSION

Provides automated Command and Control (C2) to integrate Air and Missile Defense (AMD) planning and operations for Air Defense Airspace Management (ADAM) systems in Brigade Combat Teams (BCTs) and at every Air Defense Artillery (ADA) echelon, Battery through Theater.

DESCRIPTION

The Air/Missile Defense Planning and Control System (AMDPCS) is an Army Objective Force system that provides integration of AMD operations at all echelons. AMDPCS systems are deployed with ADAM systems, ADA brigades, and Army Air and Missile Defense Commands (AAMDCs).

ADAM provides the commanders at BCTs, fires brigades, combat aviation brigades, and division and corps tactical operations systems with situational awareness (SA) of the airspace. ADAM provides collaboration and staff planning capabilities through the Army Mission Command System and operational links for airspace coordination with Joint, interagency, multinational, and coalition forces.

The AMDPCS in ADA brigades and AAMDCs provides expanded staff planning and coordination capabilities for integrating defense of the air battlespace. The AMDPCS includes shelters, automated data processing equipment, tactical communications, standard vehicles, tactical power, and the following two software systems for force operations/engagement operations: Air and Missile Defense Workstation (AMDWS) and Air Defense System Integrator (ADSI).

AMDWS is a staff planning and battlespace SA tool that provides commanders with a common tactical and operational air picture. It is the ADA component of Army Mission Command. ADSI is a Joint multi-communications processor that provides external Joint messaging for operations by subordinate or attached units.

SYSTEM INTERDEPENDENCIES

None

PROGRAM STATUS

- **2QFY12:** Completed FY11 reset of 32 ADAMs
- **4QFY12:** Urgent Materiel Release of AMDWS 6.5.1
- **4QFY12:** Fielded AMDPCS to the 10th AAMDC (Germany)

PROJECTED ACTIVITIES

- **1QFY13:** Participate in NIE 13.1
- **1QFY13:** Transition ADAM integration to Tobyhanna Army Depot
- **2QFY13:** Complete FY12 reset of 23 ADAMs from OEF
- **4QFY13:** AMDWS Software V6.6 Full Materiel Release
- **4QFY13:** Field AMDPCS to 94th AAMDC (Hawaii/Pacific)

Air/Missile Defense Planning and Control System (AMDPCS)

FOREIGN MILITARY SALES
Netherlands (AMDWS)
Germany (AMDPCS)

CONTRACTORS
Northrop Grumman (Huntsville, AL)
Ultra, Inc. (Austin, TX)

Airborne and Maritime/Fixed Station Joint Tactical Radio System (AMF JTRS)

Modernization

Recapitalization

Maintenance

MISSION

Provides tactical radio capability to enable a wide variety of Army Aviation platforms to meet networking interoperability requirements.

DESCRIPTION

Airborne and Maritime/Fixed Station Joint Tactical Radio System (AMF JTRS) is a family of software-defined networking radios designed to meet Army Aviation requirements for secure user voice, data and video communications.

The AMF JTRS program of record is being restructured around a family-of-radios approach to procure industry-developed tactical radios. The proposed acquisition strategy will authorize procurement of Non-Developmental Items (NDIs) to meet platform needs.

SYSTEM INTERDEPENDENCIES

None

PROGRAM STATUS

- **3QFY12:** Received Under Secretary of Defense for Acquisition, Technology and Logistics (USD(AT&L)) direction to close out the AMF JTRS System Development and Demonstration (SDD) contract
- **4QFY12:** Received USD(AT&L) direction to pursue an NDI acquisition approach to meet Army rotary wing aircraft requirements
- **4QFY12:** Release NDI Request for Information (RFI)
- **4QFY12:** AMF JTRS analysis on RFI

PROJECTED ACTIVITIES

- **1QFY15:** Milestone C

Technology Development | Engineering & Manufacturing Development | Production & Deployment | Operations & Support

Airborne and Maritime/Fixed Station Joint Tactical Radio System (AMF JTRS)

FOREIGN MILITARY SALES
None

CONTRACTORS*
Prime:
Lockheed Martin (San Diego, CA; Alexandria, VA)
Subcontractors:
BAE Systems (Wayne, NJ)
Northrop Grumman (San Diego, CA)
General Dynamics C4 Systems, Inc. (Scottsdale, AZ)
Raytheon (Ft. Wayne, IN)

*Contractors listed above are through 3QFY12.

Airborne Reconnaissance Low (ARL)

INVESTMENT COMPONENT

Modernization

Recapitalization

Maintenance

MISSION

Provides tactical commanders with day/night, near-all-weather, real-time airborne communications intelligence/imagery intelligence (COMINT/IMINT) collection and designated area surveillance system.

DESCRIPTION

Airborne Reconnaissance Low (ARL) is a self-deploying, multisensor, day/night, all-weather reconnaissance, intelligence system. It consists of a modified DeHavilland DHC-7 fixed-wing aircraft equipped with COMINT/IMINT, Ground Moving Target Indicator/Synthetic Aperture Radar (GMTI/SAR), and electro-optical (EO)/infrared (IR) full-motion video capability. Four on-board operators control the payloads via on-board open-architecture, multifunction workstations and communication directly with ground units. Intelligence collected on the

ARL can be analyzed, recorded, and disseminated on the aircraft workstations in real time and stored on board for post-mission processing. During multi-aircraft missions, data can be shared between cooperating aircraft via ultra-high-frequency air-to-air data links allowing multiplatform COMINT geolocation operations. The ARL system includes a variety of communications subsystems to support near-real-time dissemination of intelligence and dynamic retasking of the aircraft. ARL provides real-time down-link of MTI data to the Common Ground Station (CGS) at the Brigade Combat Team through echelon-above-corps level. Eight aircraft are configured as ARL–Multifunction (ARL-M), equipped with a combination of IMINT, COMINT, and SAR/MTI payload and demonstrated hyperspectral imager applications and multi-intelligence (multi-INT) data-fusion capabilities. Four mission workstations are on board the aircraft and are remote operator-capable. The Intelligence and Security Command (INSCOM) operates all ARL systems and currently supports Southern Command (SOUTHCOM) with one to four ARL-M aircraft, United States

Forces Korea (USFK) with three ARL-M aircraft, and U.S. Central Command (CENTCOM) with one aircraft. Future sensor enhancements are focused on upgrades to the COMINT, IMINT, and radar payloads to support emerging threats. Capabilities include:

Endurance/ceiling: 8 hours/20,000 feet
Speed/gross weight: 231 knots/47,000 pounds
Range with max payload: greater than 1,400 nautical miles
Mission completion rate: greater than 90 percent

ARL will continue to support current operations until a future system is fielded.

SYSTEM INTERDEPENDENCIES

None

PROGRAM STATUS

- **1QFY12:** Installation of Phoenix Eye Radar on ARL
- **3QFY12:** Completed workstation and beyond-line-of-sight (BLOS) upgrades on ARL system M8

PROJECTED ACTIVITIES

- **3QFY13:** Workstation and BLOS upgrades on ARL system M7
- **FY13-FY14:** Continue imagery, radar, COMINT, system interoperability, and workstation architecture upgrades

ACQUISITION PHASE

Technology Development	Engineering & Manufacturing Development	Production & Deployment	Operations & Support

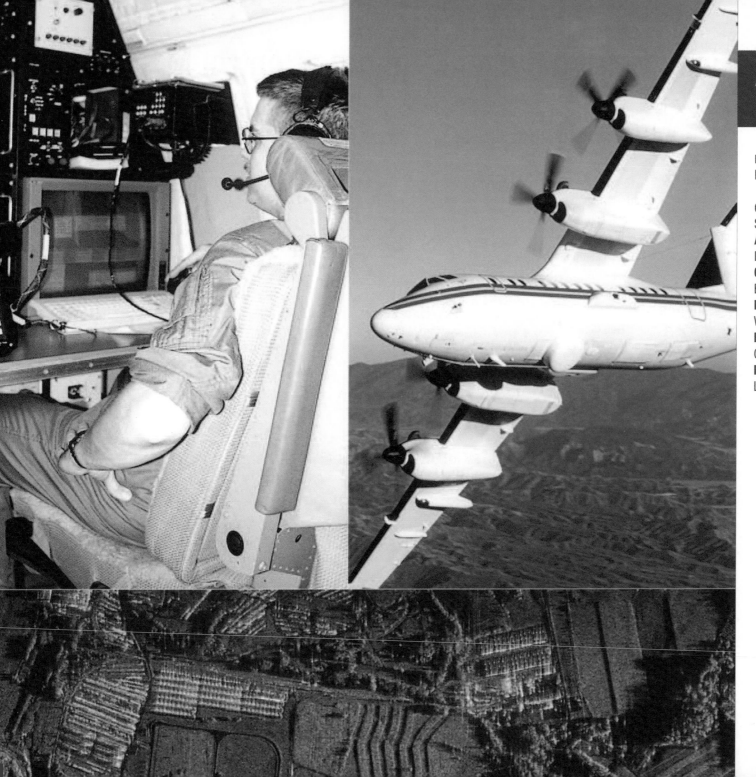

Airborne Reconnaissance Low (ARL)

FOREIGN MILITARY SALES
None

CONTRACTORS
Sierra Nevada Corp. (Hagerstown, MD)
Aircraft Survivability:
Litton Advanced Systems (Gaithersburg, MD)
COMINT Subsystem:
BAE Systems (Manchester, NH)
EO/IR Subsystem:
WESCAM (Hamilton, Ontario, Canada)
Engineering Support:
CACI (Berryville, VA)
Radar Subsystem:
Lockheed Martin (Phoenix, AZ)

Anti-Personnel Mine Clearing System, Remote Control M160

INVESTMENT COMPONENT

Modernization

Recapitalization

Maintenance

MISSION

Provides line-of-sight, unmanned mine clearing operations and real-time control of the mine clearing vehicle from either a mounted or dismounted position.

DESCRIPTION

The Anti-Personnel Mine Clearing System, Remote Control M160 is a tele-operated Light Flail system designed to provide line-of-sight (LOS) unmanned mine clearing operations. The remote control system provides real-time control of the mine-clearing vehicle and allows the operator to control the vehicle from either a mounted or stand-off dismounted position. The system is equipped with a communication system that transfers operating status and video feedback to the operator. The M160 was designed to be suitable for mine clearing in urban areas, fields, forests, bushes, forest roads, riverbanks, and muddy areas. The M160's hand-held, stand-off, remote control feature allows the operator to remain outside the range of exploding mines during the clearing process. The engine and vital components of the machine are protected by steel armor plates. The M160 is designed to be protected against mine-explosion fragments. Mine clearing is conducted by using the flailing motion of high-speed, rotating chained hammers. The machine digs and powders the soil up to a depth of 2 inches (threshold)/6 inches (objective), depending on soil type. The digging action may result in the detonation or shattering of AP mines. The M160 system consists of the following major components: engine, hydraulic system, flail head assembly, remote control system and drive train.

SYSTEM INTERDEPENDENCIES

In this Publication

None

Other Major Interdependencies

Trailers for transporting

PROGRAM STATUS

- **1QFY12:** Type Classification Standard
- **1QFY12:** Conduct Performance Based Logistics BCA and Core Logistics Analysis (CLA)
- **4QFY12:** 81 systems produced with 64 systems fielded in support of Operation Enduring Freedom (OEF)

PROJECTED ACTIVITIES

- Support ONS Request to develop/ integrate CREW communication systems for M160 systems operating in OEF
- **FY15:** Prepare for RECAP of 41 systems returning from theater after OEF drawdown for fielding of POR systems

FOREIGN MILITARY SALES
Zagreb, Croatia

CONTRACTORS
Subcontractor:
SAIC (Huntsville, AL)

AN/TPQ-53 (formerly known as the Enhanced AN/TPQ-36)

INVESTMENT COMPONENT

Modernization

Recapitalization

Maintenance

MISSION
Detects, classifies, tracks, and locates the point of origin of projectiles fired from mortar, artillery, and rocket systems.

DESCRIPTION
The AN/TPQ-53 detects, tracks, and locates the point of origin of projectiles fired from mortar, artillery, and rocket systems. It will replace the current AN/TPQ-36 and AN/TPQ-37 Firefinder radar systems. The AN/TPQ-53 provides accurate targetable data against mortars, cannon, and rockets in a 90-degree search sector or a continuous 360-degree search sector.

SYSTEM INTERDEPENDENCIES
In this Publication
Counter-Rocket, Artillery and Mortar (C-RAM)

PROGRAM STATUS
• **FY12:** Continued Low-rate initial production (LRIP) (Lot 2)

PROJECTED ACTIVITIES
• **FY13:** Continue LRIP (Lot 3)
• **2QFY13:** Limited User Test (LUT)
• **1QFY14:** Initial Operational Test & Evaluation (IOT&E)
• **3QFY14:** Full-rate production (FRP) Decision
• **4QFY14:** Follow-on Test and Evaluation (FOT&E)

ACQUISITION PHASE

Technology Development | Engineering & Manufacturing Development | Production & Deployment | Operations & Support

UNITED STATES ARMY

AN/TPQ-53

FOREIGN MILITARY SALES
None

CONTRACTORS
Integration/ICS:
Lockheed Martin (Syracuse, NY)
Technical Support:
JB Management (Alexandria, VA)

Armored Multi-Purpose Vehicle (AMPV)

INVESTMENT COMPONENT

Modernization

Recapitalization

Maintenance

MISSION

Provides continuous and simultaneous support of unified land operation (offense, defense, and stability operations), across the spectrum of conflict.

DESCRIPTION

The Armored Multi-Purpose Vehicle (AMPV) enables the Heavy Brigade Combat Team (HBCT) commander to control a relentless tempo that overwhelms the threat with synchronized and integrated assaults that transition rapidly to the next engagement. The AMPV replaces M113s in five mission roles within the HBTCs: General Purpose (GP), Mortar Carrier (MC), Mission Command (MCmd), Medical Evacuation (ME), and Medical Treatment (MT).

GP: Vehicle will include two crew and six passengers. It will be reconfigurable to carry one litter, mounted crew-served weapons, and will integrate two Joint Tactical Radio System (JTRS) Handheld, Manpack and Small Form Factor (HMS) or two Single Channel Ground and Airborne Radio System (SINCGARS), Vehicle Intercom (VIC)-3, Warfighter Information Network – Tactical (WIN-T), Driver's Vision Enhancer (DVE), DUKE v3, and Force XXI Battle Command Brigade and Below (FBCB2)/ Blue Force Tracker (BFT).

MC: Vehicle will include two crew, two mortar crew, mounted 120mm mortar and 69 rounds of 120mm ammunition. It will integrate two JTRS HMS, VIC-3, DVE, DUKE v3, FBCB2/BFT, M95 Mortar Fire Control Systems (MFCS), and Advanced Field Artillery Tactical Data System (AFATDS).

MCmd: Vehicle will include two crew, two operators, and mounted crew-served weapons. It will integrate one JTRS Ground Mobile Radio (GMR), two JTRS HMS, VIC-3, WIN-T, DVE, DUKE v3, FBCB2/BFT, MFCS and Army Battle Command Systems (AFATDS, Distributed Common Ground System – Army (DCGSA), etc.).

ME: Vehicle will include three crew, six ambulatory patients or four litter patients, or three ambulatory and two litter patients. It will integrate two JTRS HMS, VIC-3, DVE, DUKE v3, FBCB2/BFT, and the storage for Medical Equipment Sets (MES).

MT: Vehicle will include four crew, one litter patient and will integrate two JTRS HMS, VIC-3, DVE, DUKE v3, and FBCB2/BFT.

SYSTEM INTERDEPENDENCIES
In this Publication
Advanced Field Artillery Tactical Data System (AFATDS), Force XXI Battle Command Brigade and Below (FBCB2)

Other Major Interdependencies
Blue Force Tracker (BFT), Others to be determined

PROGRAM STATUS
- **2QFY12:** Materiel Development Decision
- **2QFY12:** Market surveys completed
- **3QFY12:** First industry day event
- **3QFY12:** Completed Analysis of Alternatives

PROJECTED ACTIVITIES
- **1QFY13:** Second industry day event
- **2QFY13:** Release request for proposal
- **2QFY14:** Engineering & Manufacturing Development awards
- **3QFY14:** Preliminary Design Review (PDR)

FOREIGN MILITARY SALES
None

CONTRACTORS
Prime:
To be determined
Engine:
To be determined
Transmission:
To be determined

Army Integrated Air and Missile Defense (AIAMD)

INVESTMENT COMPONENT

Modernization

Recapitalization

Maintenance

MISSION
Provides the full combat potential of an Integrated Air and Missile Defense capability through a network-centric "plug and fight" architecture at the component level (e.g., launchers and sensors) and a common mission command (MC) system.

DESCRIPTION
Army Integrated Air and Missile Defense (AIAMD) will enable the integration of modular components (current and future AMD sensors, weapons, and MC) with a common MC capability in a networked and distributed "plug and fight" architecture. This common MC system, called the IAMD Battle Command System (IBCS), will provide standard configurations and capabilities at each echelon. This allows Joint, Interagency, Intergovernmental, and Multinational (JIIM) AMD forces to organize based on mission, enemy, terrain and weather, troops and support available, time available, and civil considerations (METT-TC). Shelters and vehicles may be added to enable broader missions and a wider span of control executed at higher echelons. A network-enabled "plug and fight" architecture and common MC system will enable dynamic defense design and task force reorganization, and provide the capability for interdependent, network-centric operations that link Joint IAMD protection to the supported force scheme of operations and maneuver.

This Army IAMD system-of-systems architecture will enable extended range and non-line-of-sight engagements across the full spectrum of aerial threats, providing fire control quality data to the most appropriate weapon to successfully complete the mission. Furthermore, it will mitigate the coverage gaps and the single points of failure that have plagued AMD defense design in the past as well as reduce manpower, enhance training, and reduce operation and support costs.

SYSTEM INTERDEPENDENCIES
In this Publication
PATRIOT Advanced Capability–Three (PAC-3), Early Infantry Brigade Combat Team (E-IBCT) Capabilities IBCT Increment I, Joint Land Attack Cruise Missile Defense Elevated Netted Sensor System (JLENS), Joint Tactical Ground Stations (JTAGS)

Other Major Interdependencies
Army Battle Command System, AEGIS, Airborne Warning and Control System, BCS, Ballistic Missile Defense System, Common Aviation Command and Control System, Command, Control, Battle Management, and Communications Planner, DD(X), E-2C, Terminal High Altitude Area Defense

PROGRAM STATUS
- **2QFY12:** Program Restructure Acquisition Decision Memorandum
- **3QFY12:** AIAMD Critical Design Review
- **4QFY12:** Defense Acquisition Board In Process Review

PROJECTED ACTIVITIES
- **1QFY14:** AIAMD Demonstration

ACQUISITION PHASE

| Technology Development | Engineering & Manufacturing Development | Production & Deployment | Operations & Support |

Army Integrated Air and Missile Defense (AIAMD)

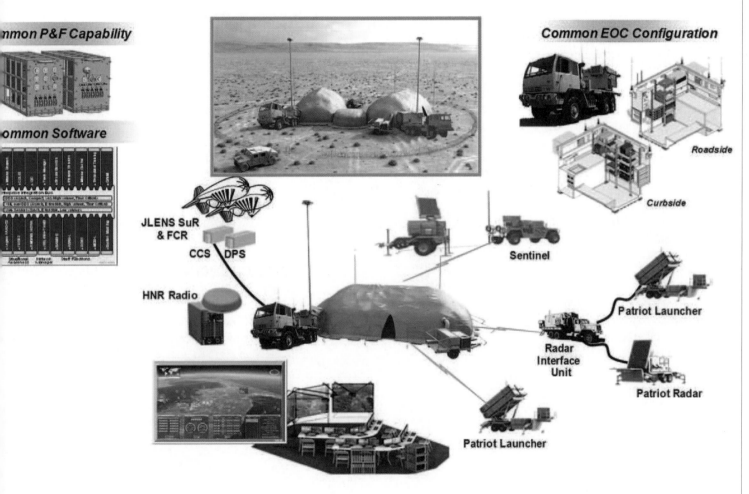

Common P&F Capability

Common Software

Common EOC Configuration

Roadside

Curbside

JLENS SuR & FCR

CCS DPS

HNR Radio

Sentinel

Patriot Launcher

Radar Interface Unit

Patriot Radar

Patriot Launcher

FOREIGN MILITARY SALES
None

CONTRACTORS
IBCS Development:
Northrop Grumman (Huntsville, AL)
A-Kit Design and Development:
Raytheon (Andover, MA; Tewksbury, MA)
SETA Support:
DMD (Huntsville, AL)

Army Key Management System (AKMS)

MISSION

Automates the functions of communication securities (COMSEC) key management, control, and distribution; electronic protection generation and distribution; and signal operating instruction management to provide planners and operators with automated, secure communications at theater/tactical and strategic/sustaining base levels.

DESCRIPTION

The Army Key Management System (AKMS) is a fielded system composed of three subsystems: Local COMSEC Management Software (LCMS), Automated Communications Engineering Software (ACES), and the Data Transfer Device/Simple Key Loader (SKL). Under the umbrella of the objective National Security Agency Electronic Key Management System, AKMS provides tactical units and sustaining bases with an organic key generation capability and an efficient secure electronic key distribution means. AKMS provides a system for distribution of COMSEC, electronic protection, and signal operating instructions (SOI) information from the planning level to the point of use in support of current, interim, and objective force at division and brigade levels.

The LCMS (AN/GYK-49) workstation provides automated key generation, distribution, and COMSEC accounting. The ACES (AN/GYK-33), which is the frequency management portion of AKMS, has been designated by the Military Communications Electronics Board as the Joint standard for use by all services in development of frequency management and cryptographic net planning and SOI generation. The SKL (AN/PYQ-10) is the associated support equipment that provides the interface between the ACES workstation, the LCMS workstation, the Warfighters' End Crypto Unit, and the Soldier. It is a small, ruggedized hand-held key loading device.

AKMS supports the Army transition to NSA's Key Management Infrastructure (KMI), which will replace the current EKMS infrastructure to provide increased security. PD COMSEC has been involved in the transition planning and will procure and field the KMI Management Client (MGC) workstations, provide New Equipment Training (NET), and total life cycle management support for the system.

SYSTEM INTERDEPENDENCIES
In this Publication
None

Other Major Interdependencies
AKMS systems are considered enabling systems for equipment/systems to receive key and frequency allotments.

PROGRAM STATUS
- **FY12-FY17:** Continue to procure and field SKLs for Air Force, Navy, Foreign Military Sales, and other government organizations
- **2QFY12:** Complete refresh of ACES hardware
- **2QFY12:** Complete LCMS software upgrade version v5.1.0.5
- **3QFY12:** KMI Initial Operational Capability
- **4QFY12:** ACES software upgrade version 3.1; SKL software upgrade version 8.0

PROJECTED ACTIVITIES
- **FY13:** ACES software upgrade version 3.2
- **FY14:** SKL upgrade to SKLv3.1
- **FY14:** ACES software upgrade version 3.3; SKL software upgrade version 9.0

ACQUISITION PHASE

| Technology Development | Engineering & Manufacturing Development | Production & Deployment | Operations & Support |

Army Key Management System (AKMS)

FOREIGN MILITARY SALES
Australia, Belgium, Bulgaria, Canada, Czech Republic, Estonia, Germany, Greece, Hungary, Latvia, Lithuania, Luxembourg, NATO, Netherlands, New Zealand, Norway, Poland, Portugal, Slovenia, Spain, Turkey, United Kingdom

CONTRACTORS
Sierra Nevada Corp. (Sparks, NV)
Mantech Sensors Technology, Inc.
  (Aberdeen, MD)
Science Applications International Corp.
  (San Diego, CA)
Websec Corp. (Murrieta, CA)
CACI (Eatontown, NJ)
CSS (Augusta, GA)

Artillery Ammunition

MISSION

Provides field artillery forces with modernized munitions to destroy, neutralize, or suppress the enemy by cannon fire.

DESCRIPTION

The Army's artillery ammunition program includes 75mm (ceremonies and simulated firing), 105mm, and 155mm projectiles and their associated fuzes and propelling charges.

Semi-fixed ammunition for short and intermediate ranges, used in 105mm howitzers, is characterized by adjusting the number of multiple propelling charges. Semi-fixed ammunition for long ranges contains a single bag of propellant optimized for obtaining high velocity and is not adjustable. The primer is an integral part of the cartridge case, and is located in the base. All 105mm cartridges are issued in a fuzed or unfuzed configuration. Both cartridge configurations are packaged with propellant.

Separate-loading ammunition, used in 155mm howizters, has separately issued projectiles, fuzes, propellant charges, and primers.

After installing the appropriate fuze on the projectile, the fuzed projectile is loaded into the cannon along with the appropriate amount of propellant charges and a primer.

The artillery ammunition program includes fuzes for cargo-carrying projectiles, such as smoke and illumination, and bursting projectiles, such as high explosives.

This program also includes bag propellant for the 105mm semi-fixed cartridges and a modular artillery charge system for 155mm howitzers.

SYSTEM INTERDEPENDENCIES

None

PROGRAM STATUS

- **3QFY12:** Full Materiel Release of the M1122, 155mm high-explosive projectile

PROJECTED ACTIVITIES

- **3QFY13:** Type Classification of the M1123/M1124 155mm illumination projectiles

Artillery Ammunition

FOREIGN MILITARY SALES
Australia, Canada, Israel, Lebanon

CONTRACTORS
General Dynamics Ordnance and Tactical
Systems (Fort Lauderdale, FL)
McAlester Army Ammunition Plant
(McAlester, OK)
General Dynamics Ordnance and
Tactical Systems-Scranton Operations
(Scranton, PA)
American Ordnance (Middletown, IA)
ARMTEC (Coachella, CA)

Assault Breacher Vehicle (ABV)

Modernization

Recapitalization

Maintenance

MISSION
Provides deliberate and in-stride minefield and complex obstacle breaching capability to the mounted maneuver force.

DESCRIPTION
The Assault Breacher Vehicle (ABV) is a highly mobile and heavily armored minefield and complex obstacle breaching system. It consists of an M1A1 Abrams tank hull, a unique turret with two Linear Demolition Charge Systems (employing two Mine Clearing Line Charges (MICLIC) and rockets), a Lane Marking System (LMS), Integrated Vision System, and a High Lift Adapter that interchangeably mounts a Full Width Mine Plow (FWMP) or a Combat Dozer Blade (CDB). The ABV requires a crew of two Soldiers. It improves the mobility and survivability of Combat Engineers and has the speed to keep pace with the maneuver force.

The ABV creates a tank-width cleared lane through a minefield by launching and detonating one of its MICLIC systems across the minefield, then proofing the lane with its FWMP while marking the cleared lane with its LMS. The ABV is fielded to the Combat Engineer company organic to the Brigade Special Troops Battalion of Armored Brigade Combat Teams. Each ABCT receives six ABVs, four FWMP and two CDBs. The ABV is air-transportable by C-17 and larger aircraft.

SYSTEM INTERDEPENDENCIES
None

PROGRAM STATUS
- **2QFY12:** 3d Brigade 3d Infantry Division
- **2QFY12:** 2d Brigade 3d Infantry Division
- **3QFY12:** 1st Brigade 3d Infantry Division

PROJECTED ACTIVITIES
- **2QFY13:** Type Classification Standard and Full Materiel Release
- **2QFY13:** Fielding 1st Brigade 34th Infantry Division (Minnesota ARNG)
- **3QFY13:** Fielding 1st Brigade 2d Infantry Division
- **4QFY13:** Fielding 4th Brigade 1st Armored Division
- **1QFY14:** Fielding 2d Brigade 1st Armored Division
- **2QFY14:** Fielding 3d Brigade 1st Cavalry Division
- **3QFY14:** Fielding 4th Brigade 1st Cavalry Division
- **FY15:** Further fielding based on Army priorities

FOREIGN MILITARY SALES
None

CONTRACTORS
Integration:
Anniston Army Depot (ANAD)
   (Anniston, AL)
Front End Equipment:
URS (Albuquerque, NM)
Line Demolition Charge System:
DRS (Los Angeles, CA)
Technical Manuals:
XMCO (Warren, MI)

Assembled Chemical Weapons Alternatives (ACWA)

INVESTMENT COMPONENT

Modernization

Recapitalization

Maintenance

MISSION

Enhances national security by destroying chemical weapons stockpiles at the U.S. Army Pueblo Chemical Depot (Colorado) and Blue Grass Army Depot (Kentucky) in a safe and environmentally sound manner.

DESCRIPTION

Established by Congressional legislation in 1996, Program Executive Officer, Assembled Chemical Weapons Alternatives (PEO ACWA) reports directly to the Under Secretary of Defense (Acquisition, Technology & Logistics) through the Assistant Secretary of Defense (Nuclear, Chemical & Biological Defense Programs) and is responsible for pilot testing selected alternative technologies and accelerating destruction of the chemical weapons stockpiles located at U.S. Army Pueblo Chemical Depot (PCD) Pueblo, CO and Blue Grass Army Depot (BGAD), Richmond, KY. PEO ACWA is specifically responsible for managing the construction, systemization, operation and closure, and any contracting related of the Pueblo

Chemical Agent-Destruction Pilot Plant (PCAPP) and the Blue Grass Chemical Agent-Destruction Pilot Plant (BGCAPP). Contractor of the PCAPP facility is 96 percent complete while the BGCAPP facility is 54 percent complete.

SYSTEM INTERDEPENDENCIES

None

PROGRAM STATUS

- **2QFY12:** ACWA Program Milestone B and new Acquisition Program Baseline approved by USD (AT&L)
- **4QFY12:** BGCAPP systemization task award

PROJECTED ACTIVITIES

- **3QFY13:** PCAPP construction complete

Assembled Chemical Weapons Alternatives (ACWA)

FOREIGN MILITARY SALES
None

CONTRACTORS
PCAPP
Systems Contractor:
Bechtel National, Inc. (Pueblo, CO)
Teaming Subcontractors:
URS Corp. (Pueblo, CO)
Parsons Government Services, Inc.
 (Pueblo, CO)
Battelle Memorial Institute (Pueblo, CO)
GP Strategies Corp. (Pueblo, CO)
BGCAPP
Systems Contractor:
Bechtel National, Inc. (Pueblo, CO)
Parsons Government Services, Inc.
 (Joint Venture) (Richmond, KY)
Teaming Partners:
URS Corporation (Richmond, KY)
Battelle Memorial Institute (Richmond, KY)
General Atomics (Richmond, KY)
GP Strategies Corporation (Richmond, KY)

Aviation Combined Arms Tactical Trainer (AVCATT)

INVESTMENT COMPONENT

Modernization

Recapitalization

Maintenance

MISSION

Provides a collective training system to meet aviation training requirements and to support institutional, organizational, and sustainment training for Active and Reserve Army aviation units worldwide in combined arms training and mission rehearsal in support of Unified Land Operations.

DESCRIPTION

The Aviation Combined Arms Tactical Trainer (AVCATT) is a mobile, transportable, multi-station, virtual simulation device designed to support unit collective and combined arms training. AVCATT provides six manned modules reconfigurable to any combination of attack, reconnaissance, lift, and/or cargo helicopters. There are four role player stations for battalion/squadron staff, combined arms elements, integrated threat, or friendly semi-automated Forces (SAF). Exercise

record/playback with simultaneous AAR capability is provided. The Non-Rated Crew Member Manned Module (NCM3), a sub-system of AVCATT, is a mobile, transportable, multi-station virtual simulation device designed to support training of non-rated crew members in crew coordination, flight, aerial gunnery, hoist, and sling-load related tasks.

The AVCATT single suite of equipment consists of two mobile trailers that house six reconfigurable networked simulators to support the Apache Longbow, Kiowa Warrior, Chinook, and Black Hawk. An after-action review theater and a battle master control station are also provided as part of each suite.

AVCATT builds and sustains training proficiency on mission-essential tasks through crew and individual training by supporting aviation collective tasks, including armed reconnaissance (area, zone, route), deliberate attack, covering force operations, downed aircrew recovery operations, Joint air attack team, hasty attack, and air assault operations. The system also

has multiple Synthetic Environment Core (SE Core) correlated visual databases.

AVCATT is fully mobile, capable of using commercial and generator power, and is transportable worldwide. The system is interoperable via local area network/wide area network with other AVCATT suites and the Close Combat Tactical Trainer (CCTT).

SYSTEM INTERDEPENDENCIES
In this Publication
None

Other Major Interdependencies
SE Core, One Semi-Automated Forces (OneSAF)

PROGRAM STATUS
- **1QFY12:** SE Core and OneSAF integration completed
- **2QFY12:** Production and fielding of the 2nd NCM3 to Fort Campbell, KY
- **2QFY12:** UH60M and CH47F Concurrency Upgrade contract awarded

PROJECTED ACTIVITIES
- **2QFY13:** Longbow Lot 13.1 Concurrency Upgrade contract awarded
- **4QFY13:** Fielding of NCM3s #3-5
- **1Q-4QFY14:** Production and fielding of NCM3s #6-16

ACQUISITION PHASE

| Technology Development | Engineering & Manufacturing Development | Production & Deployment | Operations & Support |

NCM3 Trailer Diagram

IOS & AAR

Manned Modules

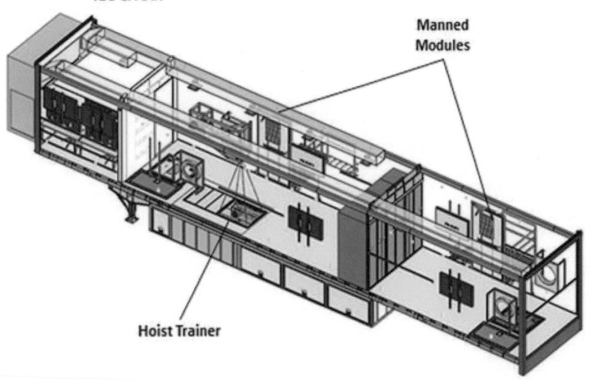

Hoist Trainer

Aviation Combined Arms Tactical Trainer (AVCATT)

FOREIGN MILITARY SALES
None

CONTRACTORS
Prime
AVCATT:
L-3 Communications (Arlington, TX)
Prime
NCM3:
SAIC (Orlando, FL)
Prime
Technology Refresh:
Daedalus Technologies (Oviedo, FL)

Biometric Enabling Capability (BEC)

INVESTMENT COMPONENT

Modernization

Recapitalization

Maintenance

MISSION
Serves as the DoD authoritative biometric repository enabling identity superiority.

DESCRIPTION
Biometric Enabling Capability (BEC), using an Enterprise System-of-Systems service-oriented architecture, will serve as DoD's authoritative biometric repository, enabling multimodal matching, storing, and sharing in support of identity superiority across the Department, Federal Agencies, and International Partners. DoD Automated Biometric Identification System (DoD ABIS), a quick reaction capability, will transition into BEC Increment 0 upon receiving a Full Deployment Decision (FDD).

This will be followed by BEC Increment 1, which will enter the Engineering and Manufacturing Development (EMD) Phase of acquisition with a Milestone B (MS-B).

SYSTEM INTERDEPENDENCIES
In this Publication
Joint Personnel Identification Version 2 (JPIv2)

Other Major Interdependencies
Automated Identity Management System, Department of Homeland Security IDENT, Federal Bureau of Investigation Integrated Automated Fingerprint Identification System, U.S. Navy Personnel Identification Version 1 Program, Special Operations Identity Dominance

PROGRAM STATUS
- **2QFY12:** NG-ABIS Capability Production Document (CDD) approved
- **4QFY12:** FDD for BEC Increment 0

PROJECTED ACTIVITIES
- **1QFY13:** Biometrics BEC Increment 1 CDD approved
- **1QFY14:** Pre-EMD review for BEC Increment 1, i.e., permission to release final request for proposal for the EMD phase
- **3QFY14:** MS-B for BEC Increment 1

ACQUISITION PHASE

| Technology Development | Engineering & Manufacturing Development | Production & Deployment | Operations & Support |

UNITED STATES ARMY

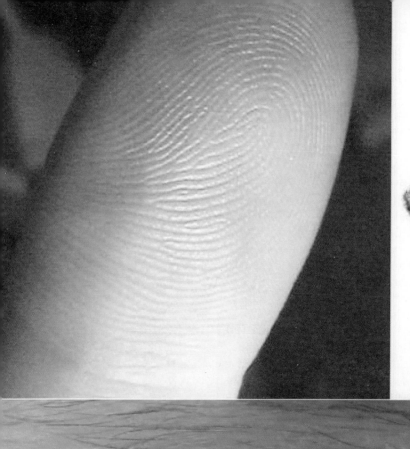

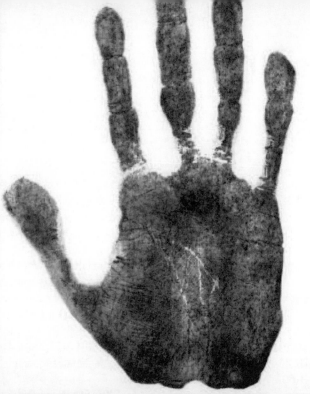

Biometric Enabling Capability (BEC)

FOREIGN MILITARY SALES
None

CONTRACTORS
Program Management Support Services:
CACI (Arlington, VA)
Strategic Support Services:
The Research Associates (New York, NY)
PM Technical Support:
MITRE (Alexandria, VA)
Systems Integration:
Northrop Grumman Information Systems
    (Fairmont, WV)

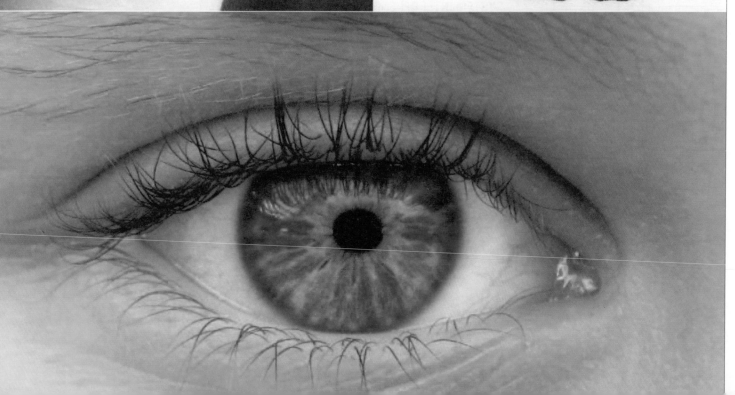

Black Hawk/UH/HH-60

INVESTMENT COMPONENT

Modernization

Recapitalization

Maintenance

MISSION

Provides air assault, general support, aeromedical evacuation, command and control, and special operations support to combat, stability, and support operations.

DESCRIPTION

The Black Hawk (UH-60) is the Army's utility tactical transport helicopter. The versatile Black Hawk has enhanced the overall mobility of the Army due to dramatic improvements in troop capacity and cargo lift capability. It will serve as the Army's utility helicopter in the Future Force.

There are multiple versions of the UH-60 Black Hawk: the original UH-60A; the UH-60L, which has greater gross weight capability, higher cruise speed, rate of climb, and external load; and the UH-60M, which includes the improved GE-701D engine and provides greater cruising speed, rate of climb, and internal load than the

UH-60A and L versions. During FY10 the Army decided to continue only with developmental testing of the UH-60M P3I Upgrade components, including Common Avionics Architecture System, fly-by-wire flight controls, and full authority digital engine control upgrade to the GE-701D Engine.

There are also dedicated Medical Evacuation (MEDEVAC) versions of the UH-60 Black Hawk: the HH-60A, HH-60L, and HH-60M each include an integrated MEDEVAC Mission Equipment Package (MEP) kit, providing day/night and adverse weather emergency evacuation of casualties.

On the asymmetric battlefield, the Black Hawk enables the commander to get to the fight quicker and to mass effects throughout the battlespace across the full spectrum of conflict. A single Black Hawk can transport an entire 11-person, fully equipped infantry squad faster than predecessor systems and in most weather conditions. The aircraft's critical components and systems are armored or redundant, and its airframe is

designed to crush progressively on impact, thus protecting crew and passengers. The UH-60M is a digital networked platform with greater range and lift to support maneuver Commanders through air assault, general support command and control, and aeromedical evacuation. Full-rate production for the new-build UH-60M began in 2007 and the UH-60M and HH-60M MEDEVAC aircraft continue to be deployed in combat rotations.

SYSTEM INTERDEPENDENCIES
In this Publication
None

Other Major Interdependencies
Blue Force Tracker (BFT)

PROGRAM STATUS
- **Current:** Production and fielding of UH-60M and HH-60M aircraft
- **FY12:** Multi-Year/Multi-Service VIII contract award
- **FY12:** Materiel Development Decision (MDD) for Improved Turbine Engine Program (ITEP)

PROJECTED ACTIVITIES
- **Continue:** Production and fielding of UH-60M and HH-60M aircraft
- **FY13:** Milestone A for ITEP

Black Hawk/UH/HH-60

FOREIGN MILITARY SALES
UH-60M:
Bahrain, Jordan, Mexico, United Arab
Emirates, Taiwan, Thailand, Sweden
UH-60L:
Brazil, Colombia, Egypt, Saudi Arabia,
Thailand

CONTRACTORS
UH-60M:
Sikorsky (Stratford, CT)
701D Engine:
General Electric (Lynn, MA)
Multi-Function Displays:
Rockwell Collins (Cedar Rapids, IA)
Flight Controls:
Hamilton Sundstrand (Windsor Locks, CT)

	UH-60A	UH60L	UH60M
MAX GROSS WEIGHT (pounds):	20,250	22,000	22,000
CRUISE SPEED (knots):	149	150	152
RATE CLIMB (feet per minute):	814	1,315	1,646
ENGINES (2 each):	GE-700	GE-701C	GE-701D
EXTERNAL LOAD (pounds):	8000	9,000	9,000
INTERNAL LOAD (troops/pounds):	11/2, 640	11/2, 640	11/3, 190
CREW:	two pilots, two crew chiefs		
ARMAMENT:	two 7.62mm machine guns		

Bradley Fighting Vehicle Systems Upgrade

INVESTMENT COMPONENT

Modernization

Recapitalization

Maintenance

MISSION

Provides infantry and cavalry fighting vehicles with digital command and control capabilities, significantly increased situational awareness, enhanced lethality and survivability, and improved sustainability and supportability.

DESCRIPTION

The Bradley M2A3 Infantry/M3A3 Cavalry Fighting Vehicle (IFV/CFV) features two second-generation, forward-looking infrared (FLIR) sensors—one in the Improved Bradley Acquisition Subsystem (IBAS), the other in the Commander's Independent Viewer (CIV). These systems provide "hunter-killer target handoff" capability with ballistic fire control. The Bradley A3 also has embedded diagnostics and an Integrated Combat Command and Control (IC3) digital communications suite hosting a Force

XXI Battle Command Brigade-and-Below (FBCB2) package with digital maps, messages, and friend/foe information. These systems provide the vehicle with increased shared battlefield situational awareness (SA). The Bradley's position navigation with GPS, inertial navigation, and enhanced squad SA includes a squad leader display integrated into vehicle digital images and IC3.

Speed: 40 mph
Range: 250 miles
Payload: 6,000 pounds
Vehicle weapons: 25mm, TOW II, 7.62mm
M2/M3A3 MMBF required/actual: 400/681
Deployable aircraft: C17, C5

SYSTEM INTERDEPENDENCIES
In this Publication
None

Other Major Interdependencies Army Battle Command System (ABCS), Blue Force Tracker (BFT), FM Voice-Advanced SINCGARS Improvement Program (ASIP) Radio, Forward Observer Systems (BFIST Only),

Ground Mobile Radio System (GMRS), System of Systems Common Operating Environment (SOSCOE)

PROGRAM STATUS
- **1QFY12:** Fielded Bradley A3s and BFIST with FS3 to 1st Brigade, 3rd Infantry Division and to APS-4 Korea, as well as A3s to TSS Korea
- **2QFY12:** Fielded Bradley A3s from LBE to 1st Brigade, 4th Infantry Division and to 4th Brigade, 1st Cavalry Division; Fielded ODS from LBE to 11 ACR
- **3QFY12:** Fielded A3s from LBE to 3rd and 2nd Brigade, 1st Cavalry Division; Fielded A3s and BFIST with FS3 to 2nd Brigade, 1st Infantry Division and to 116th HBCT, ID NG
- **4QFY12:** In process of fielding Bradley A3s and BFIST with FS3 to 155th HBCT, MS NG and to 1st Brigade, 1st Infantry Division and ODS-SAs to 278th HBCT, TN NG; Planning for fieldings to A3s to 1st Brigade, 1st Cavalry Division

PROJECTED ACTIVITIES
- **1QFY13:** Field Bradley A3s to 2nd Brigade, 4th Infantry Division from LBE
- **1QFY13:** Field Bradley A3s to 1st Brigade, 1st Cavalry Division some from LBE and 65 from production
- **2QFY13:** Field ODS-SAs to 1-34th
- **3QFY13:** Field ODS-SAs to 145 CAB
- **2QFY14:** Field ODS-SAs to 118th CAB and 55/28 HBCT
- **3QFY14:** Field ODS-SAs to 137th CAB

ACQUISITION PHASE

Technology Development

Engineering & Manufacturing Development

Production & Deployment

Operations & Support

Bradley Fighting Vehicle Systems Upgrade

FOREIGN MILITARY SALES
None

CONTRACTORS
Prime:
BAE Systems (York, PA; Santa Clara, CA)
DRS Technologies (Palm Bay, FL)
Raytheon (McKinney, TX)
L-3 Communications (Muskegon, MI)
Curtiss-Wright (Littleton, MA)
Elbit Systems of America (Fort Worth, TX)

Calibration Sets Equipment (CALSETS)

INVESTMENT COMPONENT

Modernization

Recapitalization

Maintenance

MISSION

Provides the capability to test, adjust, synchronize, repair, and verify the accuracy of Army test, measurement, and diagnostic equipment across all measurement parameters. Precision weapons require precision maintenance systems that only the calibration program provides.

DESCRIPTION

Calibration Sets Equipment (CALSETS) consists of calibration instrumentation housed in fixed facilities or contained within tactical shelters with accompanying power generation equipment. CALSETS provides support to maintenance units and area support organizations from brigade to multi-theater sustainment operations and ensures a cascading transfer of precision accuracy originating from the U.S. National Institute of Standards.

CALSETS is designed to plug into Army enterprise and battle networks. CALSETS tactical shelters are 100 percent mobile and transportable by surface mode or aircraft (C-130, C-5, and C-17). CALSETS is designed to calibrate 90 percent of the Army's test, measurement, and diagnostic equipment workload with an objective of 98 percent. CALSETS is configured in several set configurations.

Secondary Transfer Standards Basic, AN/GSM-286: This set consists of baseline instruments and components capable of supporting precision maintenance equipment in the physical, dimensional, electrical, and electronic parameters.

Secondary Transfer Standards Augmented, AN/GSM-287: This set consists of baseline instruments and augmented components with expanded capability to support a wider variety of precision maintenance equipment. It is capable of supporting precision maintenance equipment in the physical, dimensional, electrical, electronic, radiological, electro-optical, and microwave frequency parameters.

Secondary Transfer Standards, AN/GSM-705/440: This set configuration contains baseline instruments and augmented components designed for a tactical support mission. The platform applies a network-centric approach to precision maintenance support operations and data handling via an integrated data network, capable of sending calibration management system data to higher Army headquarters and obtaining calibration software updates. The set of instruments is contained in a 37-foot semi-trailer with a M1088A1 Medium Tactical Vehicle Tractor and an integrated 15-kilowatt power generator.

Secondary Transfer Standards, AN/GSM-421/439: This set is a subset of the baseline instruments designed to support up to 70 percent of the Army's high-density precision measurement equipment in forward areas. The system is modular and configurable to meet mission requirements and can operate in a true split-based mission posture. Designed for rapid deployment by surface or air, AN/GSM-421/439 set will not radiate or be disrupted by electromagnetic interference.

SYSTEM INTERDEPENDENCIES
None

PROGRAM STATUS
- **Current:** Sustainment of CALSETS Secondary Transfer Standards Basic, AN/GSM-286; Secondary Transfer Standards Augmented, AN/GSM-287; Secondary Transfer Standards, AN/GSM-421 and AN/GSM-705.
- **Current:** Fielding of CALSETS Secondary Transfer Standards, AN/GSM-705 to 756 Component Repair Company, Fort Hood, ARNG.
- **4QFY12:** AN/GSM-421A(v2) first unit equipped

PROJECTED ACTIVITIES
- **2QFY13:** AN/GSM-421A(v2) Logistics Demonstration
- **2QFY13-2QFY15:** AN/GSM-421A(v2) continues production and fielding

ACQUISITION PHASE

Technology Development | Engineering & Manufacturing Development | Production & Deployment | Operations & Support

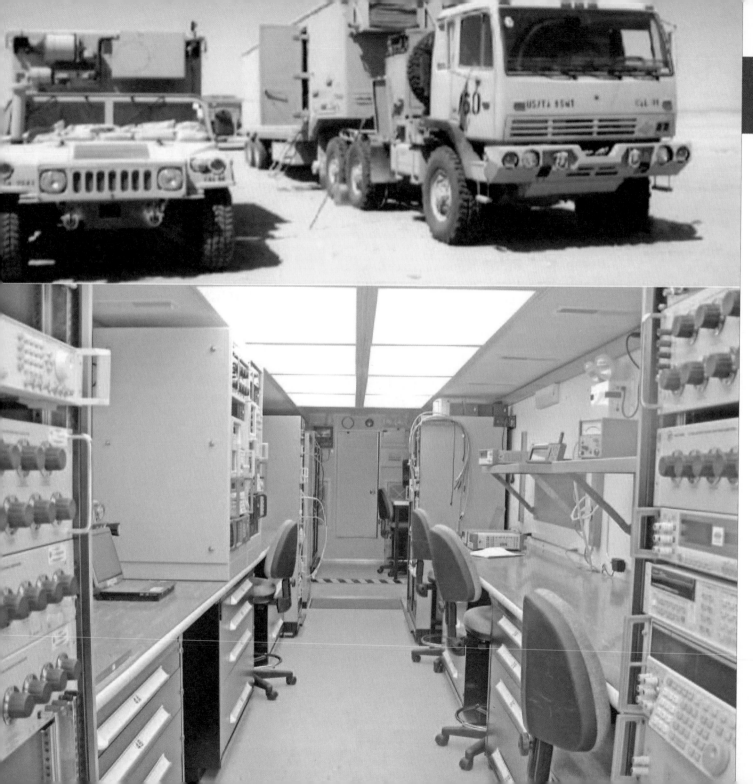

Calibration Sets Equipment (CALSETS)

FOREIGN MILITARY SALES
Afghanistan, Egypt, Japan, Lithuania, Saudi Arabia, Taiwan, United Arab Emirates

CONTRACTORS
Dynetics, Inc. (Huntsville, AL)
Agilent Technologies, Inc. (Santa Clara, CA)
Fluke Corp. (Everett, WA)

Capability Set 13 (CS 13)

INVESTMENT COMPONENT

Modernization

Recapitalization

Maintenance

MISSION

Delivers an unprecedented integrated network solution capable of supporting mission command requirements for the full range of Army operations, and an integrated voice and data capability throughout the entire Brigade Combat Team (BCT) formation.

DESCRIPTION

A Capability Set is an entire package of network components, associated equipment, and software that provides an integrated network capability from the static Tactical Operations Center (TOC) to the dismounted Soldier. Capability Set 13 (CS 13) is the first fully-integrated suite of network components fielded out of the Army's new Agile Process. CS 13 delivers an unprecedented integrated network solution capable of supporting mission command requirements for the full range of Army operations, and an integrated voice and data capability throughout the entire BCT formation.

As part of the Agile Capabilities Life Cycle Process, all Network Capability Sets will be operationally evaluated and technically integrated using a Heavy BCT (HBCT) conducting Network Integration Evaluations (NIEs) at Fort Bliss, TX, and White Sands Missile Range, NM. NIE brings Soldiers, developers, and engineers together in a realistic operational environment to guide the selection of mature capabilities for inclusion in Capability Sets.

CS 13 addresses 11 critical Operational Need Statements, giving commanders and Soldiers vastly increased abilities to communicate and share information. Enhancements include Mission Command on the Move, allowing leaders access to network capabilities found in TOCs while mounted in vehicles, and delivering the network to Soldiers at the squad level.

Capability Sets will be distributed throughout a combat formation and its supporting elements, from the brigade command post, to the commander on the move and the dismounted Soldier. This process, known as Capability Set Management, is a significant departure from the previous practice of fielding systems individually and often to only one element of the operational force at a time.

Following the fielding of CS 13, the Army will program the fielding of up to six BCT sets of network equipment per year for the FY14-18 Program Objective Memorandum (POM) in order to better synchronize its platform and network modernization efforts.

SYSTEM INTERDEPENDENCIES
In this Publication
Warfighter Information Network Tactical (WIN-T) Increment 2, Nett Warrior (NW), Joint Tactical Radio System Handheld, Manpack, Small Form Fit (JTRS HMS)

Other Major Interdependencies
Joint Capabilities Release Blue Force Tracker 2, Joint Battle Command–Platform (JBC-P), 117G (ANW2), Joint Tactical Radio System (JTRS) Rifleman Radio

PROGRAM STATUS
- **FY12-13:** Field CS 13 to 8 Infantry Brigade Combat Teams (OEF deployers and 8A Korea); Three brigades are deploying to OEF, three are next deployers, one is an 8th Army Korea forward deployed BCT and the last is the 2nd HBCT, 1st Armored Division (2/1 AD), which is the NIE brigade

PROJECTED ACTIVITIES
- **FY14-18:** Field up to six BCT sets of network equipment per year

ACQUISITION PHASE

| Technology Development | Engineering & Manufacturing Development | Production & Deployment | Operations & Support |

Capability Set 13 (CS 13)

FOREIGN MILITARY SALES
None

CONTRACTORS
To be determined

CH-47F Chinook

INVESTMENT COMPONENT

Modernization

Recapitalization

Maintenance

MISSION

Supports a full spectrum of operations including disaster relief, homeland defense and security, and current overseas contingency operations with a Future Force system design.

DESCRIPTION

The CH-47F is the Army's only heavy-lift cargo helicopter supporting critical combat and non-combat operations. The CH-47F aircraft has a suite of improved features such as an upgraded digital cockpit featuring the Common Avionics Architecture System (CAAS), a new monolithic airframe with vibration reduction, and the Digital Automatic Flight Control System (DAFCS), which provides coupled controllability for operations in adverse environments (reduced visibility, brown out, high winds). The CH-47F's common cockpit enables multiservice digital compatibility and interoperability for improved situational awareness, mission performance, and survivability, as well as future growth potential. The CH-47F has an empty weight of 24,578 pounds and a maximum gross weight

of 50,000 pounds. The CH-47F can lift intra-theater payloads up to 16,000 pounds in high/hot environments.

Max gross weight: 50,000 pounds
Max cruise speed: 160 knots
Troop capacity: 36 (33 troops plus three crew members)
Litter capacity: 24
Sling-load capacity: 26,000 pounds center hook, 17,000 pounds forward/aft hook, 25,000 pounds tandem
Minimum crew: three (pilot, copilot, and flight engineer)

SYSTEM INTERDEPENDENCIES
In this Publication
None

Other Major Interdependencies
ARC-231, BFT, CXP (APX-118), CXP (APX-123), IDM, AMPS

PROGRAM STATUS
- **2QFY07:** Completed initial operational testing
- **4QFY07:** First Unit Equipped
- **1QFY08:** Multi-year procurement contract awarded
- **3QFY12:** 11 units equipped

PROJECTED ACTIVITIES
- **FY13:** Multi-year II contract award
- **1QFY18:** Complete CH-47F fielding

Technology Development | Engineering & Manufacturing Development | Production & Deployment | Operations & Support

CH-47F Chinook

FOREIGN MILITARY SALES
None

CONTRACTORS
Aircraft and Recap:
Boeing (Philadelphia, PA)
Engine:
Honeywell (Phoenix, AZ)
Software:
Rockwell Collins (Cedar Rapids, IA)
Engine Controls:
Goodrich (Danbury, CT)

Chemical Biological Medical Systems (CBMS)–Prophylaxis

Modernization

Recapitalization

Maintenance

MISSION

Delivers safe, effective, and robust medical products that protect U.S. forces against validated chemical, biological, radiological, and nuclear (CBRN) threats by applying government and industry best practices to develop or acquire Food and Drug Administration (FDA) approved products.

DESCRIPTION

Chemical Biological Medical Systems (CBMS)–Prophylaxis consists of eight components:

Anthrax Vaccine Absorbed (AVA):
AVA is the only FDA-licensed anthrax vaccine in the United States that provides protection against cutaneous, gastrointestinal and aerosol infection by battlefield exposure to Bacillus anthracis.

Recombinant Plague Vaccine (rF1V): rF1V is a highly purified polypeptide produced from bacterial cells transfected with a recombinant vector from Yersinia pestis to prevent pneumonic plague.

Recombinant Botulinum Toxin Vaccine A/B (rBV A/B): rBV A/B is comprised of nontoxic botulinum toxin heavy chain fragments of serotypes A and B formulated with an aluminum hydroxide adjuvant and delivered intramuscularly prior to potential exposure to botulinum toxin.

Bioscavenger (BSCAV): The BSCAV program fills an urgent capability gap in the warfighter's defense against nerve agents by development of a nerve agent prophylactic that significantly reduces or eliminates the need for post-exposure antidotal therapy.

Smallpox Vaccine System (SVS):
The SVS Program provides both the ACAM2000™ smallpox vaccine and the Vaccinia Immune Globulin, Intravenous (VIGIV) to vaccinate and protect the warfighter from potential exposure to smallpox. Both products are FDA approved.

Filovirus Vaccine (Vac Filo):
The Vac Filo program covers an essential capability gap for protecting warfighters against aerosolized filovirus for which there is no current therapeutic. Target filovirus strains include Ebola Sudan, Ebola Zaire, and Marburg.

Ricin Vaccine (Vac Ricin): The Vac Ricin program will develop a vaccine against the A and B chains of this threat agent and validate performance against aerosolized material.

Western, Eastern Venezuelan Equine Encephalitis Vaccine (Vac WEVEE):
The Vac WEVEE program will develop a vaccine against three arboviruses with the goal of a single product protecting against all three threats.

SYSTEM INTERDEPENDENCIES
None

PROGRAM STATUS
- **1QFY12:** rBV A/B consistency lot manufacturing began
- **2QFY12:** Vac Ricin and Vac WEVEE Medical Device Directive (MDD)
- **3QFY12:** BSCAV MDD
- **3QFY12:** rF1V manufacturing process validation completed

PROJECTED ACTIVITIES
- **2QFY13:** Vac Ricin and Vac WEVEE Milestone A
- **3QFY13:** rFV1 Milestone C
- **4QFY13:** rBV A/B Milestone C
- **4QFY13:** rF1V Phase 3 clinical trial begins
- **4QFY13:** rBV A/B Milestone C
- **1QFY15:** rBV rF1V Initial Operational Capability (IOC)

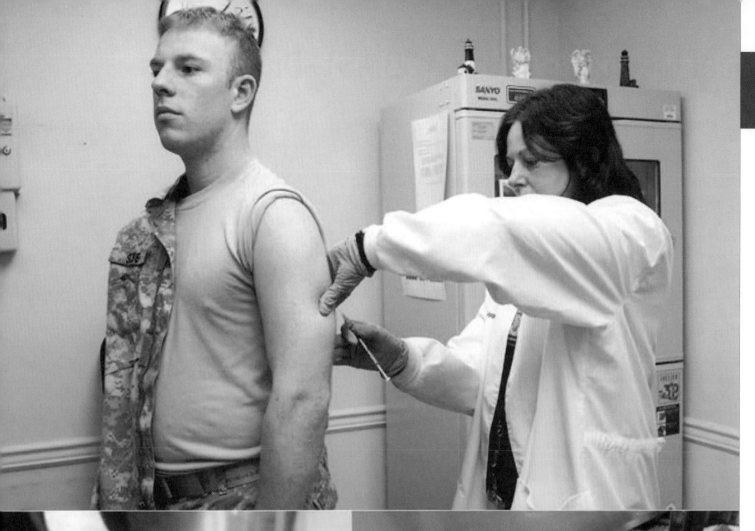

Chemical Biological Medical Systems (CBMS)–Prophylaxis

FOREIGN MILITARY SALES
Plague Vaccine:
Canada, United Kingdom

CONTRACTORS
AVA:
Emergent BioSolutions (Bioport)
  (Lansing, MI)
rF1V:
DynPort Vaccine (Frederick, MD)
rBV A/B:
DynPort Vaccine (Frederick, MD)
BSCAV:
In source selection
SVS:
Acambis PLC (Cambridge, MA)
Cangene, Corp. (Winnipeg, Manitoba,
  Canada)

Chemical Biological Medical Systems (CBMS)–Diagnostics

INVESTMENT COMPONENT

Modernization

Recapitalization

Maintenance

MISSION

Delivers safe, effective, and robust medical products that protect U.S. forces against validated chemical, biological, radiological, and nuclear (CBRN) threats by applying government and industry best practices to develop or acquire Food and Drug Administration (FDA) approved products.

DESCRIPTION

Components of the Next Generation Diagnostic System (NGDS) Family of Systems (FoS) will be deployed to all Roles (I through III) of combat health support for use during peace time and combat across the full spectrum of military operations. NGDS will be used within the continental U.S. (CONUS) and outside the continental U.S. (OCONUS) by forward deployed units, mobile hospitals and laboratories. NGDS is intended to mitigate the effects of personnel exposure to biological warfare agents and endemic infectious disease through rapid near real-time diagnostics.

SYSTEM INTERDEPENDENCIES

None

PROGRAM STATUS

- **Current:** NGDS – Materiel Solution Analysis

PROJECTED ACTIVITIES

- **2QFY14:** Conplete NGDS Materiel Solution Analysis

ACQUISITION PHASE

| Technology Development | Engineering & Manufacturing Development | Production & Deployment | Operations & Support |

Chemical Biological Medical Systems (CBMS)–Diagnostics

FOREIGN MILITARY SALES
None

CONTRACTORS
To be determined

Chemical Biological Medical Systems (CBMS)–Therapeutics

Modernization

Recapitalization

Maintenance

MISSION

Provides the warfighter and the nation with innovative medical solutions to protect against and treat emerging, genetically-engineered, or unknown biothreats.

DESCRIPTION

Chemical Biological Medical Systems (CBMS)–Therapeutics consists of the following components:

Advanced Anticonvulsant System (AAS): The AAS will consist of the drug midazolam in an autoinjector, which will replace the fielded Convulsant Antidote for Nerve Agents (CANA) that contains diazepam. Midazolam, injected intramuscularly, will treat seizures and prevent subsequent neurological damage caused by exposure to nerve agents.

AAS will not eliminate the need for other protective and therapeutic systems.

Improved Nerve Agent Treatment System (INATS): INATS is an enhanced treatment regimen against the effects of nerve agent poisoning. The new oxime component of INATS will replace 2-pyridine aldoxime methyl chloride (2-PAM) in the Antidote Treatment Nerve Agent Autoinjector (ATNAA). In addition U.S. Food and Drug Administration (FDA) approval will be obtained for use of pyridostigmine bromide (PB), the component of Soman Nerve Agent Pretreatment Pyridostigmine (SNAPP) for use against additional nerve agents.

Medical Radiation Countermeasure (MRADC): Acute Radiation Syndrome (ARS) manifests primarily as hematopoietic (bone marrow), gastrointestinal, and cerebrovascular sub-syndromes depending on the dose of radiation received. The portfolio of MRADC will, when used as a system, provide a robust capability to the warfighter. The current lead

MRADC, Protectan CBLB502, is under investigation for reducing the risk of death following whole body irradiation.

Hemorrhagic Fever Viruses (HFV): HFV medical countermeasures will mitigate the threat of illness or death, as well as lessen issues with performance degradation resulting from exposure to hemorrhagic fever viruses (Ebola and Marburg). Due to the general severity of these diseases, HFV therapeutics will be administered to infected warfighters while under direct medical observation.

Emerging Infectious Disease Influenza (EID Flu): EID Flu is an FDA-approved, broad-spectrum medical countermeasure that will protect against naturally occurring or biologically-engineered influenza viruses. This therapeutic will mitigate the threat of pandemic and drug resistant influenza viruses and will mitigate performance degradation issues associated with exposure to this organism.

SYSTEM INTERDEPENDENCIES
None

PROGRAM STATUS
- **1QFY12:** INATS Phase 1 Clinical Trial began
- **2QFY12:** Contract awarded for EID Flu
- **3QFY12:** MRADC expanded cooperative efforts with Biomedical Advanced Research and Development Authority (BARDA)

PROJECTED ACTIVITIES
- **1QFY13:** INATS Milestone B
- **1QFY13:** EID Flu Milestone B
- **1QFY13:** AAS Milestone C
- **3QFY13:** HFV Milestone B
- **3QFY14:** INATS initiates animal efficacy and clinical trials
- **4QFY14:** AAS Initial Operational Capability (IOC)

Chemical Biological Medical Systems (CBMS)–Therapeutics

FOREIGN MILITARY SALES
None

CONTRACTORS
AAS:
Meridian Medical Technologies
  (Columbia, MD)
Battelle Biomedical Research Center
  (Columbus, OH)
INATS:
Southwest Research Institute
  (San Antonio, TX)
Battelle Memorial Institute (Columbus, OH)
MRADC:
Osiris Therapeutics (Columbia, MD)

Chemical Biological Protective Shelter (CBPS) M8E1

INVESTMENT COMPONENT

Modernization

Recapitalization

Maintenance

MISSION

Provides medical personnel with a collective protection capability to perform their mission in a toxic free area without the encumbrance of individual protective equipment in the forward battle area.

DESCRIPTION

The Chemical Biological Protective Shelter (CBPS) M8E1 is a mobile, self-contained, rapidly deployable, chemically and biologically protected shelter that provides a contamination free, environmentally controlled medical treatment area for U.S. Army medical units. CBPS provides the operating crew with a Chemical, Biological and Radiological (CBR) protected Toxic Free Area (TFA) to execute their mission without the encumbrance of individual protective clothing/equipment. The 400 square foot CBR protected, decontaminable, air beam supported fabric shelter allows for rapid deployment and strike. It maintains an internal temperature of 60° F to 90° F in environments from 40° F to +125° F. It includes onboard primary and fully redundant auxiliary power for uninterrupted power during CBR medical operations. Armor equipped M1085A1R2 prime-mover provides crew protection during mobile/convoy operations. It provides 400 cubic-feet-per-minute of CBR-filtered air to maintain the TFA. CBPS will be assigned to trauma treatment teams/squads of maneuver battalions, medical companies of forward and division support battalions, nondivisional medical treatment teams/squads, division and corps medical companies, and forward surgical teams and can provide a dual- use medical capability for homeland defense.

SYSTEM INTERDEPENDENCIES

In this Publication

Family of Medium Tactical Vehicles (FMTV), Single Channel Ground and Airborne Radio System (SINCGARS), Joint Chemical Agent Detector (JCAD)

Other Major Interdependencies

The shelter system is integrated onto an armored MTV.

PROGRAM STATUS

- **1QFY12-4QFY12:** First Article Testing (FAT)

PROJECTED ACTIVITIES

- **2QFY13:** Type Classification/ Materiel Release (TC/MR) Decision
- **2QFY14:** Full-rate production

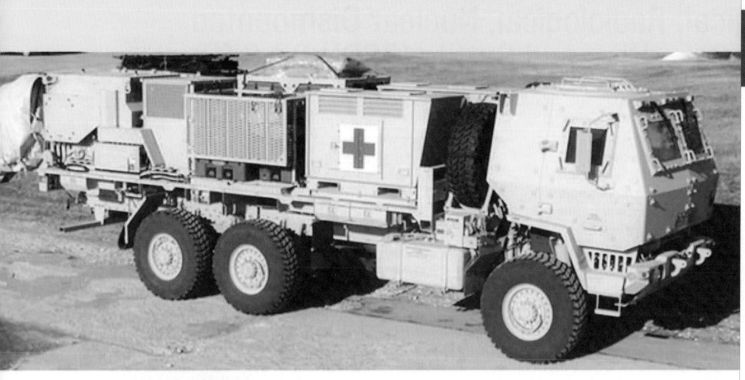

Chemical Biological Protective Shelter (CBPS) M8E1

FOREIGN MILITARY SALES
None

CONTRACTORS
Smiths Detection, Inc. (Edgewood, MD)

Chemical, Biological, Radiological, Nuclear Dismounted Reconnaissance Sets, Kits, and Outfits (CBRN DR SKO)

INVESTMENT COMPONENT

Modernization

Recapitalization

Maintenance

MISSION

Provides Chemical, Biological, Radiological, and Nuclear (CBRN) reconnaissance in confined spaces and terrain inaccessible by traditional CBRN reconnaissance mounted platforms/vehicles.

DESCRIPTION

The CBRN Dismounted Reconnaissance Sets, Kits, and Outfits system will consist of commercial and government off-the-shelf equipment that will provide detection, identification, sample collection, decontamination, marking, and hazard reporting of CBRN threats, as well as personnel protection from CBRN hazards.

CBRN DR SKO is composed of handheld, man-portable detectors that detect and identify potential Weapons of Mass Destruction (WMD) and/ or WMD precursors and determine levels of protection required to assess a sensitive site. The system supports dismounted reconnaissance, surveillance, and CBRN site-assessment missions to enable more

detailed CBRN information reports for commanders. These site locations may be enclosed or confined, and are therefore inaccessible by traditional CBRN reconnaissance-mounted platforms. CBRN site assessments help planners determine if more thorough analysis is required to mitigate risks or gather intelligence on adversaries' chemical warfare agents, biological warfare agents, or toxic industrial material capabilities.

From 2008 through 2012, 46 DR SKO-like systems were fielded in support of Joint Urgent Operational Needs Statements (JUONS) to U.S. Central Command and active Army units.

SYSTEM INTERDEPENDENCIES

None

PROGRAM STATUS

- **Current:** Developmental Testing

PROJECTED ACTIVITIES

- **1QFY13-2QFY13:** Multi Service operational testing
- **4QFY13:** Milestone C
- **4QFY13:** Full-rate production decision

Notional Equipment Set

Sabre 4000

CDS Kit Draeger Tubes

IdentiFINDER

Draeger AirBoss PSS 100 Plus

Kestrel 4500

AHURA First Defender

XM328 CBRN Marking Kit

Universal Pressure Kit

AN/UDR-14 RADIAC

Motorola XTS2500i

MultiRae Plus

HazMat ID

HazMaster G3

Yanman Generator 5.5kw

QuickSilver Analytics (QSA) 102

Individual Protective Equipment - Shower

Protective Gear with JCAD

Collapsible Cart

Bio Capture 650

FOREIGN MILITARY SALES
None

CONTRACTORS
FLIR Systems (Elkridge, MD)

Clip-on Sniper Night Sight (CoSNS), AN/PVS-30

INVESTMENT COMPONENT

Modernization

Recapitalization

Maintenance

MISSION

Enables the sniper to acquire and engage targets using the M110 Semi-Automatic Sniper System (SASS) and XM2010 Enhanced Sniper Rifle (ESR) during periods of limited visibility and at low-light levels.

DESCRIPTION

The AN/PVS-30 Clip-on Sniper Night Sight (CoSNS) is a lightweight, in-line weapon-mounted sight used in conjunction with the day optic sight on the M110 SASS and XM2010 (ESR). It employs a variable gain image intensification tube that can be adjusted by the sniper depending on ambient light levels. When used in conjunction with the M110 or XM2010 day optic sight, it provides for personnel-sized target recognition at quarter moon illumination in clear air to a range of 600 meters. The CoSNS has an integrated rail adapter that interfaces directly to the MIL-STD-1913 rail for quick and easy mounting to or dismounting from the weapon.

The CoSNS allows a sniper to maintain the current level of accuracy with the M110 and XM2010 and to deliver precise fire within one minute of angle. Use of the CoSNS does not affect the zero of the day optic sight and allows the M110 SASS and XM2010 to maintain bore sight throughout the focus range of the CoSNS and both the M110 and XM2010 day optic sights.

Weight: < 3.5 pounds
Focus range: 25 meters to infinity
Power: One AA battery

SYSTEM INTERDEPENDENCIES

None

PROGRAM STATUS

- **FY12:** Continued to field in accordance with Headquarters Department of the Army guidance

PROJECTED ACTIVITIES

- **FY13:** Continue to field in accordance with Headquarters Department of the Army guidance

ACQUISITION PHASE

Technology Development | Engineering & Manufacturing Development | Production & Deployment | Operations & Support

Clip-on Sniper Night Sight (CoSNS), AN/PVS-30

FOREIGN MILITARY SALES
None

CONTRACTORS
Knight's Armament Co. (Titusville, FL)

Close Combat Tactical Trainer (CCTT)

INVESTMENT COMPONENT

Modernization

Recapitalization

Maintenance

MISSION

Provides collective training for infantry, armor, mechanized infantry, and cavalry units, using manned module simulators to improve readiness and provide more realistic collective training.

DESCRIPTION

The Close Combat Tactical Trainer (CCTT) is a virtual, collective training simulator that allows Soldiers to operate in simulators representing dismounted infantry, mechanized infantry/tank, company teams, armored cavalry troops, or combat service support weapon systems. Crewed simulators, such as the Abrams Main Battle Tank, the Bradley Fighting Vehicle, and other vehicles, offer sufficient fidelity for collective mission training. Modular components include the Reconfigurable Vehicle Simulator/ Reconfigurable Vehicle Tactical Trainer

(RVS/RVTT), which simulates other vehicles. The newest simulator is the Dismounted Soldier Training System (DSTS), which enables individual Soldiers to train on infantry and improvised explosive device-defeat tasks in a fully immersive, virtual environment.

Soldiers simulate the battle direction of artillery, mortar, combat engineers, and logistics units to support their training mission. A semi-automated forces (SAF) workstation provides supporting units and all opposing forces. All battlefield operating systems are represented, ensuring an effective simulation of a combat environment that encompasses various conditions. CCTT's virtual terrain databases cover 100 by 150 kilometers, 3.5 kilometers of active visual terrain, and eight kilometers of extended range for M1A2 System Enhancement Program (SEP) Abrams tanks and M2A3 Bradley Fighting Vehicles. CCTT supports the training of both active Army and Army National Guard units at locations in the U.S., Europe, and South Korea.

SYSTEM INTERDEPENDENCIES

In this Publication
None

Other Major Interdependencies
CCTT requires Synthetic Environment Core (SE Core) to provide terrain databases and virtual models. The One Semi-Automated Force (OneSAF) will provide a common SAF through SE Core in the future.

PROGRAM STATUS

- **1QFY12:** Production and fielding of the RVTT to Fort Leonard Wood, MO
- **2QFY12:** Production and fielding of RVS/RVTT to Fort Jackson, SC; Camp Atterbury, IN; Camp Roberts, CA; Fort McCoy, WI; Fort Hood, TX; successful completion of the Developmental Test of the DSTS
- **2QFY12-4QFY12:** Production and fielding of RVTT to Fort Bragg, NC (Special Operations Forces); Gowen Field, ID; Camp Bullis, TX
- **3QFY12:** Award of the Post Deployment Software Support contract. Successful completion of User Assessment of DSTS at Fort Benning, GA. Approval to proceed with the DSTS fielding

- **4QFY12:** Award of the CCTT Concurrency Contract. Production and fielding of the DSTS to Fort Bragg, NC: Fort Leonard Wood, MO; Fort Bliss, TX; Fort Campbell, KY; Camp Bullis, TX; Fort Hood, TX; and Fort Lewis, WA

PROJECTED ACTIVITIES

- **1QFY13:** Production and fielding of the DSTS to Gowen Field, ID; Fort Pickett, VA; Fort McCoy, WI; Camp Grayling, FL; Camp Casey, ROK; Fort Stewart, GA; Fort Carson, CO; Fort Riley, KS; Fort Hunter Liggett, CA; Schofield Barracks, HI; Camp Atterbury, IN; Fort Knox, KY; and Fort Drum, NY
- **2QFY13:** Production and fielding of the DSTS to Fort Dix, NJ; Fort Indiantown Gap, PA; Fort Wainwright, AK; Fort Richardson, AK; Camp Shelby, MS; and Grafenwoehr, GDR
- **3QFY13-4QFY13:** Production and fielding of the CCTT Concurrency M1A2 SEP V2 to Fort Hood, TX and Gowen Field, ID. Production and fielding of the M2A3 Chassis Modernization Embedded Diagnostic (CMED) System to Gowen Field, ID

ACQUISITION PHASE

| Technology Development | Engineering & Manufacturing Development | Production & Deployment | Operations & Support |

Close Combat Tactical Trainer (CCTT)

FOREIGN MILITARY SALES
None

CONTRACTORS
RVTT:
Lockheed Martin Global Training and
   Logistics (Orlando, FL)
Dismounted Soldier Training System:
Intelligent Decisions (Ashburn, VA)
Post Deployment Software Support:
AVT (Orlando, FL)
Image Generator:
Rockwell Collins (Cedar Rapids, IA)
Dismounted Soldier Training System:
Quantum 3D (San Jose, CA)

Combat Service Support Communications (CSS Comms)

INVESTMENT COMPONENT

Modernization

Recapitalization

Maintenance

MISSION

Provides a worldwide commercial satellite communications network, engineering services, integrated logistics support, infrastructure, and portable remote terminal units in support of Army Combat Service Support (CSS) logistics management information systems operating from garrison or while deployed.

DESCRIPTION

Combat Service Support Communications (CSS Comms) includes the Combat Service Support Automated Information Systems Interface (CAISI) and the CSS Satellite Communications (CSS SATCOM) system. CAISI allows current/emerging battlefield CSS automation devices to electronically exchange information via tactical networks. CAISI provides unit commanders and logistics managers an interface device to support CSS doctrine for full-spectrum operations. CAISI allows deployed Soldiers to connect CSS automation devices to a secure wireless network and electronically exchange information via tactical or commercial communications. CAISI employs a deployable wireless LAN infrastructure linking Army logistics information system computers in a seven square-kilometer area. It is certified in accordance with Federal Information Processing Standards 140-2 Level 2. CSS SATCOM includes commercial-off-the-shelf, Ku-band, auto-acquire satellite terminals, called CSS Very Small Aperture Terminals (CSS VSATs), repackaged in fly-away transit cases, along with a fixed infrastructure of four primary and three Contingency Operating Point teleports and high-speed terrestrial links that provide a highly effective, easy-to-use, transportable, SATCOM-based solution to CSS nodes. CSS SATCOM supports information exchange up to the sensitive information level, is rapidly deployable anywhere in the world and is fully integrated into the Non-secure Internet Protocol Router Network (NIPRNET). CSS SATCOM eliminates the often dangerous need for Soldiers to hand-deliver requisitions via convoys in combat areas.

SYSTEM INTERDEPENDENCIES

In this Publication

Global Combat Support System–Army (GCSS-Army)

PROGRAM STATUS

- **1QFY12:** Provided support to the Global Combat Support System Army (GCSS-Army) Initial Operational Test and Evaluation (IOT&E)
- **2QFY12:** Awarded CSS SATCOM Network Services Contract
- **2QFY12:** Awarded contract for CSS SATCOM airtime
- **3QFY12:** Initial reset of Southwest Asia theater provided equipment
- **4QFY12:** Awarded contract to replace modem for CSS VSAT terminals

PROJECTED ACTIVITIES

- **1QFY13:** Award Integrated Network Operations Center Contract
- **4QFY14:** Complete fielding of CAISI systems

ACQUISITION PHASE

Technology Development | Engineering & Manufacturing Development | Production & Deployment | Operations & Support

Combat Service Support Communications (CSS Comms)

FOREIGN MILITARY SALES
None

CONTRACTORS
Equipment:
Telos Corp. (Ashburn, VA)
LTI DataComm Inc. (Reston, VA)
Juniper Networks (Herndon, VA)
L-3 Global Communications Solutions Inc. (Victor, NY)
Artel Inc. (Reston, VA)
Inmarsat Government (Herndon, VA)
Project support/training:
Systems Technologies (Systek) Inc. (West Long Branch, NJ)
Tobyhanna Army Depot (Tobyhanna, PA)
Software Engineering Center-Belvoir (SEC-B) (Fort Belvoir, VA)
U.S. Army Information Systems Engineering Command (USAISEC) (Fort Huachuca, AZ)
Defense Information System Agency (DISA) Satellite Transmission Services-Global NETCOM (Fort Huachuca, AZ)

Command Post Systems and Integration (CPS&I) Standardized Integrated Command Post Systems (SICPS)

INVESTMENT COMPONENT

Modernization

Recapitalization

Maintenance

MISSION

Provides commanders standardized and mobile command posts with a tactical, fully integrated, and digitized physical infrastructure to execute Networked enabled Mission Command (NeMC) and achieve information dominance.

DESCRIPTION

The Command Post Systems and Integration (CPS&I) Standardized Integrated Command Post Systems (SICPS) product office provides commanders with standardized, mobile, and fully integrated command posts for the modular expeditionary force, including support for Future Force capabilities and Joint and coalition forces. The SICPS-based command post is where commanders and their staffs collaborate, plan, and execute NeMC, maintain situational awareness using the Common Operational Picture (COP) and make decisions based on available information. Per the SICPS Capabilities Production Document (CPD), a family of Command Post Platforms (CPP) with Standardized Shelters, Command Center Systems (CCS), Command Post

Communications Systems (CPCS), and Trailer Mounted Support Systems (TMSS), is currently being fielded to the Army's active component, Army National Guard, and Army Reserve units.

SICPS provides the integrated NeMC platform and infrastructure to allow shared situational understanding of the COP based on the various Army and Joint command and control, communications, and network systems in the command post. Scalable and modular, SICPS supports echelons from battalion through Army Services Component Command providing tactical flexibility to support all phases of operations. By integrating the tactical internet with current and future mission command capabilities, command post operations are revolutionized through a combination of state-of-the-art data processing, communications, and information transport methods to achieve information dominance.

SYSTEM INTERDEPENDENCIES

In this Publication

Warfighter Information Network–Tactical (WIN-T), Distributed Common Ground System–Army (DCGS-A)

Other Major Interdependencies

Battle Command Common Services (BCCS) Server, Mobile Electric Power (MEP), Command Post of the Future (CPOF)

PROGRAM STATUS

- **1QFY12-4QFY12:** Fielded SICPS to 38 brigade level or higher units, 70 separate battalions and six Mission Command Training Centers

PROJECTED ACTIVITIES

- **2QFY13-1QFY15:** Continue SICPS fielding and New Equipment Training (NET) to 21 brigade level units, and 60 separate battalions and two Mission Command Training Centers

ACQUISITION PHASE

| Technology Development | Engineering & Manufacturing Development | Production & Deployment | Operations & Support |

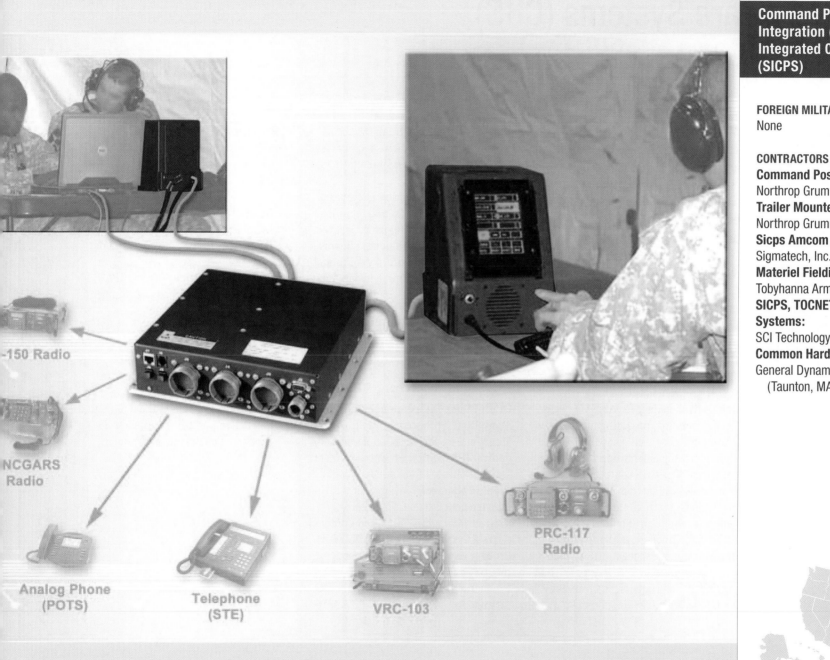

FOREIGN MILITARY SALES
None

CONTRACTORS
Command Post Platform/NET:
Northrop Grumman (Huntsville, AL)
Trailer Mounted Support System:
Northrop Grumman (Huntsville, AL)
Sicps Amcom Express (SETA):
Sigmatech, Inc. (Huntsville, AL)
Materiel Fielding:
Tobyhanna Army Deport (Tobyhanna, PA)
SICPS, TOCNET® Intercommunications Systems:
SCI Technology, Inc. (Huntsville, AL)
Common Hardware Systems:
General Dynamics C4 Systems, Inc. (Taunton, MA)

Common Hardware Systems (CHS)

INVESTMENT COMPONENT

Modernization

Recapitalization

Maintenance

MISSION

Provides state-of-the-art, fully qualified, interoperable, compatible, deployable, and survivable hardware for command, control, and communications at all echelons of command for the Army and other DoD services with world-wide repair, maintenance, and logistics support through contractor-operated CHS Regional Support Centers and management of a comprehensive warranty program.

DESCRIPTION

The Common Hardware Systems (CHS) program is the command and control enabler for Army Transformation, providing modularity, interoperability, and compatibility to support implementation of net-centricity. The CHS contract includes a technology insertion capability to continuously refresh the network-centric architectural building blocks, add new technology, and prevent hardware obsolescence. CHS products can be procured in four versions: version 1 (non ruggedized), version 1+ (moderate ruggedization of v1), version 2 (ruggedized), and versions 3 (fully rugged, military specification (MIL-SPEC) Rugged Handheld Unit).

CHS also provides worldwide repair, maintenance, logistics, and technical support through strategically located, contractor-operated regional support centers for tactical military units and management of a comprehensive five-year warranty.

SYSTEM INTERDEPENDENCIES

In this Publication

Warfighter Information Network–Tactical (WIN-T)

Other Major Interdependencies

Army Tactical Systems (ATS), Product Manager Fire Support Command and Control (PM FSC2), Product Manager Tactical Mission Command (PM TMC), Product Manager Training Integration Management System (PM TIMS), Product Manager Distributed Common Ground System–Army (PM DCGS-A), Product Manager Tactical Airspace Integration System (PM TAIS), Product Manager Terminal High-Altitude Area Defense (PM THAAD)

PROGRAM STATUS

- **FY12:** Establishing Repair Service Center in Bagram, Afghanistan.
- **FY12:** National Training Center (NTC)/Joint Readiness Training Center (JRTC)/Network Integration Evaluation (NIE) support at Fort Irwin, CA; Fort Polk, LA; and Fort Bliss, TX
- **FY12:** First Article Testing (FAT) for CHS-4 for eight pieces of equipment

PROJECTED ACTIVITIES

- **FY13:** High-Altitude Electro-Magnetic Pulse (HEMP) and Nuclear, Biological, and Chemical (NBC) testing
- **FY13-FY15:** Manage the acquisition and delivery of CHS equipment in support of customer requirements

ACQUISITION PHASE

| Technology Development | Engineering & Manufacturing Development | Production & Deployment | Operations & Support |

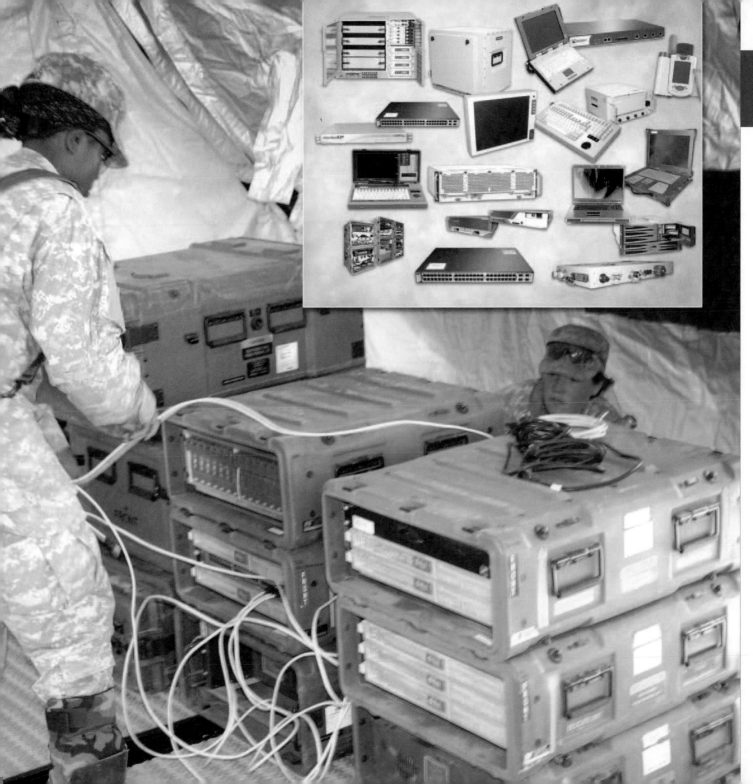

Common Hardware Systems (CHS)

FOREIGN MILITARY SALES
None

CONTRACTORS
CHS-3 Production Contract:
General Dynamics (Taunton, MA)
CHS-4 Production Contract:
General Dynamics (Taunton, MA)
Engineering:
Engineering Solutions and Products (ESP)
   (Oceanport, NJ)
CACI (Eatontown, NJ)
Sensor Technologies (Red Bank, NJ)
Logistics, Ordering:
Engineering Solutions and Products (ESP)
   (Oceanport, NJ)
Lab/Tech Support:
Northrop Grumman (Eatontown, NJ)
Consultant:
Sensor Technologies (Red Bank, NJ)

Common Remotely Operated Weapon Station (CROWS)

INVESTMENT COMPONENT

Modernization

Recapitalization

Maintenance

MISSION
Enables Soldiers to acquire and engage targets with precision while protected inside an armored vehicle.

DESCRIPTION
Common Remotely Operated Weapon Station (CROWS) is a stabilized mount that contains a sensor suite and fire control software, allowing on-the-move target acquisition and first-burst target engagement. Capable of target engagement under day and night conditions, the CROWS sensor suite includes a daytime video camera, thermal camera, and laser rangefinder. CROWS is designed to mount on any tactical vehicle and supports the MK19 Grenade Machine Gun, M2 .50 Caliber Machine Gun, M240B Machine Gun, and M249 Squad Automatic Weapon.

CROWS also features programmable target reference points for multiple locations, programmable sector surveillance scanning, automatic target ballistic lead, automatic target tracking, and programmable no-fire zones.

Potential enhancements include integration of other weapons, escalation-of-force systems, sniper detection, integrated 360-degree situational awareness, increased weapon elevation, Javelin integration, and commander's display.

SYSTEM INTERDEPENDENCIES
In this Publication
None

Other Major Interdependencies
CROWS mounts the MK19, M2, M240B, or M249 machine guns.

PROGRAM STATUS
- **2QFY12:** Initial CROWS Fixed Site fielding in support of Operation Enduring Freedom (OEF)
- **2QFY12:** Basis of Issue Plan (BOIP) approved for 11,269 systems
- **2QFY12:** Program designated Acquisition Category IC
- **3QFY12:** Full Materiel Release

PROJECTED ACTIVITIES
- **Ongoing:** Fielding against the BOIP
- **Ongoing-1QFY14:** Field and sustain CROWS in support of OEF
- **4QFY12:** Full-rate production decision
- **2QFY13:** Transition to organic field support

ACQUISITION PHASE

| Technology Development | Engineering & Manufacturing Development | Production & Deployment | Operations & Support |

UNITED STATES ARMY

Common Remotely Operated Weapon Station (CROWS)

FOREIGN MILITARY SALES
United Arab Emirates

CONTRACTORS
Prime:
Kongsberg Defense & Aerospace
   (Johnstown, PA)
Subcontractors:
Vingtech (Biddeford, ME)
BAE Systems (Austin, TX)
Smith's Machine (Cottondale, AL)
JWF Defense Systems (Johnstown, PA)

Computer Hardware, Enterprise Software and Solutions (CHESS)

Modernization

Recapitalization

Maintenance

MISSION

Serves as the Army's primary source in supporting Soldier's information dominance objectives by developing, implementing and managing commercial information technology contracts that provide enterprise-wide, net-centric hardware, software and support services for the Army.

DESCRIPTION

Computer Hardware, Enterprise Software and Solutions (CHESS) provides a no-fee flexible procurement strategy through which an Army user may procure commercial-off-the-shelf (COTS) information technology (IT) hardware, software and services that adhere to DoD and Army standards. CHESS contracts provide continuous vendor competition and consolidation of requirements to leverage the Army's buying power. CHESS is also the Army's Enterprise Software Initiative (ESI) Software Product Manager (SPM). Responsibilities in this area include: consolidating software requirements and business cases and assisting contracting officers in negotiating, and administering

the resulting agreements. CHESS offers the twice yearly Consolidated Buy (CB) with discounts of up to 62 percent off already discounted desktops, notebooks, tablets, slates and peripherals.

SYSTEM INTERDEPENDENCIES

In this Publication

None

Other Major Interdependencies

Army Chief Information Officer/G-6, Army Network Enterprise Technology Command/9th Signal Command, Information Systems Engineering Command, and Army Contracting Command-Rock Island partner with CHESS in implementing and promoting standards of COTS products Army-wide.

PROGRAM STATUS

- **2QFY12:** CB 14 yielded a cost avoidance of $7.1 million
- **3QFY12:** Oracle agreement consolidated 225 software agreements and provided unlimited use of 11 products. The cost avoidance is estimated to be between $10 million and $50 million
- **4QFY12:** Cost avoidance between $31 million - $65 million
- **4QFY12:** Pilot of a portal for management of software enterprise license agreements (ELAs)

PROJECTED ACTIVITIES

- **1QFY13:** Completion of portals for software ELAs
- **1QFY13:** Execution of software ELAs
- **4QFY13:** Information Technology Enterprise Solutions-3H (ITES-3H) which is a follow on the ITES-2H contract will offer state-of-the-art commercial IT equipment and related services
- **1QFY15:** Execution of software ELAs

Computer Hardware, Enterprise Software and Solutions (CHESS)

FOREIGN MILITARY SALES
None

CONTRACTORS
CHESS, in partnership with a host of large and small prime contract holders, subcontractors, resellers and original equipment manufacturers, provide architecturally sound, cost-effective, standards and policy-compliant IT enterprise solutions to all Army activities and organizations.

Counter Defilade Target Engagement (CDTE)–XM25

INVESTMENT COMPONENT

Modernization

Recapitalization

Maintenance

MISSION

Provides the Soldier with a "smart" revolutionary weapon system that breaks the current small arms direct fire parity and dramatically increases our forces' lethality and range with a family of 25mm programmable ammunition.

DESCRIPTION

The XM25 Counter Defilade Target Engagement (CDTE) enables the small unit and individual Soldier to engage defilade targets by providing a 25mm air bursting capability that can be used in all operational environments. The CDTE is an individually fired, semi-automatic, man-portable weapon system. An individual Soldier employing basic rifle marksmanship skills can effectively engage exposed or defilade targets in just seconds out to 700 meters.

The system allows the individual Soldier to quickly and accurately engage targets by producing an adjusted aimpoint based on range, environmental factors, and user inputs. The target acquisition/fire control integrates thermal capability with direct-view optics, laser rangefinder, compass, fuze setter, ballistic computer, and an internal display.

The CDTE System reduces the reliance of small units on non-organic assets (mortars, artillery, and air support) and the need to compete for priority of fires when time is critical. In addition to air bursting ammunition, a family of ammunition is being developed to support other missions, which could include armor-piercing and nonlethal scenarios.

SYSTEM INTERDEPENDENCIES

None

PROGRAM STATUS

- **1QFY11-4QFY11:** Prototype units engaged in Operation Enduring Freedom (OEF) Forward Operational Assessment (FOA)
- **2QFY11:** Capabilities Development Document approved
- **2QFY11:** Engineering & Manufacturing Development (EMD) contract awarded
- **3QFY11:** Integrated Baseline Review
- **3QFY11:** System Requirements Review

PROJECTED ACTIVITIES

- **2QFY11-2QFY14:** Conduct EMD phase
- **1QFY13:** Conduct Government Development Testing
- **2QFY13:** FOA continuation
- **4QFY13:** Milestone C decision

ACQUISITION PHASE

Technology Development

Engineering & Manufacturing Development

Production & Deployment

Operations & Support

UNITED STATES ARMY

Counter Defilade Target Engagement (CDTE)–XM25

FOREIGN MILITARY SALES
None

CONTRACTORS
Prime:
Alliant Techsystems (Plymouth, MN)
25mm Airburst Weapon:
H&K Gmbh (Oberdorf, Germany)
Target Acquisition/Fire Control:
L-3 Communications/Brashear
 (Pittsburgh, PA)
25mm Ammunition:
Alliant Techsystems (Plymouth, MN)

Countermine

Modernization

Recapitalization

Maintenance

MISSION

Provides Soldiers and maneuver commanders with a full range of countermine capabilities plus immediate solutions to counter improvised explosive devices (IEDs) and other explosive hazards, allowing the maneuver commander to achieve assured mobility on the battlefield.

DESCRIPTION

The Countermine product line comprises several different systems.

- The AN/VSS-6 Husky Mounted Detection System (HMDS) is a rapid acquisition system developed to support multiple Joint Urgent Operational Needs (JUONS). It is a ground penetrating radar (GPR) that provides the Vehicle Mounted Mine Detection (Husky) platform with the capability to detect and mark buried, low-metallic, and metallic-cased explosive hazards and their trigger mechanisms.
- The HMDS Program of Record (POR) enhances the JUONS GPR capability and adds both a deep buried detection sensor and semi-autonomous control capability.

- The AN/PSS-14 Mine Detecting Set is a POR hand held multi-sensor mine detector undergoing significant enhancements beginning in FY13.
- The VMR-2 Minehound is a rapid acquisition hand held multi-sensor explosive hazard detector developed to support multiple JUONS.
- The DSP-27 is a rapid acquisition hand held IED detection system developed to support multiple JUONS.
- The VMC-1 Gizmo is a rapid acquisition hand held metal detector developed to support multiple JUONS.

SYSTEM INTERDEPENDENCIES
In this Publication
None

Other Major Interdependencies
Mine Resistant Ambush Protected (MRAP) Vehicle (Husky)

PROGRAM STATUS
- **Current:** The AN/VSS-6 HMDS (JUONS), VMR-2 Minehound, DSP-27, and VMC-1 Gizmo are fielded in support of Overseas Contingency Operations
- **1QFY12:** The Product Management Office transitioned the Area

Mine Clearance System and the Interrogation Arm products to Program Executive Office Combat Support and Combat Services Support
- **2QFY12:** The Vehicle Optics Sensor Suite product transitioned to Program Executive Office Intelligence Electronic Warfare and Sensor
- **2QFY12:** The HMDS POR acquired a Materiel Development Decision and is preparing for a request for proposal in 4QFY12
- **4QFY12:** The AN-PSS-14 is undergoing an Engineering Change Proposal to add improvised explosive hazard detection capabilities

PROJECTED ACTIVITIES
- **2QFY13:** HMDS POR Increment A Milestone B
- **3QFY14:** HMDS POR Increment B Milestone B
- **4QFY14:** HMDS POR Increment A Milestone C
- **4QFY15:** HMDS POR Increment A Full-rate production (FRP)

Technology Development

Engineering & Manufacturing Development

Production & Deployment

Operations & Support

Countermine

FOREIGN MILITARY SALES
AN/VSS-6 HMDS:
Australia, Canada

CONTRACTORS
AN/PSS-14:
L-3 CyTerra Corp. (Waltham, MA;
  Orlando, FL)
AN/VSS-6 HMDS (JUONS):
NIITEK (Sterling, VA)
VMR-2 Minehound and VMC-1 Gizmo:
Vallon GmbH (Germany)
DSP-27 Goldie:
Gill Technology (United Kingdom)

Counter-Rocket, Artillery, Mortar (C-RAM) / Indirect Fire Protection Capability (IFPC)

INVESTMENT COMPONENT

Modernization

Recapitalization

Maintenance

MISSION

Integrates multiple Army and DoD managed systems and commercial off-the-shelf (COTS) systems with a command and control (C2) system, to provide protection of fixed and semi-fixed sites from rockets and mortar rounds.

DESCRIPTION

The Counter-Rocket, Artillery, Mortar (C-RAM) / Indirect Fire Protection Capability (IFPC) system-of-systems (SoS) was developed in response to a Multi-National Force-Iraq Operational Needs Statement (ONS) validated in September 2004. An innovative SoS approach was implemented in which multiple DoD acquisition program systems were integrated with COTS items to provide seven C-RAM functions: sense, warn, respond, intercept, C2, shape, and protect.

C-RAM component systems include the following: Forward Area Air Defense Command and Control (FAAD C2) and Air and Missile Defense Workstation (AMDWS) for C2; Lightweight Counter Mortar Radars (LCMR), Firefinder Radars, and Ka-band Multi-Function Radio Frequency Systems (MFRFS) for sense; Land-based Phalanx Weapon System (LPWS) and emerging Accelerated Improved Intercept Initiative (AI3) for intercept; Wireless warn; and a wireless Local Area Network (LAN) for comms integration of all components. Response is provided thru C-RAM integration with Army/Joint mission command systems.

C-RAM sense and warn is currently deployed to forward operating bases (FOBs) in Iraq in support of Department of State/Office of Security Cooperation-Iraq (DoS/OSC-I) operations and in Afghanistan in support of Operation Enduring Freedom (OEF). C-RAM's sense and warn performance has been extremely successful, providing timely warning for more than 2,500 rocket and mortar attacks against C-RAM equipped FOBs, with minimum false warnings. Current ONS-based capability

enhancements include development/deployment of Ka and Ku-band MFRFS sensors for detection of high/low quadrant elevation rockets and improvised rocket assisted munitions, and an enhanced interceptor for improved mobility and extended range.

SYSTEM INTERDEPENDENCIES
In this Publication
None

Other Major Interdependencies
Army and Marine Corps Battle Command Systems, Sentinel Radar

PROGRAM STATUS
- **Current:** Sustainment of fielded C-RAM SoS capability in OEF and Iraq
- **1QFY12:** Completed C-RAM sense and warn fielding to OEF
- **1QFY12:** Initial Ka MFRFS declared fully operational in OEF
- **2QFY12:** Initiated support for C-RAM sense and warn for DoS/OSC-I
- **2QFY12:** Approval of a Directed Requirement for the AI3
- **4QFY12:** Ka MFRFS fielding complete

PROJECTED ACTIVITIES
- **1QFY13:** Deploy initial Ku MFRFS to OEF
- **1QFY14:** Conduct AI3 live fire testing

ACQUISITION PHASE

| Technology Development | Engineering & Manufacturing Development | Production & Deployment | Operations & Support |

UNITED STATES ARMY

Counter-Rocket, Artillery, Mortar (C-RAM) / Indirect Fire Protection Capability (IFPC)

FOREIGN MILITARY SALES
Australia, United Kingdom

CONTRACTORS
Hardware/Integration/Fielding/ Contractor Logistics Support:
Northrop Grumman (Huntsville, AL)
Land-based Phalanx Weapon System (LPWS):
Raytheon Missile Systems (Tucson, AZ)
Software Development/Maintenance:
Northrop Grumman (Redondo Beach, CA)
Lightweight Counter Mortar Radars (LCMR):
SRCTec Inc. (Syracuse, NY)
Common Hardware, Software:
General Dynamics (Taunton, MA)

Cryptographic Systems

Modernization

Recapitalization

Maintenance

MISSION

Provides Army users strategic and tactical advantages through communication security (COMSEC) superiority by modernizing and fielding cryptographic equipment and systems that protect against cyber threats, increase battlefield survivability/lethality, and enable critical mission command activities.

DESCRIPTION

Cryptographic Systems procures and fields solutions to secure the National Network Enterprise. New and emerging architectures are driving the need to replace current inventory of stove pipe systems with technologically advanced [network centric/Global Information Grid (GIG) compliant)] devices that incorporate Chairman of the Joint Chiefs of Staff and Joint Requirements Oversight Council directed cryptographic modernization, advanced

key management and network centric performance capabilities. This program enables DoD to equip the force with critical cryptographic solutions and services during peacetime, wartime, and contingency operations.

The In-Line Network Encryptor (INE) family of network encryption devices provides network communications security in support of the movement to Everything over Internet Protocol (EoIP). These systems are used in both tactical and strategic networks and support multiple bandwidth configurations.

The Link & Trunk Encryptor Family (LEF) is used to multiplex and encrypt numerous signals into wideband data streams to be transmitted over fiber, cable, or satellites. The wide-band circuits require systems with rapid encryption capabilities.

Finally, the Secure Voice (SV) family uses security tokens and/or public key encryption to provide secure voice communication. There is a drive

towards substitution in preference from wide-bandwidth to narrow-bandwidth communication channels.

Cryptographic Systems is modernizing legacy devices in order to accomplish cryptographic standardization of the Network for the Army, effectively countering the emerging cyber threat while supporting decisive full spectrum operations.

The Army-wide Cryptographic Network Standardization (ACNS) project, a multi-year effort which commenced in May 2012, ensures that legacy items with NSA mandated cease-key dates are replaced with fully modernized COMSEC equipment in support of the Army Modernization Strategy and Plan.

SYSTEM INTERDEPENDENCIES
In this Publication
None

Other Major Interdependencies
Cryptographic Systems are considered enabling systems which provide required COMSEC capabilities.

PROGRAM STATUS
- **1QFY12:** Started depot balancing efforts providing operational cost avoidance to PMs/Units
- **2QFY12:** Accelerated process of obtaining Standard Line Item Numbers (LINs)
- **3QFY12:** Decreased time in depot receiving process from 57 to 20 days
- **3QFY12:** Commencement of ACNS effort

PROJECTED ACTIVITIES
- **FY13-FY14:** Pursue and execute upon the Army Airborne Secure Voice requirement
- **FY13-FY15:** Continue modernization of INE, LEF and Secure Voice devices
- **FY13-FY15:** Continue the ACNS effort

Technology Development

Engineering & Manufacturing Development

Production & Deployment

Operations & Support

Cryptographic Systems

FOREIGN MILITARY SALES
None

CONTRACTORS
General Dynamics Communication
   Systems (Needham, MA)
Harris Corp. (Palm Bay, FL)
L3 Communications (Camden, NJ)
SafeNet (Columbia, MD)
ViaSat (Carlsbad, CA)

Defense Enterprise Wideband SATCOM System (DEWSS)

INVESTMENT COMPONENT

Modernization

Recapitalization

Maintenance

MISSION

Provides combatant commanders, deployed warfighters, and senior leadership with secure, high-capacity satellite connectivity, enabling reachback for voice, video, and data communications and transfer of intelligence information.

DESCRIPTION

The Defense Enterprise Wideband SATCOM System (DEWSS) provides strategic Army and DoD satellite communications (SATCOM) infrastructure, enabling national and senior leader communications; Joint Chiefs of Staff-validated command, control, communications, and intelligence (C3I) requirements; tactical reachback to sustaining base for deployed warfighters; and transport for critical intelligence information transfer to deployed forces worldwide. DEWSS is modernizing the enterprise satellite terminals, baseband systems, and payload and network control systems required to support warfighters use of the high-capacity Wideband Global SATCOM (WGS) satellite constellation, which

DoD launched in October 2007. DEWSS capabilities include super high frequency (SHF), beyond-line-of-sight communications; tactical reachback via DoD Teleport and Standardized Tactical Entry Point (STEP) sites; survivable communications for critical nuclear command and control; and an anti-jam, High-Altitude Electromagnetic Pulse (HEMP) hardened, anti-scintillation capability for key strategic forces. Management capabilities include the Common Network Management System (CNPS), Wideband Global Spectrum Monitoring System (WGSMS), Wideband Remote Monitoring Sensor (WRMS), Remote Monitoring and Control Equipment (RMCE), Joint Management Operations System (JMOS), and the Replacement Frequency Management Orderwire (RFMOW).

SYSTEM INTERDEPENDENCIES

None

PROGRAM STATUS

- **1QFY12:** Began Modernization of Earth Terminals (MET) First Article Test (FAT) and RMCE DoD Information Assurance Certification and Accreditation Process certification
- **2QFY12:** Began DEWSS transmission security modernization; MET large fixed FAT (non-HEMP) installation; WGS-4 launch and RMCE installation at Wahiawa/ Wideband SATCOM Operational Centers (WSOCs); began WSOC upgrade
- **3QFY12:** RMCE onsite acceptance test/on orbit test and RMCE Phase I installation in East/West Australia
- **4QFY12:** MET FAT terminal HEMP installation

PROJECTED ACTIVITIES

- **1QFY13:** Support WGS-5 launch
- **1QFY13:** WRMS installation
- **2QFY13:** MET FAT and commissioning, begin earth terminal relocation/terminal installation; conduct WRMS installation
- **3QFY13:** Second MET installation/ earth terminal relocation
- **4QFY13:** MET installation, terminal installation, and RMCE system integration test

Technology Development | Engineering & Manufacturing Development | **Production & Deployment** | Operations & Support

Defense Enterprise Wideband SATCOM System (DEWSS)

FOREIGN MILITARY SALES
None

CONTRACTORS
Johns Hopkins University Applied Physics
    Laboratory (Laurel, MD)
Northrop Grumman (Winter Park, FL)
ITT (Colorado Springs, CO)
Harris Corp. (Melbourne, FL)
Computer Sciences Corp. (Eatontown, NJ)

Distributed Common Ground System–Army (DCGS-A)

INVESTMENT COMPONENT

Modernization

Recapitalization

Maintenance

MISSION

Provides distributed intelligence, surveillance, and reconnaissance (ISR) planning, management, control, and tasking; multi-intelligence fusion; and robust Joint, allied, and coalition forces interoperability.

DESCRIPTION

The Distributed Common Ground System–Army (DCGS-A) provides unprecedented timely, relevant, and accurate targetable data to the warfighter. DCGS-A will be fully interoperable with the Army's Unified Mission Command System and provide access to information and intelligence to support battlefield visualization and ISR management in accordance with the Army Common Operating Environment. It provides a flattened network enabling information discovery, collaboration, production, and dissemination to commanders and staffs in seconds and minutes versus hours and days. This system enables the commander to achieve situational understanding by leveraging multiple sources of data, information, and intelligence, and to synchronize Joint and combined arms combat power to see, understand and act first, and finish decisively.

DCGS-A will incrementally assume lifecycle management responsibility and consolidate/replace the operational capabilities provided by several post-Milestone C Programs of Record (PORs) and fielded quick reaction capabilities. The Army will produce and field DCGS-A capability on various hardware (HW) platforms using a consolidated DCGS-A Software Baseline (DSB). HW platforms will range from single laptops to multi-server transportable configurations to large cloud-based computing nodes able to process and store the enormous volumes of data that DCGS-A must manage. DCGS-A's modular, open systems architecture and heavy emphasis on "design for change" allows rapid adaptation to changing circumstances. DCGS-A will support three primary roles: DCGS-A enables the user to collaborate, synchronize, and integrate organic and non-organic direct and general-support collection elements with operations; DCGS-A can discover and use all relevant threat, noncombatant, weather, geospatial, and space data and evaluate technical data and information on behalf of a commander; and DCGS-A provides organizational elements the ability to control select sensor platforms/payloads and process the collected data.

DCGS-A leverages commercial products from both large and small businesses creating a level playing field for industry through an open architecture design. DCGS-A is capable of mutli-intelligence processing and is built to intelligence community framework standards.

SYSTEM INTERDEPENDENCIES
In this Publication
Battle Command Sustainment Support System (BCS3), Distributed Common Ground System–Army (DCGS-A), Enhanced Medium Altitude Reconnaissance and Surveillance System (EMARSS), Extended Range/Multiple Purpose (ER/MP) Unmanned Aircraft System (UAS), Guardrail Common Sensor (GR/CS)

Other Major Interdependencies
DCGS Family of Systems (Services), Global Information Grid (GIG)

PROGRAM STATUS
- **1QFY12:** DCGS-A DSB 1.0 Milestone C
- **2QFY12:** DCGS-A DSB 1.0 Initial Operational Test & Evaluation
- **4QFY12:** DCGS-A DSB 1.0 Full Deployment Decision

PROJECTED ACTIVITIES
- **1QFY13:** Complete DCGS-A DSB 1.1 Build
- **FY13:** Field DCGS-A DSB 1.1
- **FY13:** Complete DCGS-A DSB 1.2 Build
- **FY14:** Field DCGS-A DSB 1.2
- **FY14:** Complete DCGS-A DSB 1.3 Build
- **FY15:** Field DCGS-A DSB 1.3

ACQUISITION PHASE

Technology Development

Engineering & Manufacturing Development

Production & Deployment

Operations & Support

Distributed Common Ground System–Army (DCGS-A)

FOREIGN MILITARY SALES
None

CONTRACTORS
System Integration and Design:
Northrop Grumman (Linthicum, MD)
Software Engineering:
Azimuth Inc. (Morgantown, WV)
All Source Integration:
Lockheed Martin (Denver, CO)
GMTI Integration:
General Dynamics (Scottsdale, AZ)
Program Support:
CACI (Tinton Falls, NJ)
Engineering Support:
MITRE (Eatontown, NJ)
Battle Command Integration and Interoperability:
OverWatch Systems (Austin, TX)
Program Support, System Engineering & Architecture:
Booz Allen Hamilton (Eatontown, NJ), MITRE
(Eatontown, NJ)
DCGS Integrated Backbone (DIB):
Raytheon (Garland, TX)
Other Support:
NetApp (CA), Cloudera (CA), Vmware (CA), Esri
(CA), Tucson Embedded Systems (AZ), L3 Comm
(AZ), Dell (TX), Potomac Fusion (TX), Overwatch
(TX), Ringtail Design (TX), Redhat (NC), Digital
Reasoning (TN), IBM (MD), NetCentric (VA), Pixia
(VA), Data Tactics (VA), HP (CA), Cogility (CA),
EMC2 (VA), CSC (MD), Informatica (CA)

Distributed Learning System (DLS)

INVESTMENT COMPONENT

Modernization

Recapitalization

Maintenance

MISSION
Acquires, deploys, and maintains a worldwide distributed learning system to ensure our nation's Soldiers receive critical training for mission success.

DESCRIPTION
The Distributed Learning System (DLS) provides a worldwide information technology infrastructure that innovatively combines hardware, software, and telecommunications resources with training facilities and web-based applications to electronically deliver course content for training of Soldiers and Department of the Army (DA) Civilians anytime, anywhere. DLS uses technology to increase training efficiencies, increase individual and unit readiness, support Soldiers' career advancement, and improve their quality of life.

DLS provides:
- Access to Army e-Learning, web-based training consisting of more than 5,400 commercial businesses and information technology courses

- Globally located Digital Training Facilities (DTFs) capable of delivering multimedia courseware for individual or group training via computer or video tele-training (VTT)
- Enterprise management of the DLS infrastructure, with customer support for training applications
- The Army Learning Management System (ALMS), for web-based delivery of more than 1,500 multimedia training courses and streamlined, automated training management functions
- Deployed Digital Training Campuses (DDTCs) to deliver multimedia courseware to Soldiers deployed in-theater

SYSTEM INTERDEPENDENCIES
In this Publication
None

Other Major Interdependencies
Army Knowledge Online/Defense Knowledge Online (AKO/DKO) and Army Training Requirements and Resources System (ATRRS)

PROGRAM STATUS
- **1QFY12:** Upgraded ALMS to ALMS 3.1 software
- **1QFY12-4QFY12:** Delivered DDTC to 24 brigade combat teams deploying in Germany, Kuwait, Iraq, and Afghanistan
- **1QFY12-4QFY12:** Delivered full-time-equivalent training to more than 300,000 Soldiers using DTFs
- **1QFY12-4QFY12:** 10 million ALMS registrations
- **1QFY12-4QFY12:** 5.5 million Course completions by ALMS users
- **3QFY12:** Completed installation of Virtual Battle Space 2 simulations in 73 DTF sites
- **4QFY12:** Produced two DDTCs

PROJECTED ACTIVITIES
- **1QFY13:** Transition ALMS to Acquisition, Logistics and Technology Enterprise Systems and Services Data Center
- **4QFY13:** Full DDTC deployment
- **4QFY14:** Complete production of DDTC

Technology Development | Engineering & Manufacturing Development | Production & Deployment | Operations & Support

Distributed Learning System (DLS)

FOREIGN MILITARY SALES
None

CONTRACTORS
IBM (Fort Eustis, VA)
CRGT (Fort Eustis, VA)
Saba (Fort Eustis, VA)
Lockheed Martin (Fort Eustis, VA)
Engility (Newport News, VA)
CACI (Fort Eustis, VA; Newport
  News, VA)
eScience & Technology Solutions Inc.
  (Fort Eustis, VA; Newport News, VA)
Bryant & Associates (Newport News, VA)
N-Link (Bend, OR)
CRGT (Washington, DC)
EIBOT (Peachtree City, GA)
JLMI (Petersburg, VA)
MILVETS Systems Technology Inc.
  (Orlando, FL)
Skillsoft Corp. (Nashua, NH)

Dry Support Bridge (DSB)

INVESTMENT COMPONENT

Modernization

Recapitalization

Maintenance

MISSION

Supports military load classification 100 (wheeled)/80 (tracked) vehicles over 40-meter gaps via a modular military bridge.

DESCRIPTION

The Dry Support Bridge (DSB) is a mobile, rapidly erected, modular military bridge system. DSB is fielded to Multi-Role Bridge Companies (MRBCs) and requires a crew of eight Soldiers to deploy a 40-meter bridge in fewer than 90 minutes (daytime). DSB sections have a 4.3-meter road width and can span a 40-meter gap or two 20-meter gaps at military load classification (MLC) 100 (wheeled)/80 (tracked) normal crossing and MLC 110 (W) caution crossing. The system includes a DSB bridge, a launcher mounted on a dedicated Palletized Load System (PLS) chassis that deploys the modular bridge sections, and seven M1077 Flatracks to transport the bridge sections. The bridge modules are palletized onto seven flat racks and transported by equipment organic to the MRBC. DSB is designed to replace the M3 Medium Girder Bridge.

DSB modular structure allows launch and retrieval from either end without a dedicated or special training area and can be placed directly over pavement to reinforce damaged sections, bridges, or spans. Air transport for the DSB system is accomplished by C-130 if divided (bridge: one flat-rack per a/c; launcher vehicle: split into three loads, five hours work), or by C-17 and C-5 intact.

This system has been fielded since 2003.

SYSTEM INTERDEPENDENCIES
In this Publication
None

Other Major Interdependencies
DSB operations rely and are interdependent upon fully mission-capable M1977 Common Bridge Transporters and M1076 PLS trailer assets within a fully Modified Table of Organization and Equipment-equipped MRBC.

PROGRAM STATUS
- **2QFY12:** Fielded 50th MRBC
- **2QFY12:** Fielded 341st MRBC
- **3QFY12:** Fielded 250th MRBC
- **4QFY12:** Fielding 401st MRBC

PROJECTED ACTIVITIES
- **2QFY13:** Field 189th MRBC
- **4QFY13:** Field 2225th MRBC
- **2QFY14:** Field 132nd MRBC
- **3QFY14:** Field 551st MRBC
- **1QFY15:** Field 361st MRBC

ACQUISITION PHASE

Technology Development | Engineering & Manufacturing Development | Production & Deployment | Operations & Support

Dry Support Bridge (DSB)

FOREIGN MILITARY SALES
None

CONTRACTORS
Manufacturer:
Williams Fairey Engineering, Ltd.
(Stockport, United Kingdom)
PLS Chassis:
Oshkosh Corp. (Oshkosh, WI)
Logistics:
XMCO (Warren, MI)

Enhanced Medium Altitude Reconnaissance and Surveillance System (EMARSS)

INVESTMENT COMPONENT

Modernization

Recapitalization

Maintenance

MISSION

Provides a persistent multi-intelligence capability to detect, locate, classify/identify, and track surface targets in day/night, near-all-weather conditions with a high degree of timeliness and accuracy.

DESCRIPTION

The Enhanced Medium Altitude Reconnaissance and Surveillance System (EMARSS) contributes to filling critical gaps in the Airborne Intelligence, Surveillance and Reconnaissance (AISR) coverage Brigade Combat Teams (BCTs) require to be successful across the Range of Military Operations and especially in Irregular Warfare (IW) operations. The capabilities include an electro-optical/infrared high definition full motion video sensor, communications intelligence (COMINT) sensor, and an

aerial precision geo-location (APG) sensor, all supported by line-of-sight and beyond line-of-sight (LOS/BLOS) communications and hosted on a manned, medium-altitude derivative of the commercial Hawker-Beechcraft King Air 350ER aircraft.

EMARSS contains a tailored set of Distributed Common Ground System–A (DCGS-A) enabled software and ISR processing software functionalities to process, exploit, and rapidly disseminate the intelligence derived from the imagery sensor. The imagery and APG operators release time-sensitive information directly to the supported BCT and subordinate units, and to the DCGS-A, enabling tactical ground forces to operate at their highest potential. EMARSS complies with the DoD Information Technology Standards Registry and the Defense Information Systems Network (DISN). This architecture permits interoperability with any multi-service or Joint system that complies with DoD-standard formats for data transfer and dissemination.

EMARSS is an improvement over the existing MARSS in that it hosts an on board DCGS-A capability, improved satellite communications, improved aircraft performance, and life cycle logistics sustainment capability.

SYSTEM INTERDEPENDENCIES
In this Publication
Distributed Common Ground System–Army (DCGS-A)

Other Major Interdependencies
DCGS-A is the EMARSS ground station supporting pre-mission, mission, and post-mission operations. DCGS-A software is on-board the EMARSS work stations and will be updated as DCGS-A provides incremental software builds.

PROGRAM STATUS
- **1QFY12:** Completed System Design Review
- **1QFY12:** Conducted Integrated Baseline Review
- **2QFY12:** Took Receipt of "Green" Aircraft

- **3QFY12:** Completed Engineering Risk Reduction Prototype (ERRP) Aircraft

PROJECTED ACTIVITIES
- **2QFY13:** Contractor Testing and Developmental Testing (CT/DT)
- **3QFY13:** Joint Requirements Oversight Council consideration of the CPD
- **3QFY13:** DD250 of EMARSS EMD Systems to Government
- **4QFY13:** Limited User Testing (LUT) and Logistics Demo
- **4QFY13:** Milestone C

Enhanced Medium Altitude Reconnaissance and Surveillance System (EMARSS)

FOREIGN MILITARY SALES
None

CONTRACTORS
Prime:
The Boeing Co. (Ridley Park, PA)
Airframe:
Hawker-Beechcraft (Wichita, KA)
SATCOM:
L-3 Communications West (Salt Lake City, UT)
COMINT Hardware/Software:
BAE Systems (Nashua, NH)
Training and Operational Testing:
Avenge (Dulles, VA)
Cockpit Avionics:
Rockwell Collins (Cedar Rapids, IA)
SETA Support:
CACI (Tinton Falls, NJ)
Booz Allen Hamilton (Eatontown, NJ)
Engineering/Program Management:
MITRE (Eatontown, NJ)
Aircraft Engineering:
CAS, Inc. (Huntsville, AL)
Science Applications International Corp. (SAIC)
    (Huntsville, AL)
Information Assurance:
Sensor Technologies (Red Bank, NJ)
Program Support:
CACI (Arlington, VA)
Software Engineering Support:
Lockheed Martin (Tinton Falls, NJ)

Enterprise Email (EE)

Modernization

Recapitalization

Maintenance

MISSION

Improves Soldier communication resulting in improved operations by providing access to email from any location, a complete global address list, calendar sharing, and management capability across the enterprise.

DESCRIPTION

Enterprise Email (EE) services will increase functionality, improve mission effectiveness, unify services, heighten security, and reduce cost. The Army is obtaining EE capabilities as a managed service through an interagency acquisition with the Defense Information Systems Agency (DISA). The DoD's EE service provides secure, cloud-based email to the DoD enterprise that is designed to increase operational efficiency and facilitate collaboration across organizational boundaries. It provides secure access to email anywhere, at any time, from any place, whether stationary or mobile. EE supports coordination efforts by sharing individual, organizational, and resource calendars across the DoD and its mission partners. EE reduces the cost of

email by eliminating unnecessary administration and inefficient network configurations, enabling resources to focus on other priorities. It provides user organizations with the level of assurance required to know that their communications are secure.

SYSTEM INTERDEPENDENCIES

In this Publication
None

Other Major Interdependencies
EE utilizes Microsoft Outlook 2010 and accommodates use with many Microsoft products.

PROGRAM STATUS

- **Current:** Migrated 459,904 Non-Classified Internet Protocol Route (NIPR) users/41,594 Blackberries to date
- **2QFY12:** Established Product Director EE
- **2QFY12:** Successfully resumed migrations after 30-day pause
- **3QFY12:** Army/DISA Metrics Integrated Product Team (IPT)
- **3QFY12:** DCIO, G6, and Army CYBER quarterly update

- **3QFY12:** EE acquisition strategy approved
- **3QFY12:** Army National Guard (ARNG) Requirements Working Groups
- **3QFY12:** DoD EE portal requirements finalized
- **3QFY12:** ARNG requirements analysis
- **3QFY12:** End-to-End user experience analysis
- **4QFY12:** Service Level Agreement (SLA) signed between Program Executive Office Enterprise Information Systems and DISA
- **4QFY12:** Secure Internet Protocol Router (SIPR) schedule to be finalized (200,000 users)
- **4QFY12:** Army continues to successfully migrate users

PROJECTED ACTIVITIES

- **1QFY13:** ARCENT NIPR migration
- **1QFY13:** Planned migration for rest of ARNG states
- **2QFY13:** Complete NIPR migrations
- **2QFY13:** Complete SIPR migrations

Technology Development	Engineering & Manufacturing Development	Production & Deployment	Operations & Support

Enterprise Email (EE)

FOREIGN MILITARY SALES
None

CONTRACTORS
DISA maintains and manages multiple contracts, all of which provide a component(s) of EE.

PRODUCT DIRECTOR
ENTERPRISE EMAIL

Excalibur (M982)

MISSION

Provides improved fire support to the maneuver force commander through a precision-guided, extended-range, artillery projectile that increases lethality and reduces collateral damage.

DESCRIPTION

Excalibur (M982) is a 155mm, Global Positioning System (GPS)-guided, extended-range artillery projectile, in use as the Army's next-generation cannon artillery precision munition. The target, platform location, and GPS-specific data are entered into the projectile's mission computer through an Enhanced Portable Inductive Artillery Fuze Setter (EPIAFS).

Excalibur uses a jam-resistant internal GPS receiver to update the inertial navigation system, providing precision in-flight guidance and dramatically improving accuracy regardless of range. Excalibur has three fuze options (height-of-burst, point-detonating, and delay/penetration) and is employable in all weather conditions and terrain.

The program is using an incremental approach to provide a combat capability to the Soldier as quickly as possible, and to deliver advanced capabilities and lower costs as technology matures. The initial variant (Increment Ia-1) includes a unitary high-explosive warhead capable of penetrating urban structures and is also effective against personnel and light materiel targets. Increment Ia-2 provides increased range (up to 37.5 kilometers) and reliability improvements. The third variant (Increment Ib) will maintain performance and capabilities while reducing unit cost and increasing reliability.

Excalibur is designed for fielding to the Lightweight 155mm Howitzer (M777A2), the 155mm M109A6 self-propelled Howitzer (Paladin), and the Swedish Archer Howitzer. Excalibur is an international cooperative program with Sweden, which contributes resources toward the development in accordance with established cooperative development and production agreements.

SYSTEM INTERDEPENDENCIES

In this Publication
Advanced Field Artillery Tactical Data System (AFATDS)

Other Major Interdependencies
Enhanced Portable Inductive Artillery Fuze Setter (EPIAFS), Modular Artillery Charge System (MACS)

PROGRAM STATUS

- **Current:** Army and Marine Corps units in Afghanistan are Excalibur capable with Increments Ia-1 and Ia-2
- **1QFY12:** Initial Operational Capability for Increment Ia-2
- **1QFY12-4QFY12:** Continued deliveries of Increment Ia-2

PROJECTED ACTIVITIES

- **1QFY13:** Milestone C Low-rate initial production decision for Increment Ib
- **3QFY13:** Initial Operational Test & Evaluation for Increment Ib
- **4QFY14:** Increment Ib Initial Operational Capability

Excalibur (M982)

FOREIGN MILITARY SALES
Australia, Canada, Sweden, United Kingdom

CONTRACTORS
Raytheon (Tucson, AZ)
General Dynamics Ordnance and Tactical Systems (Healdsburg, CA)
General Dynamics Ordnance and Tactical Systems (Niceville, FL)
M/A-COM (Lowell, MA)
L3 Communications (Cincinnati, OH)

Family of Medium Tactical Vehicles (FMTV)

Modernization

Recapitalization

Maintenance

MISSION
Provides unit mobility/resupply, equipment/personnel transportation, and key ammunition distribution, using a family of vehicles based on a common chassis.

DESCRIPTION
The Family of Medium Tactical Vehicles (FMTV) is a system of strategically deployable vehicles that performs general resupply, ammunition resupply, maintenance and recovery, and engineer support missions. The FMTV also serves as weapon systems platforms for combat, combat support, and combat service support units in a tactical environment.

The Light Medium Tactical Vehicle (LMTV) has a 2.5-ton capacity (cargo, van, and chassis models) and has a companion trailer.

The Medium Tactical Vehicle (MTV) has a 5-ton capacity (cargo, long-wheelbase-cargo with and without materiel handling equipment, tractor, van, wrecker, 8.8-ton Load Handling System (LHS), 8.8-ton LHS trailer, and 10-ton dump truck models). Three truck variants and two companion trailers, with the same cube and payload capacity as their prime movers, provide air drop capability. MTV also serves as the platform for the High Mobility Artillery Rocket System (HIMARS) and a resupply vehicle for PATRIOT and HIMARS. MTV operates worldwide in all weather and terrain conditions.

FMTV enhances crew survivability through the use of armor capable cabs, three-point seat belts, automatic braking system, and central tire inflation capability.

FMTV enhances tactical mobility and is strategically deployable in C-5, C-17, and C-130 aircraft (excluding wrecker). It reduces the Army's logistical footprint by providing commonality of parts and components, reduced maintenance downtime, high reliability, and high operational readiness rate.

FMTV incorporates a vehicle data bus and class V interactive electronic technical manual, significantly lowering operating and support costs as compared with older trucks. Units are equipped with FMTVs at more than 68 locations worldwide; 65,172 trucks and 16,752 trailers are in field units as of June 2012. The Army developed, tested, and installed add-on-armor and enhanced add-on-Protection kits, and a Low Signature Armored Cab (LSAC) for Southwest Asia. The newest armored version, the Long-term Armor Strategy (LTAS) A-Cabs are integral to new production and are being fielded. The LTAS B-kit is also available.

SYSTEM INTERDEPENDENCIES
In this Publication
Chemical Biological Protective Shelter (CBPS), High Mobility Artillery Rocket System (HIMARS)

Other Major Interdependencies
AGSE, CBDP-CP, AN/TPQ-53, HMMWV Replacement Interchange, LMS-788 Ops Shelter and Sensor Pallet, Other Interchange, P/M CAP, USAF AN/TPS-75 Radar

PROGRAM STATUS
- **3QFY12:** Government First Article Test (FAT) of new Oshkosh FMTV completed
- **3QFY12:** Government Follow on Production Test (FPT) of Oshkosh FMTV variants begins

PROJECTED ACTIVITIES
- **Continue:** Full production and fielding to support Army transformation

Technology Development	Engineering & Manufacturing Development	Production & Deployment	Operations & Support

FMTV with Armor Kit

Family of Medium Tactical Vehicles (FMTV)

FOREIGN MILITARY SALES
Afghanistan, Canada, Djibouti, Greece, Iraq, Jordan, Macedonia, Saudi Arabia, Taiwan, Thailand, United Arab Emirates

CONTRACTORS
Prime:
Oshkosh Corp. (Oshkosh, WI)
Axles:
Meritor (Troy, MI)
Transmission:
Allison Transmission (Indianapolis, IN)
Engine:
Caterpillar (Greenville, SC)

	LMTV A1 Cargo	MTV A1 Cargo
Payload:	5,000 pounds	10,000 pounds
Towed load:	12,000 pounds	21,000 pounds
Engine:	Caterpillar 6-cylinder diesel	Caterpillar 6-cylinder diesel
Transmission:	Allison Transmission Automatic	Allison Transmission Automatic
Horsepower:	275	330
Drive:	4 x 4	6 x 6

Fixed Wing

INVESTMENT COMPONENT

Modernization

Recapitalization

Maintenance

MISSION

Performs operational support and focused logistics missions for Army, Joint services, national agencies, and multinational users in support of reconnaissance and surveillance, homeland security, transportation of key personnel, medical evacuation, and movement of critical, time-sensitive logistical support for battle missions.

DESCRIPTION

Army fixed wing aviation units serve as intelligence and electronic warfare assets, provide timely movement of key personnel to critical locations throughout the theater of operations, and support worldwide peacetime contingencies and humanitarian relief efforts. The fixed wing fleet consists of over 370 aircraft comprised of 11 missions, 40 designs, and 73 series. All Army fixed wing aircraft are either commercial-off-the-shelf (COTS) products or commercial derivative aircraft.

The EO-5, RC-12, and B-300 are classified as special electronic mission aircraft (SEMA) and provide real-time intelligence collection in peace and wartime environments. The C-12, C-23, C-26 and UC-35 are classified as transport aircraft and provide direct fixed wing support to warfighting combatant commanders worldwide. The C-20 and C-37 are classified as Executive Transport for the Chiefs of Staff and the service secretaries. The remaining mission support aircraft including the C-208 and T-34 are used for Research, Development, Test & Evaluation, the CE-182 Engineering Flight Laboratory United States Military Academy (USMA), the UV-18 and C-31 for the U.S. Army Golden Knights Parachute Team, and the C-41 and UV-20 for Special Operations Training.

SYSTEM INTERDEPENDENCIES

None

PROGRAM STATUS

- **FY12:** C-12, RC-12, and UC-35 aircraft sustained using a Life Cycle Contractor Support (LCCS) maintenance contract (L-3 Vertex)
- **FY12:** C-23 and C-26 aircraft sustained using a LCCS maintenance contract (M-7 Aerospace)
- **FY12:** C-20 and C-37 aircraft sustained using LCCS maintenance contracts (Northrop Grumman Technical Services (NGTS) and Gulfstream)
- **FY12:** EO-5 aircraft sustained using a LCCS maintenance contract (King Aerospace)
- **FY12:** Contract awarded to procure one replacement aircraft for the Air Traffic Services Command (ATSCOM)

PROJECTED ACTIVITIES

- **FY13:** Deliver three specially modified UV-18C (DHC-6 Model 400) replacement aircraft for the U.S. Army Golden Knights Parachute Team
- **FY15:** Procure four T-6 Texan II aircraft as a participating service in the Joint Primary Aircraft Training System (JPATS) ACAT 1C Program to replace four T-34C aircraft for the Army Test and Evaluation Command (ATEC)

ACQUISITION PHASE

| Technology Development | Engineering & Manufacturing Development | Production & Deployment | Operations & Support |

C-20/37

C-26

RC-12

EO-5 (ARL)

C-12

C23

Force Protection Systems

Modernization

Recapitalization

Maintenance

MISSION

Detects, assesses, and responds to unauthorized entry or attempted intrusion into installations or facilities.

DESCRIPTION

Force Protection Systems consist of the following components:

Automated Installation Entry (AIE) is a software and hardware system designed to read and compare vehicles and personnel identification media. The results of the comparison are used to permit or deny access to an installation in accordance with installation commanders' criteria. AIE will use a database of personnel and vehicles that have been authorized entry onto an Army installation and appropriate entry lane hardware to permit/deny access to the installation. The system will validate the authenticity of credentials presented by a person with data available from defense personnel and vehicle registration databases. AIE will have the capability to process permanent personnel and enrolled visitors and to present a denial barrier to restrict unauthorized personnel. The system will also be capable of

adapting to immediate changes in threat conditions and apply restrictive entrance criteria consistent with the force protection condition.

The Battlefield Anti-Intrusion System (BAIS) is a compact, modular, sensor-based warning system that can be used as a tactical stand-alone system. The system consists of a handheld monitor and three seismic/acoustic sensors and provides coverage across a platoon's defensive front (450 meters). It delivers early warning and situational awareness information, classifying detections as personnel, vehicle, wheeled, or tracked intrusions.

The Lighting Kit, Motion Detector (LKMD) is a simple, compact, modular, sensor-based early-warning system providing programmable responses of illumination and sound. The LKMD enhances unit awareness during all types of operations and environments, including those in urban terrain.

SYSTEM INTERDEPENDENCIES

None

PROGRAM STATUS

- **1QFY12-4QFY12:** BAIS production systems delivery and fielding
- **1QFY12-4QFY12:** LKMD production and fielding
- **3QFY12:** LKMD follow on procurement contract awarded

PROJECTED ACTIVITIES

- **1QFY13-4QFY14:** BAIS production and fielding
- **1QFY13-4QFY14:** LKMD production and fielding

Technology Development | Engineering & Manufacturing Development | Production & Deployment | Operations & Support

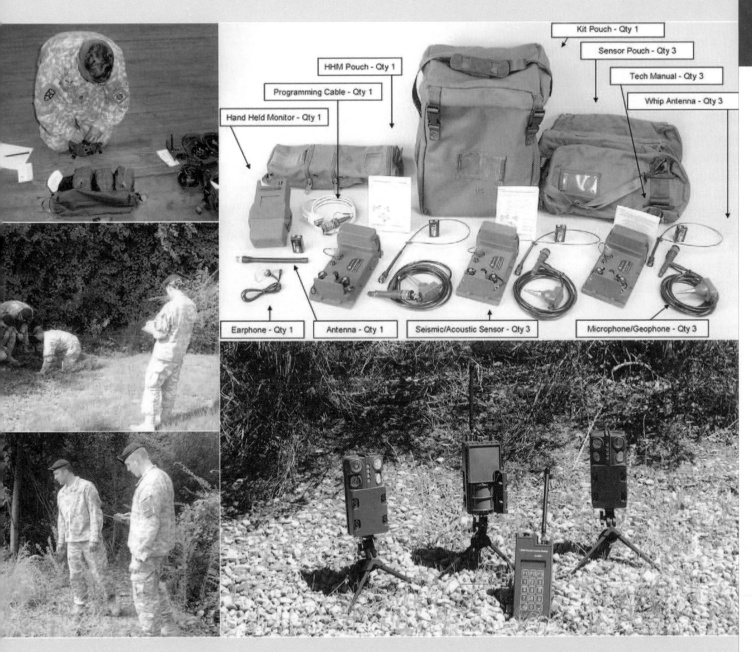

Kit Pouch - Qty 1

Sensor Pouch - Qty 3

HHM Pouch - Qty 1

Tech Manual - Qty 3

Programming Cable - Qty 1

Whip Antenna - Qty 3

Hand Held Monitor - Qty 1

Earphone - Qty 1

Antenna - Qty 1

Seismic/Acoustic Sensor - Qty 3

Microphone/Geophone - Qty 3

Force Protection Systems

FOREIGN MILITARY SALES
None

CONTRACTORS
BAIS:
L-3 Communications-East (Camden, NJ)
LKMD:
URS Corp. (San Francisco, CA)

Force Provider (FP)

Modernization

Recapitalization

Maintenance

MISSION

Provides Army personnel and other forces as directed by the Secretary of Defense with a high quality deployable base camp to support expeditionary missions; develops, integrates, acquires, fields, sustains, and modernizes base camp support systems to improve the warfighter's fighting capabilities, performance, and quality of life.

DESCRIPTION

Each Force Provider (FP) is a high-quality deployable base camp that provides billeting, laundry, shower, latrine, food service, shower water reuse, and morale, welfare, and recreation (MWR) kits to support 600 Soldier camps. Additionally, FP can be configured to support 150 base camps. FP includes 75 deployable triple container (TRICON) systems, with eight latrine systems, eight shower systems, four kitchen systems, containerized batch laundry systems, four TRICON refrigerated containers, 26 60-kilowatt tactical quiet generators, 26 modular personnel tents (air supported), four 400,000

British thermal unit water heaters, four improved fuel distribution systems, two wastewater evacuation tank/trailers, 26 mobile electric power distribution replacement systems, and 56 environmental control units. FP is prepositioned in Army Prepositioned Stocks (APS) 1, 3, and 4 to support combatant commanders' requirements. All system components weigh less than 10,000 pounds and are prepacked for rapid transport via air (C-130, C-141, C-5, and C-17), sea, road, or rail.

Additional operational add-on kits include: a cold-weather kit that allows operation to -15 degrees Fahrenheit; prime-power kit; large-scale electric kitchen; and resource efficiency add-ons to include a shower water reuse system and energy saving shelter shade and insulating liner systems. New modules use an Airbeam Shelter technology which reduces set up time from days to hours.

Operational energy upgrades reduce fuel, power, and water requirements.

SYSTEM INTERDEPENDENCIES
In this Publication
None

Other Major Interdependencies
60-kilowatt Tactical Quiet Generator

PROGRAM STATUS
- **1QFY12:** Additional funding was received to provide Operational Energy upgrades (solar shields, tent liners) for 20 FP modules

PROJECTED ACTIVITIES
- **2QFY13:** Capability Production Document for the Force Provider Expeditionary approval
- **2QFY14:** Incorporate Black Water Treatment capability into Force Provider Modules
- **4QFY14:** Milestone C approval for Force Provider Expeditionary

Technology Development | Engineering & Manufacturing Development | Production & Deployment | Operations & Support

Force Provider (FP)

FOREIGN MILITARY SALES
None

CONTRACTORS
Force Provider Assembly:
Sotera Defense Systems (Easton, MD)
Letterkenny Army Depot
  (Chambersburg, PA)
**Expeditionary TRICON Kitchen System
and FP Electric Kitchen:**
Tri-Tech USA Inc. (South Burlington, VT)
Airbeam TEMPER Tent:
Vertigo Inc. (Lake Elsinore, CA)
Environmental Control:
Hunter Mfg. (Solon, OH)
TRICON Container:
Charleston Marine Containers
(Charleston, SC)
Waste Water Evacuation Tank/Trailer:
Marsh Industrial (Kalkaska, MI)
Cold Weather Kit Assembly:
Berg Companies, Inc. (Spokane, WA)
**Mobile Electric Power Distribution
System Replacement:**
Lex Products Corp. (Stamford, CT)
**Expeditionary TRICON Systems
(shower, laundry, latrine):**
To be determined

Force XXI Battle Command Brigade and Below (FBCB2)

INVESTMENT COMPONENT

Modernization

Recapitalization

Maintenance

MISSION

Provides integrated, on-the-move, timely, relevant battle command information to tactical combat leaders and Soldiers from brigade to platform and across platforms within the brigade task force and other Joint forces.

DESCRIPTION

The Force XXI Battle Command Brigade and Below (FBCB2) forms the principal digital command and control system for the Army at brigade levels and below. It provides increased situational awareness (SA) on the battlefield by automatically disseminating throughout the network timely friendly force locations, reported enemy locations, and graphics to visualize the commander's intent and scheme of maneuver.

FBCB2 is a key component of the Army Battle Command System (ABCS). Appliqué hardware and software are integrated into the various platforms at brigade-and-below, as well as at appropriate division and corps slices necessary to support brigade operations.

The system features platform interconnections through two communication systems: FBCB2-Enhanced Position Location Reporting System (EPLRS), supported by the tactical Internet; and FBCB2-Blue Force Tracking, supported by L-Band satellite. The Joint Capabilities Release (JCR) is the next software release and addresses Joint requirements, database simplification, Type 1 encryption, a product line software approach, and enables the transition to the Blue Force Tracking II (BFT II) transceiver, allowing a tenfold increase in data throughput. FBCB2 is the primary platform-level digital Battle Command (BC) for the Army and Marine Corps at brigade-and-below, consisting of computer hardware and software integrated into tactical vehicles and aircraft. The system distributes SA data and BC messages within/between platforms and command posts using the Lower Tactical Internet EPLRS or L-Band satellite as its means of communication.

The Joint Capabilities Release (JCR), which acts as a bridge between FBCB2 and Joint Battle Command–Platform, is part of Capability Set 13 (CS 13), the Army's first package of network components, associated equipment, and software that provides integrated connectivity from the static tactical operations center to the commander on-the-move to the dismounted Soldier. CS 13 begins fielding to Brigade Combat Teams in October 2012.

SYSTEM INTERDEPENDENCIES

In this Publication

Advanced Field Artillery Tactical Data System (AFATDS), Battle Command Sustainment Support System (BCS3), Distributed Common Ground System–Army (DCGS-A), Movement Tracking System (MTS), Nett Warrior (NW), Warfighter Information Network Tactical (WIN-T) Increment 2, Warfighter Information Network Tactical (WIN-T) Increment, Warfighter Information Network–Tactical (WIN-T) Increment 1

Other Major Interdependencies

AMDWS, ASAS, BFT-AVN, DTSS, CPOF, JTCW, JSTARS, MCS, JC2C

PROGRAM STATUS

- **Current:** FBCB2 hardware and software have transitioned to sustainment

PROJECTED ACTIVITIES

- **Continue:** Deployment of FBCB2/JCR and BFT-2

ACQUISITION PHASE

| Technology Development | Engineering & Manufacturing Development | Production & Deployment | Operations & Support |

Force XXI Battle Command Brigade and Below (FBCB2)

FOREIGN MILITARY SALES
Australia

CONTRACTORS
Software, Encryption, and Installation Kits Prime:
Northrop Grumman (Carson, CA)
Field Service Representatives, Trainers, Installers:
Engineering Solutions and Products (ESP)
    (Eatontown, NJ)
Hardware:
DRS Technologies (Palm Bay, FL)
ViaSat Inc. (Carlsbad, CA)
Program Management Support:
CACI (Eatontown, NJ)
Test Support:
MANTECH (Killeen, TX)
Aviation Hardware:
Prototype Integration Facility
    (Huntsville, AL)

Forward Area Air Defense Command and Control (FAAD C2)

INVESTMENT COMPONENT

Modernization

Recapitalization

Maintenance

MISSION

Collects, processes, and disseminates real-time target tracking and cuing information to all short-range air defense weapons and provides command and control (C2) for the Counter-Rocket, Artillery, Mortar (C-RAM) System-of-Systems (SoS).

DESCRIPTION

Forward Area Air Defense Command and Control (FAAD C2) software provides critical C2, situational awareness (SA), and automated air track information by integrating engagement operations software for multiple systems, including:
- Avenger
- Sentinel
- Army Mission Command
- C-RAM SoS

FAAD C2 supports air defense and C-RAM weapon systems engagement operations by tracking friendly and enemy aircraft, cruise missiles, unmanned aerial systems, mortar and rocket rounds as identified by radar systems, and by performing C2 engagement operations for Short Range Air Defense (SHORAD) and C-RAM SoS.

FAAD C2 uses the following communication systems:
- Enhanced Position Location Reporting System (EPLRS)
- Multifunctional Information Distribution System (MIDS)
- Single Channel Ground and Airborne Radio System (SINCGARS)
- Wireless Local Area Network (LAN)

FAAD C2 provides Joint C2 interoperability and horizontal integration with all Army C2 and air defense artillery systems, including, but not limited to:
- PATRIOT
- Avenger
- Theater High Altitude Area Defense (THAAD)
- Airborne Warning and Control System (AWACS)
- C-RAM SoS
- Army Mission Command

SYSTEM INTERDEPENDENCIES

In this Publication
None

Other Major Interdependencies
Radar systems providing input data such as Sentinel, Firefinder, Lightweight Counter-Mortar Radar (LCMR), and Joint external sensors (e.g. AWACS)

PROGRAM STATUS
- **1QFY12:** Completed fielding of Sensor C2 nodes to all COMPO 1 divisions
- **4QFY12:** Urgent Materiel Release of FAAD C2 V5.5A.11.4
- **4QFY12:** Completed fielding of Sensor C2 nodes to all COMPO 2 divisions

PROJECTED ACTIVITIES
- **1QFY13:** Field upgrades to 1-204 Air Defense Artillery Battalion (Mississippi Army National Guard)
- **1QFY13:** Full Materiel Release of FAAD C2 version 5.5A
- **4QFY14:** Complete FAAD fieldings to 5-5 ADA BN and 2-44 ADA BN
- **1QFY15:** FAAD C2 Program transitions to Sustainment

ACQUISITION PHASE

| Technology Development | Engineering & Manufacturing Development | Production & Deployment | Operations & Support |

Forward Area Air Defense Command and Control (FAAD C2)

FOREIGN MILITARY SALES
Australia, Egypt, United Kingdom

CONTRACTORS
Software:
Northrop Grumman Space and Mission
    Systems Corp. (Redondo Beach, CA)
Hardware:
Tobyhanna Army Depot (Scranton, PA)
PKMM (Las Vegas, NV)
CHS 3:
General Dynamics (Taunton, MA)

General Fund Enterprise Business Systems (GFEBS)

Modernization

Recapitalization

Maintenance

MISSION

Provides a new core financial management capability compliant with Congressional mandates, administers the Army's General Fund, and improves performance, standardize processes, and meets future needs.

DESCRIPTION

The General Fund Enterprise Business Systems (GFEBS) is a commercial-off-the-shelf (COTS), Enterprise Resource Planning (ERP) system that meets the requirements of the Chief Financial Officers Act and the Federal Financial Management Improvement Act (FFMIA) of 1996, and which can support the DoD with accurate, reliable, and timely financial information. GFEBS is deployed to over 52,000 users at 227 locations worldwide and facilitates the management of nearly $140 billion in the General Fund and $80 billion in Overseas Contingency Operations funds.

SYSTEM INTERDEPENDENCIES

None

PROGRAM STATUS

- **1QFY12:** Deployed (Wave 7) to U.S. Southern Command, U.S. Army Pacific Command, Defense Finance and Accounting, Army Test and Evaluation Command
- **1QFY12:** Fielded of final development software release (Release 1.4.4)
- **3QFY12:** Completed FFMIA compliancy test
- **3QFY12:** Deployed (Wave 8a) to Redstone Arsenal, Program Executive Offices, Army Medical Command, Aberdeen Proving Grounds, Fort. Belvoir, Yuma Proving Grounds, Dugway Proving Grounds, Army Central Command Headquarters and Army Test and Evaluation Command
- **3QFY12:** Completed Standard Financial Information Structure (SFIS) compliancy test: 93 percent compliance rate with SFIS version 8.0
- **3QFY12:** FFMIA Draft Report: Test shows system remains substantially compliant meeting 96.7 percent of baseline requirements

- **4QFY12:** Final deployment (Wave 8b) to the Army Materiel Command; Army Central Command (Kuwait; Iraq; Afghanistan; Saudi Arabia) and Program Executive Offices
- **4QFY12:** Full deployment

PROJECTED ACTIVITIES

- **Continue:** Support Acquisition Phase

General Fund Enterprise Business Systems (GFEBS)

FOREIGN MILITARY SALES
None

CONTRACTORS
Accenture Federal Services (Arlington, VA)
Ilumina Solutions (California, MD)
Binary Group (Rosslyn, VA)

Global Combat Support System–Army (GCSS-Army)

INVESTMENT COMPONENT

Modernization

Recapitalization

Maintenance

MISSION

Provides commanders and staffs with a responsive and efficient automated system that provides one coherent source for accurate and timely tactical logistics and financial information to improve situational awareness (SA) and facilitate the decision-making cycle.

DESCRIPTION

Global Combat Support System–Army (GCSS-Army) is one program with two components. GCSS-Army Enterprise Resource Planning (ERP) Solution is an automation information system that serves as the primary tactical logistics enabler to support Army and Joint transformation for sustainment using an ERP system. The program re-engineers current business processes to achieve end-to-end logistics and integration with applicable command and control (C2)/Joint systems. The second component, Army Enterprise Systems Integration Program (AESIP), formerly known as Product Lifecycle Management Plus, integrates Army business functions by providing a single source for enterprise hub services, master data, and business intelligence. GCSS-Army uses commercial-off-the-shelf (COTS) ERP software products to support rapid force projection in the battlefield functional areas of arming, fixing, fueling, sustaining, and tactical logistics financial processes.

The GCSS-Army solution replaces five logistics Standard Army Management Information Systems in tactical units and will establish an interface/integration with applicable C2 and Joint systems. GCSS-Army (ERP Solution) is the primary enabler for the Army transformation vision of a technologically advanced ERP that manages the flow of logistics resources and information to satisfy the Army's modernization requirements. AESIP integrates Army business functions by providing a single source for enterprise hub services, business intelligence and analytics, and centralized master data management across the business domain. GCSS-Army will meet the Soldier's need for responsive support at the right place and time and improve the commander's SA with accurate and responsive information.

SYSTEM INTERDEPENDENCIES

In this Publication

General Fund Enterprise Business System (GFEBS)

PROGRAM STATUS

- **1QFY12:** Initial Operational Test and Evaluation

PROJECTED ACTIVITIES

- **2QFY13:** GCSS-Army is approaching a Full Deployment Decision and will begin Army-wide fielding

ACQUISITION PHASE

| Technology Development | Engineering & Manufacturing Development | Production & Deployment | Operations & Support |

What GCSS-Army Provides

GCSS-Army will field an Army automated information system as the primary tactical logistics enabler to support Army and Joint Transformation for Sustainment using an Enterprise Resource Planning (ERP) System

LOGISTICS — GCSS-Army Supports the ARFORGEN Process — **OPERATIONS**

Reduces stockpiles of supplies on the battlefield

Single Integrated database that provides status on incoming materiel and integrates AIT, RFID and a feed from an ITV feed to identify content in containers

Provides Commanders with Equipment Readiness

Allows leaders to have visibility of a ways equipment status and maintenance second prior to task organizing

Supports Task Organization

Gives the Warfighter and accurate status on parts and supplies

Reduces decision cycle time

Provides the status of all supplies ordered on the battlefield 24/7

Able to change supply and maintenance relationships while assembling subordinate and personnel equipment between organizational structures

AEIP integrated Army Business functions by providing a single source for enterprise Hub services, business intelligence and analytics, and centralized master data management

System Architecture

GCSS-Army and AESIP
Two components under one program

GFEBS
Financial ERP

GCSS-Army
Tactical Sustainment ERP

LMP
Army National
Sustainment
ERP

AESIP(PLM+)
Technical Enabler

AKO Single Sign-on

DoD Business Management Systems e.g.
- EBS-Enterprise Business System (DLA)
- DCAS-Defense Cash Accountability System (Treasury interface)
- DAAS-Defense Automatic Addressing System
- TC-AIMS II - Transportation Coordinators' Automated Information for Movements Systems - II

CAISI CSS VSAT

War Related Immediate Need

Modernization Investment

Connecting the Modular Force
GCSS-Army is designed to support current operations and adapt to future requirements

Enables the Warfighter to order, move, track, account for and maintain equipment, supplies and ammunition from factory to foxhole

Global Combat Support System–Army (GCSS-Army)

FOREIGN MILITARY SALES
None

CONTRACTORS
Prime GCSS-Army:
Northrop Grumman Information Systems (Richmond, VA)
Prime AESIP:
Systems Integrator:
SSC-Army (Picatinny, NJ)
Attain LLC (Vienna, VA)
Oakland Consulting (Lanham, MD)
Program Support:
MITRE (Alexandria, VA)
Logistics Management Institute (LMI) (Alexandria, VA)
iLuMinA Solutions Inc. (Alexandria, VA)
Cap Gemini (Alexandria, VA)
SAP (Alexandria, VA)
Engility (formerly L3 MPRI) (Alexandria, VA)

Global Command and Control System–Army (GCCS-A)

INVESTMENT COMPONENT

Modernization

Recapitalization

Maintenance

MISSION

Provides critical automated command and control (C2) tools for combatant commanders to enhance Soldier capabilities throughout the spectrum of conflict during Joint and combined operations.

DESCRIPTION

Global Command and Control System–Army (GCCS-A) is the Army's strategic, theater and tactical command, control, and communications (C3) system. It provides a seamless link of operational information and critical data from the strategic Global Command and Control System-Joint (GCCS-J) to Army theater elements and below through a common picture of Army tactical operations to the Joint and coalition communities. GCCS-A is provided by Defense Readiness Reporting System-Army (DRRS-A) and is the system of record for theater Army headquarters worldwide.

SYSTEM INTERDEPENDENCIES

In this Publication

Advanced Field Artillery Tactical Data System (AFATDS), Battle Command Sustainment Support System (BCS3), Distributed Common Ground System–Army (DCGS-A), Force XXI Battle Command Brigade and Below (FBCB2)

Other Major Interdependencies

Command Post Of the Future (CPOF), Defense Readiness Reporting System-Army (DRRS-A), Global Command and Control System-Joint (GCCS-J), Global Status of Resources and Training System (GSORTS), Joint Operation Planning and Execution System (JOPES), Defense Readiness Reporting System-Strategic (DRRS-S), Air and Missile Defense Workstation (AMDWS), Tactical Air Space Integration System

PROGRAM STATUS

- **1QFY12:** Support Operation New Dawn (OND)
- **1QFY12-4QFY12:** Support Operation Enduring Freedom (OEF)
- **1QFY12-4QFY12:** Development in support of GCCS-A modernization strategy bridge efforts

- **3QFY12:** Release DRRS-A version 2.4 to the field
- **1QFY12-4QFY12:** Support to the Joint Command and Control (JC2) Planning, Reporting, and Engineering Efforts
- **1QFY12-4QFY12:** Continue fielding hardware to support GCCS-A and DRRS-A
- **1QFY12-4QFY12:** Continue training units and supporting unit exercises for GCCS-A

PROJECTED ACTIVITIES

- **2QFY13-4QFY14:** Continue support for OEF
- **2QFY13-4QFY14:** Development of DRRS-A for continuing Global Force Management Data Initiative (GFMDI) efforts
- **2QFY13-2QFY15:** Continue development in support of GCCS-A modernization strategy bridge effort and DRRS-A 2.4 requirements
- **2QFY13-2QFY15:** Development of assigned JC2 capability requirements
- **2QFY13-4QFY15:** Continue fielding and refreshing hardware to support GCCS-A and DRRS-A

ACQUISITION PHASE

Technology Development | Engineering & Manufacturing Development | Production & Deployment | Operations & Support

Global Command and Control System–Army (GCCS-A)

FOREIGN MILITARY SALES
None

CONTRACTORS
Develop and Field Software:
Lockheed Martin (Springfield, VA; Tinton
  Falls, NJ)
System Hardware:
General Dynamics (Taunton, MA)
Systems Engineering and Support:
CACI (Aberdeen, MD)
Accenture (Reston, VA)
Field Support Representatives (FSRs):
Engineering Solutions and Products (ESP)
  (Eatontown, NJ)
General Dynamics (GDIT)(Fairfax, VA)
Program Support:
Booz Allen Hamilton (Eatontown, NJ)

Ground Combat Vehicle (GCV)

Modernization

Recapitalization

Maintenance

MISSION

Provides the infantry squad with a highly mobile, protected transport on the battlefield.

DESCRIPTION

The Ground Combat Vehicle (GCV) is a critical element of the Army's effort to transform, replace, and improve its Combat Vehicle fleet. The GCV Infantry Fighting Vehicle (IFV) will provide force protection to deliver a nine-Soldier infantry squad in an improvised explosive device (IED) threat environment. It will protect occupants from IEDs, mines, and other ballistic threats with scalable armor that provides mission flexibility for the commander.

The GCV IFV will be designed with sufficient power and space to host the Army's advanced network. The IFV will feature an open architecture to facilitate the integration of current and future communications, computers, and surveillance and reconnaissance systems. The GCV IFV will have enhanced mobility to allow it to operate effectively in a variety of complex environments, including urban and cross country terrain. The GCV IFV's organic weapons will be capable of providing both destructive fires against armored vehicle threats and direct fire support for the squad during dismounted assaults. Flexible capabilities can shape the operating environment with effects that can vary from a "shove" to a lethal overmatch.

SYSTEM INTERDEPENDENCIES

None

PROGRAM STATUS

- **FY12:** Competitive Development prior to Engineering and Manufacturing Development Phase; Non-Developmental Vehicle Assessment; AoA Dynamic Update; Conducted Knowledge Points in preparation for staffing of Draft Capabilities Development Document

PROJECTED ACTIVITIES

- **4QFY11-1QFY14:** The Technology Development phase
- **1QFY14:** Milestone B

Technology Development   Engineering & Manufacturing Development   Production & Deployment   Operations & Support

Concept of Operation: Squad Deployment

GCV Platoon

 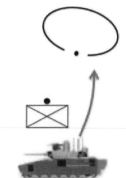 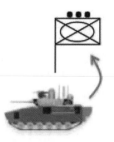

PL
Medic, RTO, FO
Interpreter, PRT, etc.

Concept of Operation: Platoon Deployment

GCV Infantry Company

 PLT
Medic/ RTO/FO
Interpreter, PRT, etc

 PLT
Medic/ RTO/FO
Interpreter, PRT, etc

 PLT
Medic/ RTO/FO
Interpreter, PRT, etc

Distributed Combat Power
Decentralized Operations —

Ground Combat Vehicle (GCV)

FOREIGN MILITARY SALES
None

CONTRACTORS
To be determined

Guardrail Common Sensor (GR/CS)

INVESTMENT COMPONENT

Modernization

Recapitalization

Maintenance

MISSION

Provides signals intelligence (SIGINT) collection and precision targeting that intercepts, collects, and precisely locates hostile communications intelligence radio frequency emitters and electronic intelligence threat radar emitters. Provides near-real-time info to tactical commanders in the Joint Task Force Area supporting full spectrum of operations (close in & deep look collections).

DESCRIPTION

The Guardrail Common Sensor (GR/CS) is a fixed-wing, airborne, SIGINT collection and precision targeting location system. It provides near-real-time information with emphasis on Indications and Warnings (I&W). It collects low-, mid-, and high-band radio signals and electronic intelligence (ELINT) signals; identifies and classifies them; determines source location; and provides near-real-time reporting, ensuring information dominance to commanders. GR/CS uses a Guardrail Mission Operations Facility (MOF) for the control, data processing, and message center for the system. GR/CS includes:

- Integrated COMINT and ELINT collection and reporting
- Enhanced signal classification and recognition and precision emitter geolocation
- Near-real-time direction finding
- Advanced integrated aircraft cockpit
- Tactical Satellite Remote Relay System

A standard system has RC-12 aircraft flying operational missions in single ship or multi-ship operations. Up to three aircraft/systems simultaneously collect communications and electronics emitter transmissions and gather lines of bearing and time-difference-of-arrival data, which is transmitted to the MOF, correlated, and supplied to supported commands via NSANet.

Enhancements include precision geo-location subsystem, the Communications High-Accuracy Location Subsystem—Compact (CHALS-C), with increased frequency coverage and a higher probability to collect targets; a modern COMINT infrastructure and core COMINT subsystem, providing a frequency extension, Enhanced Situational Awareness (ESA); a capability to process special high-priority signals through the high-end COMINT subsystems High Band COMINT (HBC) and X-Midas; and elimination of non-supportable hardware and software. Ground processing software and hardware are being upgraded for interoperability with the Distributed Common Ground System—Army (DCGS-A) architecture and Distributed Information Backbone.

SYSTEM INTERDEPENDENCIES
None

PROGRAM STATUS
- **1QFY12:** Fielded Aircraft #6 and 7 to 224th Military Intelligence Battalion
- **3QFY12:** Fielded Aircraft #8 to 224th Military Intelligence Battalion

PROJECTED ACTIVITIES
- **FY13-14:** Field the remaining six aircraft; retrofit aircraft 1-8 with enhancement; and begin defielding legacy systems

ACQUISITION PHASE

Technology Development | Engineering & Manufacturing Development | Production & Deployment | Operations & Support

Guardrail Common Sensor (GR/CS)

FOREIGN MILITARY SALES
None

CONTRACTORS
System Integrator, ESA Subsystem, and MOF Software/System Support:
Northrop Grumman (Sacramento, CA)
Data Links:
L-3 Communications (Salt Lake City, UT)
CHALS-C:
Lockheed Martin (Owego, NY)
X-MIDAS Subsystem:
ZETA (Fairfax, VA)
HBC Subsystem:
ArgonST Radix, Part of the Boeing Co.
(Mountain View, CA)

Guided Multiple Launch Rocket System (GMLRS) DPICM/Unitary/Alternative Warhead (Tactical Rockets)

INVESTMENT COMPONENT

Modernization

Recapitalization

Maintenance

MISSION
Provides an all-weather, rapidly-deployed, long-range, surface-to-surface, area-and-point precision strike capability.

DESCRIPTION
The Guided Multiple Launch Rocket System (GMLRS) is an upgrade to the M26 rocket, producing precise destructive and shaping fires against a variety of target sets. GMLRS is employed with the M270A1 upgraded Multiple Launch Rocket System (MLRS) tracked launcher and the M142 High Mobility Artillery Rocket System (HIMARS) wheeled launchers. GMLRS munitions have greater accuracy with a resulting higher probability of kill and smaller logistics footprint.

There are two fielded variants of the GMLRS: the previously produced GMLRS dual-purpose improved conventional munitions (DPICM) variant designed to service area targets; and the GMLRS Unitary variant with a single 200-pound class high-explosive charge to service point targets with low collateral damage. The development of a third variant incorporating an alternative warhead (AW) is ongoing. The GMLRS AW will replace the DPICM increment in compliance with the 2008 DoD Policy on Cluster Munitions & Unintended Harm to Civilians. The GMLRS AW rocket will service area targets without producing unexploded ordnance. It is scheduled to begin fielding in FY15.

The original GMLRS development was an international cooperative program with the United Kingdom, Germany, France, and Italy. Over 2,200 rockets were used in overseas contingency operations through June 2012.

Rocket length: 3,937 millimeters
Rocket diameter: 227 millimeters
Rocket reliability: Threshold 92 percent; objective: 95 percent
Ballistic range(s): 15 to 70+ kilometers

SYSTEM INTERDEPENDENCIES
In this Publication
Advanced Field Artillery Tactical Data System (AFATDS), High Mobility Artillery Rocket System (HIMARS), Multiple Launch Rocket System (MLRS) M270A1

Other Major Interdependencies
GPS, Joint Systems, National Systems

PROGRAM STATUS
- **4QFY11:** Selection of GMLRS AW warhead vendor for further development in Engineering and Manufacturing Development (EMD) Phase
- **2QFY12:** GMLRS AW Milestone B

PROJECTED ACTIVITIES
- **2QFY15:** GMLRS AW Milestone C

ACQUISITION PHASE

| Technology Development | Engineering & Manufacturing Development | Production & Deployment | Operations & Support |

UNITED STATES ARMY

Guided Multiple Launch Rocket System (GMLRS) DPICM/Unitary/Alternative Warhead (Tactical Rockets)

FOREIGN MILITARY SALES
Bahrain, Canada, Finland, France, Germany, Singapore, Italy, Japan, Jordan, Thailand, United Arab Emirates, United Kingdom

CONTRACTORS
Prime:
Lockheed Martin (Camden, AR; Grand Prairie, TX)
Lockheed Martin Missiles and Fire Control (Las Cruces, NM)
Guidance Set:
Honeywell (Clearwater, FL)
Rocket Motors:
Aerojet (Camden, AR)
Alternative Warhead:
Alliant Techsystems (Plymouth, MN)
Technical System Support:
Systems, Studies, and Simulation (Huntsville, AL)

Harbormaster Command and Control Center (HCCC)

INVESTMENT COMPONENT

Modernization

Recapitalization

Maintenance

MISSION

Provides Army logisticans who conduct distributed logistics a deployable and tactically mobile system that provides sensors and knowledge management tools to establish and maintain Battle Awareness (BA) and command and control (C2) of the harbor and littoral environment for all worldwide Overseas Contingency Operations (OCO).

DESCRIPTION

The Command Post Systems and Integration (CPS&I) Product Office provides a Harbormaster Command and Control Center (HCCC) System that provides the ability to facilitate safe navigation of watercraft in the harbor and littorals. The HCCC System is capable of Command, Control, and Communications operations that incorporate Local Area Network (LAN) equipment and Satellite Communications (SATCOM). The system provides sensors and management tools to collect and process environmental and asset tracking data relevant to supporting distribution in the littorals. The HCCC System possesses Non-Secure Internet Protocol (IP) Network (NIPRNET) and Secret IP Router Network (SIPRNET) technical connectivity to populate the Common Operating Picture (COP). The system provides the technical C2 connectivity to shift time and point of delivery of forces, equipment, sustainment, and support. The HCCC System is composed of a main and remote command center. Each system consists of two Command Post Platforms (CPP), two Trailer-Mounted Support Systems-Medium (TMSS-M), two Harbormaster Trailer-Mounted Sensor Platforms (HTSP), two Dual 18 kilowatt Generator Sets, and two Family of Medium Tactical Vehicles (FMTV).

SYSTEM INTERDEPENDENCIES

In this Publication
Global Command and Control System (GCCS), Movement Tracking System (MTS)

Other Major Interdependencies
SIPR/NIPR Access Point (SNAP), Battle Command Sustainment Support System (BCS3)

PROGRAM STATUS

- **1QFY12:** Delta Logistics/ Maintainability Demonstration
- **2QFY12:** Final HCCC TMs
- **2QFY12:** Harbormaster Operations Detachment (HMOD) NET and Mission Command System Integration (MCSI) Training Events I & II
- **2QFY12:** HMOD NMIB
- **3QFY12:** HCCC Received Type Classification Approval
- **4QFY12:** HMOD NMIB, NET and MCSI Training Events I, II, & III
- **4QFY12:** HMOD Joint Logistics Over The Shore (JLOTS)/MCSI Training Events I, II, & III
- **4QFY12:** HCCC Receives Conditional Materiel Release

PROJECTED ACTIVITIES

- **1QFY13:** HMOD MCSI Training Events I, II, & III
- **1QFY13:** HMOD NMIB
- **2QFY13:** HMOD MCSI Training Event III
- **2QFY13:** HCCC Receives Full Materiel Release
- **3QFY13:** HMOD NET & MCSI Events I, II, and III

ACQUISITION PHASE

Technology Development | Engineering & Manufacturing Development | Production & Deployment | Operations & Support

Harbormaster Command and Control Center (HCCC)

FOREIGN MILITARY SALES
Arab Emirates, Bahrain, Canada, Finland, France, Germany, Italy, Japan, Jordan, Thailand, United Kingdom, United Singapore

CONTRACTORS
Command Post Platform:
Northrop Grumman (Huntsville, AL)
Trailer-Mounted Support System:
Northrop Grumman (Huntsville, AL)
AMCOM EXPRESS (SETA):
Sigmatech, Inc. (Huntsville, AL)
Materiel Fielding:
Tobyhanna Army Depot (Tobyhanna, PA)
TOCNET Intercommunications Systems:
SCI Technology, Inc. (Huntsville, AL)
Common Hardware Systems:
General Dynamics C4 Systems, Inc.
   (Taunton, MA)
**Harbormaster Trailer-Mounted Sensor
   Package (HTSP):**
SPAWAR Pacific (San Diego, CA)
HP-6G 18kW Generator:
DHS Systems (Huntsville, AL)

Heavy Expanded Mobility Tactical Truck (HEMTT)/HEMTT Extended Service Program (ESP)

INVESTMENT COMPONENT

Modernization

Recapitalization

Maintenance

MISSION

Supports combat units by performing line and local haul, unit resupply, aviation refueling, tactical vehicle refueling, and related missions in a tactical environment.

DESCRIPTION

The Heavy Expanded Mobility Tactical Truck (HEMTT) 10-ton, 8-wheel drive is designed for cross-country military missions up to 11 tons to transport ammunition, petroleum, oils, and lubricants. Variants include: M977, M985, M978, M983, M984, and M1120.

The M977 is utilized for delivery of general supplies, equipment, and ammunition with an on-board crane with 4,500 pounds load capacity. The M985 cargo has an on-board crane with 5,400 pounds load capacity and is the primary transporter for Multiple Launch Rocket System (MLRS) ammunition.

The M978 tanker is a 2,500 gallon fuel transporter for field refueling of ground vehicles and aircraft. The M984 wrecker includes a crane and winch retrieval system and serves the primary role of recovery and evacuation of heavy wheel vehicles and combat systems. The M983 Tractor is the prime mover for the PATRIOT missile. The M983 Light Equipment Transporter (LET) Tractor serves as the prime mover for tactical semitrailers in engineering units to include the M870 series, Intermediate Stryker Recovery System (ISRS), and Mine-Resistant, Ambush-Protected (MRAP) vehicles. The HEMTT Load Handling System (LHS) provides NATO interoperability with standard flatrack and mission modules for delivery of general supplies, equipment, and ammunition with Palletized Load System (PLS) style load handling systems. The system is compatible with the PLS Trailer, capable of a 26,000 pound payload.

The HEMTT A4 began fielding in December 2008. Enhancements include a modern power train consisting of a Caterpillar C-15/500 horsepower engine and Allison Transmission (4500 SP/5-speed automatic), anti-lock braking system and traction control, air-ride suspension, a J-1939 data-bus providing an updated electrical system, climate control, and a larger common cab.

HEMTT ESP, known as HEMTT RECAP, is a recapitalization program that converts high-mileage, older version HEMTT trucks into the current A4 production configuration. Modernizing the fleet to one model reduces logistics footprint and operational and sustainment (O&S) cost of maintaining older vehicles.

HEMTT has several configurations:
M977: Cargo truck with light materiel handling crane
M985: Cargo truck with heavy materiel handling crane
M978: 2,500-gallon fuel tanker
M984: Wrecker
M983: Tractor
M983 LET: LET fifth wheel tractor with main recovery winch and towing capacity of 97,500 pounds
M1120: LHS transports palletized materiel and International Standards Organization (ISO) containers

SYSTEM INTERDEPENDENCIES

In this Publication
None

Other Major Interdependencies The M983 HEMTT LET Tractor paired with the Fifth Wheel Towing Device and High Mobility Recovery Trailer are together designated as the Interim Stryker Recovery System (ISRS) for Stryker and MRAP recovery. Other vehicles that utilize the HEMTT chassis are: M1142 Tactical Fire Fighting Trucks, M1158 Heavy Mobility Water Tender Truck, M1977 HEMTT Common Bridge Transporter (CBT), Theater High Altitude Area Defense Missile System (THAAD), and the M985 GMT Guided Missile Transport used in PATRIOT Battalions.

PROGRAM STATUS
- **Continue:** Production and fielding both New Production and RECAP HEMTTs to Active Army, Army National Guard, Army Reserves, and Army Pre-Position Stock (APS)

PROJECTED ACTIVITIES
- Continue to RECAP HEMTTs to support the modernization of the HEMTT fleet

ACQUISITION PHASE

Technology Development | Engineering & Manufacturing Development | **Production & Deployment** | Operations & Support

Heavy Expanded Mobility Tactical Truck (HEMTT)/HEMTT Extended Service Program (ESP)

FOREIGN MILITARY SALES
Egypt, Taiwan, United Arab Emirates

CONTRACTORS
Prime:
Oshkosh Corp. (Oshkosh, WI; Killeen, TX)
Engine:
Caterpillar (Peoria, IL)
Transmission:
Allison Transmission (Indianapolis, IN)
Tires:
Michelin (Greenville, SC)

Heavy Loader

Modernization

Recapitalization

Maintenance

MISSION
Provides engineering units the capability to perform lifting, loading, hauling, digging, and trenching operations in support of Combat Support Brigades and Brigade Combat Teams.

DESCRIPTION
The Heavy Loader is a commercial vehicle modified for military use. The military version of the loader will be armored with an A-kit (armored floor plate) on all loaders and C-kit (armored cab) on selected loaders. There are two types of loaders: the Type I–Quarry Team, with a capacity of 4.5 cubic yards; and Type II–General Use, with a capacity of 5 cubic yards. The Heavy Loader currently has state-of-the-art operator displays, on-board diagnostics and prognostics, and blackout lighting. For operator comfort, each loader is equipped with heating and air conditioning as well as an air suspension seat. Modifications include chemical-resistant coating paint, rifle rack, military standard (MIL-STD-209) lift and tie-down, and hydraulic quick coupler systems for attachments.

Heavy Loaders provide the capability to lift, move, and load a variety of materials. They are also used to perform horizontal and vertical construction tasks supporting military construction operations, including construction of roads, bridges, airfields, medical facilities, and demolition of structures, as well as loading in quarry operations.

SYSTEM INTERDEPENDENCIES
In this Publication
None

Other Major Interdependencies
M983 LET/M870 truck trailer for highway transportability

PROGRAM STATUS
• **Current:** Completed fielding

PROJECTED ACTIVITIES
None

Technology Development

Engineering & Manufacturing Development

Production & Deployment

Operations & Support

Heavy Loader

FOREIGN MILITARY SALES
Afghanistan

CONTRACTORS
OEM:
Caterpillar Defense and Federal Products
(Peoria, IL)
Armor:
BAE Systems (Rockville, MD)
Logistics:
XMCO (Warren, MI)

HELLFIRE Family of Missiles

INVESTMENT COMPONENT

Modernization

Recapitalization

Maintenance

MISSION

Engages and defeats individual moving or stationary advanced armor, mechanized or vehicular targets, patrol craft, buildings, or bunkers while increasing aircraft survivability.

DESCRIPTION

The AGM-114 HELLFIRE Family of Missiles includes the HELLFIRE II and Longbow HELLFIRE Missiles. HELLFIRE II is a precision strike, Semi-active Laser (SAL) guided missile and is the principal air-to-ground weapon for the Army AH-64 Apache, OH-58 Kiowa Warrior, Gray Eagle Unmanned Aerial System (UAS), Special Operations aircraft, Marine Corps AH-1W Super Cobra, and Air Force's Predator/Reaper UAS.

The SAL HELLFIRE II missile is guided by laser energy reflected off the target. It has three warhead variants: a dual warhead, shaped charge high-explosive anti-tank (HEAT) capability for armored targets (AGM-114K); a blast fragmentation warhead (BFWH) for urban, patrol boat and other "soft" targets (AGM-114M); and a metal augmented charge (MAC) warhead (AGM-114N) for urban structures, bunkers, radar sites, communications installations, and bridges. Beginning in 2012, a HELLFIRE II multipurpose warhead variant (AGM-114R) will allow the warfighter select warhead effects corresponding to a specific target type. The AMG-114R is capable of being launched from Army rotary-wing and UAS platforms and provides the pilot increased operational flexibility.

The Longbow HELLFIRE (AGM-114L) is also a precision strike missile using millimeter wave (MMW) radar guidance instead of the HELLFIRE II's semi-active laser (SAL). It is the principal anti-tank system for the AH-64D Apache Longbow helicopter and uses the same anti-armor warhead as the HELLFIRE II. The MMW Seeker provides beyond line-of-sight fire and forget capability, as well as the ability to operate in adverse weather and battlefield obscurants.

Diameter: 7 inches
Weight: 99.8-107 pounds
Length: 64-69 inches

Maximum Range:
HELLFIRE II: AGM-114R
Direct fire: 7 kilometers
Indirect fire: 8 kilometers
Minimum range: 0.5-1.5 kilometers

SYSTEM INTERDEPENDENCIES
None

PROGRAM STATUS
- **Current:** Laser HELLFIRE II missiles are procured annually to replace combat expenditures and war reserve requirements

PROJECTED ACTIVITIES
Laser HELLFIRE
- **Continue:** Maintain production
Longbow HELLFIRE
- **Continue:** Sustainment activities

ACQUISITION PHASE

| Technology Development | Engineering & Manufacturing Development | Production & Deployment | Operations & Support |

System Description	Production	Characteristics	Performance
AGM-114A, B, C, F, FA – HELLFIRE Weight = 45 kg Length = 163 cm	1982 – 1992	A, B, C have a Single Shaped-Charge Warhead; Analog Autopilot	• Not Capable Against Reactive Armor • Non-Programmable
		F has Tandem Warheads; Analog Autopilot	• Reactive Armor Capable • Non-Programmable
AGM-114K/K2/K2A – HELLFIRE II Weight = 45 kg Length = 163 cm	1993 – until complete	• Tandem Warheads • Electronic Safe & Arm Device • Digital Autopilot & Electronics • Improved Performance Software	• Capable Against 21st Century Armor • Hardened Against Countermeasures • K-2 adds Insensitive Munitions (IM) • K-2A adds Blast-Frag Sleeve
AGM-114L – HELLFIRE LONGBOW Weight = 49 kg Length = 180 cm	1995 – 2005	• Tandem Warheads • Digital Autopilot & Electronics • Millimeter-Wave (MMW) Seeker • IM Warheads	• Initiate on Contact • Hardened Against Countermeasures • Programmable Software • All-Weather
AGM-114M – HELLFIRE II (Blast Frag) Weight = 49 kg Length = 180 cm	1998 – 2010	• Blast-Frag Warhead • 4 Operating Modes • Digital Autopilot & Electronics • Delayed-Fuse Capability	• For Buildings, Soft-Skin Vehicles • Optimized for Low Cloud Ceilings • Hardened Against Countermeasures • WH Penetrates Target Before Detonation
AGM-114N – HELLFIRE II (MAC) Weight = 49 kg Length = 180 cm	2003 – until complete	• Metal-Augmented Charge – Sustained Pressure Wave • 4 Operating Modes • Delayed-Fuse Capability	• For Buildings, Soft-Skin Vehicles • Optimized for Low Cloud Ceilings • Hardened Against Countermeasures • WH Penetrates Target Before Detonation
AGM-114P/P+ – HELLFIRE II (for UAS) Weight = 49 kg Length = 180 cm	2003 – 2012	• Shaped-Charge or Blast-Frag • Designed for UAV Altitudes • 21st Century Armor Capability	• P-2A adds Steel Frag Sleeve • P-2B adds Tantalum Frag Sleeve • P+ adds Enhanced IMU and SW Support, Many Customizations for Varying Battlefield
AGM-114R – HELLFIRE II (Bridge to JAGM- RW/UAS) Weight = 49 kg Length = 180 cm	2012- until complete	• Multi-purpose warhead • Designed for all platforms • Health Monitoring	• For all Target Sets • P+ capabilities • Increased Lethality and Engagement Envelope

Guidance Section **Warhead Section** **Propulsion Section** **Control Section**

HELLFIRE Family of Missiles

FOREIGN MILITARY SALES
Laser HELLFIRE:
Australia, Egypt, France, Greece, Israel, Japan, Kuwait, Netherlands, Saudi Arabia, Singapore, Spain, Taiwan, Sweden, United Arab Emirates, United Kingdom
Direct commercial sale:
Netherlands, Norway, Saudi Arabia, Turkey, United Kingdom
Longbow HELLFIRE:
Israel, Japan, Kuwait, Singapore, Taiwan, United Arab Emirates
Direct commercial sale:
United Kingdom

CONTRACTORS
Prime Contractor:
Lockheed Martin (Orlando, FL)
Seeker:
Lockheed Martin (Ocala, FL)
Rocket Motor/Warhead:
Alliant Techsystems (Rocket City, WV)
Control Section:
Moog Inc. (Salt Lake, UT)
Firing Component (ESAF):
L-3 Communications (Chicago, IL)

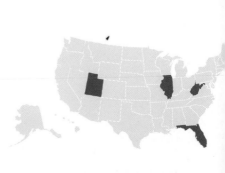

Helmet Mounted Night Vision Devices (HMNVD)

INVESTMENT COMPONENT

Modernization

Recapitalization

Maintenance

MISSION

Enhances the warfighter's visual ability and situational awareness while successfully engaging and executing operations day or night, whether in adverse weather or visually obscured battlefield conditions.

DESCRIPTION

Helmet Mounted Night Vision Devices (HMNVDs) enable the individual Soldier to see, understand, and act first during limited visibility conditions. These devices include:

The AN/PSQ-20 Enhanced Night Vision Goggle (ENVG)

The AN/PSQ-20 combines the visual detail in low light conditions that is provided by image intensification with the thermal sensor's ability to see through fog, dust, and smoke obscuration. This thermal capability makes the ENVG useful during the day as well as at night. The system is now more compact, enhancing the Soldier's mobility. The ENVG is compatible with aiming lasers currently in use, allowing for a fully integrated system of thermal, laser and image intensification.

AN/PVS-14 Monocular Night Vision Device (MNVD)

The AN/PVS-14 provides the ability to perform night time operations, while driving, walking, performing first aid, reading maps, and conducting maintenance. The device amplifies ambient light and very near infrared (IR) energy to enable night operations. The system is designed for use in conjunction with rifle-mounted aiming lights.

AN/AVS-6 Aviator's Night Vision Imaging System (ANVIS)

The AN/AVS-6 provides Army aviators the capability to support missions during periods of reduced visibility at night. The device amplifies ambient light from sources such as the moon, stars and sky glow, making the viewed scene clearly visible to the operator. The ANVIS enables the aircrew to maneuver the aircraft during low level, nap-of-the-earth flights, providing the capability to support normal and wartime missions.

SYSTEM INTERDEPENDENCIES

None

PROGRAM STATUS

- **FY12:** Production and fielding
- **FY12:** Fielded to units supporting Operation Enduring Freedom and Operation New Dawn
- **FY12:** Completed AN/PSQ-20 qualification testing, reached Full-rateproduction decision and issued two production awards
- **FY12:** Full-materiel release decisions
- **FY12:** Award new production contract(s) for AN/AVS-6(v)3
- **4QFY12:** Final Army AN/PVS-14 delivery

PROJECTED ACTIVITIES

- **FY13:** Production and fielding in accordance with Headquarters Department of the Army
- **FY13:** Issue Request for Proposal for follow-on Family of Weapons Sights (FWS) compatible ENVG contract
- **FY13-FY14:** Qualification testing for follow-on FWS compatible ENVG

ACQUISITION PHASE

| Technology Development | Engineering & Manufacturing Development | Production & Deployment | Operations & Support |

Helmet Mounted Night Vision Devices (HMNVD)

FOREIGN MILITARY SALES
PVS-7:
Oman, Romania, Thailand
PVS-14:
United Kingdom

CONTRACTORS
AN/PVS-14:
ITT-Exelis (Roanoke, VA)
L-3 Communications Electro-Optic
   Systems (Tempe, AZ; Garland, TX;
   Londonderry, NH)
AN/AVS-6(V)3:
ITT-Exelis (Roanoke, VA)
AN/PSQ-20:
ITT Exelis (Roanoke, VA)
L-3 Insight (Londonderry, NH)
DRS (Melbourne, FL)
Raytheon (Dallas, TX)

High Mobility Artillery Rocket System (HIMARS) M142

INVESTMENT COMPONENT

Modernization

Recapitalization

Maintenance

MISSION

Provides coalition ground forces with highly lethal, responsive, and precise long-range rocket and missile fires that defeat point and area targets via a highly mobile, responsive multiple launch system for Army and Marine early-entry expeditionary forces, contingency forces, and Modular Fires Brigades supporting Brigade Combat Teams.

DESCRIPTION

The M142 High Mobility Artillery Rocket System (HIMARS) is a combat-proven, wheeled artillery system that is rapidly deployable via C-130 and operable in all weather and visibility conditions. HIMARS is mounted on a five-ton modified Family of Medium Tactical Vehicles (FMTV) chassis. The wheeled chassis allows for faster road movement and lower operating costs, and requires far fewer strategic airlifts (via C-130 or C-17) to transport a firing battery than the tracked M270 Multiple Launch Rocket System (MLRS) that it replaces. The M142 provides responsive, highly accurate, and extremely lethal surface-to-surface rocket and missile fires from 15 to 300 kilometers. HIMARS can fire all munitions in the current and planned suite of the MLRS Family of Munitions (MFOM), including Army Tactical Missile System (ATACMS) missiles and Guided MLRS (GMLRS) rockets. HIMARS carries either six rockets or one missile, is self-loading and self-locating, and is operated by a three-man crew protected from launch exhaust/debris and ballistic threats by an armored man-rated cab. It operates within the MLRS command, control, and communications structure.

Ordnance options: All current and future MLRS rockets and ATACMS missiles, to include GMLRS DPICM and Unitary

Empty weight: 29,800 pounds

Max speed: 100 kilometers per hour

Max cruising range: 480 kilometers

SYSTEM INTERDEPENDENCIES

In this Publication

None

Other Major Interdependencies

C130/C-17, CNR (Combat Net Radio), GPS, JSTARS, MLRS MODS, Q36/Q37 FIREFINDER, Sensor Suite, TBMCS (Air Space Clearance)

PROGRAM STATUS

- **Current:** Fielded two Army National Guard (ARNG) battalions for total of 15 battalions fielded
- **Current:** Provide support to fielded units and units in combat
- **Current:** Field and provide sustainment and support activities for foreign military sales customers
- **2QFY12:** Completed retrofit of M142 fleet with new, Increased Crew Protection (ICP) armored cab
- **2QFY12:** Completed installation of Universal Fire Control System (UFCS) on entire M142 fleet
- **3QFY12:** Awarded Sapphire Glass Contract to Saint Gobain

PROJECTED ACTIVITIES

- **Continue:** Fielding to active and reserve components, with the last of 17 battalions fielding in FY13
- Field Long Range Communication (LRC), Blue Force Tracker (BFT), and Drivers Vision Enhancement (DVE) modifications
- **FY13:** Complete U.S. Full-rate production (FRP)

ACQUISITION PHASE

Technology Development

Engineering & Manufacturing Development

Production & Deployment

Operations & Support

High Mobility Artillery Rocket System (HIMARS) M142

FOREIGN MILITARY SALES
Jordan, Singapore, United Arab Emirates

CONTRACTORS
Prime:
Lockheed Martin (Grand Prairie, TX; Camden, AR)
Increased Crew Protection (ICP) Cab, Chassis:
BAE Systems (Sealy, TX)
Launcher Interface Unit (LIU), Weapons Interface Unit (WIU), and Power Switching Unit (PSU):
Harris Corp. (Melbourne, FL)

High Mobility Engineer Excavator (HMEE) I and III

INVESTMENT COMPONENT

Modernization

Recapitalization

Maintenance

MISSION

Provides the Army with earth-moving vehicles that support self-deployability, mobility, and speed to keep pace with the Brigade Combat Teams (BCTs).

HMEE-III Backhoe Loader (BHL) provides the Army with general excavation and earth-moving capabilities for general engineer construction units.

DESCRIPTION

The High Mobility Engineer Excavator Type I (HMEE-I) is a non-developmental, military-unique vehicle fielded to the Armys BCTs and other selected engineer units. The HMEE-I can travel up to 60 miles per hour on primary roads and up to 25 miles per hour on secondary roads. The high mobility of the HMEE-I provide earthmoving machines capable of maintaining pace with the Army's current combat systems. All HMEE-Is will be capable of accepting armor in the form of an armor cab (Crew Protection Kit), are C-130 transportable without armor, and are diesel driven. HMEE-I replaces

Small Emplacement Excavators (SEEs) in BCTs and Intern HMEEs in Stryker BCTs (SBTCs). The HMEE-I is employed in IBCTs, Heavy BCTs, SBCTs, Multi Role Bridge Companies, and Engineer Support Companies.

The HMEE-III Backhoe Loader (BHL) is a commercial-off-the-shelf backhoe loader with military modifications to include an armored cab intended for units that are relatively stationary and do not require speed and rapid deployability. Its maximum speed is 23 miles per hour on improved roads, and seven miles per hour off-road. The HMEE-III Backhoe Loader is used by Combat Support Brigades in general construction tasks. It is employed by Horizontal and Vertical Construction Units, and other non-engineer units such as Military Police and Quartermaster Units.

Tasks performed by the HMEE-I/III include repair and improvement of roads, trails, bridges, and airfields.

SYSTEM INTERDEPENDENCIES

In this Publication
HMEE III: M983 LET/M870 Trailer for highway transportability

PROGRAM STATUS

- **Current:** In fielding

PROJECTED ACTIVITIES

- **Continued:** Fielding is ongoing for both HMEE-I and III and should complete FY15

ACQUISITION PHASE

| Technology Development | Engineering & Manufacturing Development | Production & Deployment | Operations & Support |

High Mobility Engineer Excavator (HMEE) I and III

FOREIGN MILITARY SALES
HMEE-I:
Australia, Germany, New Zealand,

CONTRACTORS
HMEE-I OEM:
JCB Inc. (Pooler, GA)
Armor:
ADSI (Hicksville, NY)
Logistics:
XMCO (Warren, MI)
HMEE-III Backhoe Loader OEM:
Case New Holland (Racine, WI)
Armor:
BAE (Columbus, OH)
Logistics:
XMCO (Warren, MI)

High Mobility Multipurpose Wheeled Vehicle (HMMWV) Recapitalization (RECAP) Program

INVESTMENT COMPONENT

Modernization

Recapitalization

Maintenance

MISSION
Supports combat and combat service support units with a versatile, light, mission-configurable, tactical wheeled vehicle.

DESCRIPTION
The High Mobility Multipurpose Wheeled Vehicle (HMMWV) Recapitalization (RECAP) program supports the recapitalization of Up-Armored HMMWVs (UAH) returning from theater and Non-Armored HMMWVs (NAH) for National Guard homeland security and disaster relief missions. The recapitalization of UAHs will incorporate the latest HMMWV technical insertions common to the fleet. Current plans include the relocation of the air conditioning condenser to the engine compartment, high-load rating runflat tires, an improved geared fan drive, and electrical improvements. These adjustments will be incorporated into the depot Recapitalization (Recap) production lines under an automotive improvements program (AIP).

The HMMWV is a lightweight, highly-mobile, high-performance, diesel-powered, four-wheel drive, air transportable and air droppable family of tactical vehicles. The HMMWV uses common components to enable its reconfiguration as a troop carrier, armament carrier, shelter carrier, Tubed-Launched, Optically-Tracked, Wire-Guided missile carrier, ambulance, and scout vehicle. Since its inception, the HMMWV has undergone continuous evolution, including: improved survivability, technological upgrades; higher payload capacity; radial tires; Environmental Protection Agency emissions updates; commercial bucket seats; three-point seat belts and other safety enhancements; four-speed transmissions; and, in some cases, turbocharged engines and air conditioning.

There are numerous HMMWV variants. During RECAP the non-armored configurations are converted to the M1097R1 configuration. The HMMWV M1097R1 configuration incorporates a four-speed, electronic transmission, a 6.5-liter diesel engine, and improvements in transportability. It has a payload of 4,400 pounds.

The M1114 UAH may be converted to the M1151 during RECAP, and the M1151, M1152 and M1165 remain in their current configuration after RECAP. The M1114, M1151, M1152 and M1165 UAH configurations are based on the expanded capacity vehicle (ECV) chassis. The UAH was developed to provide increased ballistic and blast protection, primarily for military police, special operations, and contingency force use.

SYSTEM INTERDEPENDENCIES
In this Publication
None

Other Major Interdependencies
The HMMWV supports numerous data interchange customers, who mount various shelters and other systems on it. The M1101/1102 Light Tactical Trailer is the trailer designed for this vehicle.

PROGRAM STATUS
- **1QFY12-2QFY12:** Resumption of the UAH Depot Recap program at Red River Army Depot (RRAD) and Letterkenny Army Depot (LEAD)
- **1QFY12-3QFY12:** Automated Improvement Program improvements design and testing

PROJECTED ACTIVITIES
- **2QFY13:** Begin UAH Depot Recap AIP production at RRAD
- **2QFY13:** Complete UAH Depot Recap production at LEAD

ACQUISITION PHASE

Technology Development | Engineering & Manufacturing Development | Production & Deployment | Operations & Support

FOREIGN MILITARY SALES
None

CONTRACTORS
Red River Army Depot (Texarkana,TX)
Letterkenny Army Depot
   (Chambersburg, PA)
AM General (South Bend, IN)
GEP (Franklin, OH)
General Transmissions Products
   (South Bend, IN)

Improved Environmental Control Unit (IECU)

INVESTMENT COMPONENT

Modernization

Recapitalization

Maintenance

MISSION
Provides standardized environmental control capabilities to the DoD in accordance with DoDD 4120.11.

DESCRIPTION
The Improved Environmental Control Unit (IECU) program consists of three standard shelter-mounted systems in the following sizes: 9,000 BTUH (British thermal units per hour); 18,000 BTUH; 36,000 BTUH; and one skid-mounted unit of 60,000 BTUH. The IECU systems provide critical cooling to vital military electronic and support systems and equipment for the U.S. Army and the wider DoD. IECU Characteristics:
- Form, fit, and function replacement of military standard (MIL-STD) ECUs
- Ruggedized for military environments
- Increased reliability over current MIL-STD ECUs
- Reduced power consumption (results in overall fuel savings)
- Reduced weight, 10% lighter than current MIL-STD
- Fully operable up to 125 degrees Fahrenheit

- R-410A refrigerant, a commercial industry standard (compliant with all environmental legislative requirements)
- Soft start, limited inrush current (no voltage drop or breaker trip due to compressor startup)
- Electromagnetic Interference (EI) and nuclear, biological and chemical (NBC) protected
- Remote control capability; automatic safety controls
- Organically supportable

The IECU systems provide quality cooling, heating, and dehumidification for command posts; command, control, communications, computers, intelligence, surveillance, and reconnaissance (C4ISR) systems; weapon systems; and other battlefield operating systems while using a non-ozone-depleting refrigerant.

Additional improvements to the warfighter in theater are the IECUs soft start feature which limits inrush current; NBC & EI protected interface; fully embedded diagnostics; automatic safety

controls; and remote control capability for operations that require users to be out of the direct area.

SYSTEM INTERDEPENDENCIES
None

PROGRAM STATUS
- **1QFY12-4QFY12:** 60k IECU Full-rate production (FRP)
- **1QFY12-4QFY12:** 9/18/36k IECU Engineering and Manufacturing Development (EMD) Phase II

PROJECTED ACTIVITIES
- **1QFY13-3QFY15:** Continue 60k IECU Full-rate production (FRP)
- **1QFY13-2QFY14:** 9/18/36k IECU Engineering and Manufacturing Development (EMD) Phase II
- **4QFY13:** 120k IECU Milestone B Enter Engineering and Manufacturing Development (EMD)
- **3QFY14:** 9/18/36k IECU Milestone C Enter Production and Deployment
- **4QFY14-2QFY15:** 9/18/36 IECU Low-rate initial production
- **1QFY14-4QFY16:** 120K IECU Engineering and Manufacturing Development (EMD)

ACQUISITION PHASE

| Technology Development | Engineering & Manufacturing Development | Production & Deployment | Operations & Support |

Improved Environmental Control Unit (IECU)

FOREIGN MILITARY SALES
None

CONTRACTORS
9,000, 18,000, 36,000 BTUH IECU:
Mainstream Engineering (Rockledge, FL)
60,000 BTUH IECU:
DRS-ES (Florence, KY)

Improved Ribbon Bridge (IRB)

MISSION

Improves mobility by providing continuous roadway or raft configurations capable of crossing military vehicles over non-fordable wet gaps.

DESCRIPTION

The Improved Ribbon Bridge (IRB) is used to transport weapon systems, troops, and supplies over water when permanent bridges are not available. The IRB Float Ribbon Bridge System is issued to the Multi-Role Bridge Company (MRBC). The U.S. Army Modified Table of Organization and Equipment (MTOE) authorizes MRBCs to consist of: 42 IRB bridge bays (30 interior bays and 12 ramp bays), 42 Bridge Adapter Pallets (BAP), 14 Bridge Erection Boats (BEB), 14 Improved Boat Cradles (IBC), and 56 Common Bridge Transporters (CBT). These assets collectively address Tactical Float Ribbon Bridge wet gap bridging. All components are required to transport, launch, erect, and retrieve up to 210 meters of floating bridge per MRBC. The IRB can be configured as either a continuous full closure bridge or assembled and used for rafting operations. The IRB has a Military Load Capacity (MLC) of 105 wheeled/85 tracked (normal) and 110 wheeled/90 tracked (caution crossing). This MLC will support the Joint Force Commanders ability to employ and sustain forces worldwide. Bridge capabilities are provided in water currents moving at up to 10 feet per second.

The bridge system allows two-way traffic for HMMWV-width vehicles and increased MLC at all water current speeds over those of the Standard Ribbon Bridge. It is usable on increased bank heights over 2.2 meters (7.2 feet) and the improved folding/unfolding mechanism avoids cable breakage. Partially disassembled bays are C-130 transportable, and externally transportable by CH-47 and CH-53 aircraft.

SYSTEM INTERDEPENDENCIES
In this Publication
None

Other Major Interdependencies
IRB operations rely upon and are interdependent with fully mission-capable CBTs, BAPs, IBCs, and BEB assets within a fully MTOE equipped MRBC.

PROGRAM STATUS
This system has been fielded since 2002.
- **2QFY12:** Fielding 361st and 892nd MRBCs
- **3QFY12:** Fielding 125th and 502nd MRBCs
- **4QFY12:** Fielding 250th MRBC

PROJECTED ACTIVITIES
Fieldings are ongoing based on the Army Requirements Prioritization List.
- **3QFY13:** Fielding 310th MRBC
- **3QFY14:** Fielding APS

Improved Ribbon Bridge (IRB)

FOREIGN MILITARY SALES
None

CONTRACTORS
General Dynamics European Land Systems
    Germany (Kaiserslautern, Germany)
Logistic support:
AM General (Livonia, MI)
CBT manufacturer:
Oshkosh Corp. (Oshkosh, WI)
BEB manufacturer:
FBM Babcock Marine (Isle of Write,
    United Kingdom)

Improved Target Acquisition System (ITAS)

INVESTMENT COMPONENT

Modernization

Recapitalization

Maintenance

MISSION

Provides long-range sensor and anti-armor/precision assault fire capabilities, enabling the Soldier to shape the battlefield by detecting and engaging targets at long range with tube-launched, optically tracked, wire-guided (TOW) missiles or directing the employment of other weapon systems to destroy those targets.

DESCRIPTION

The Improved Target Acquisition System (ITAS) is a multipurpose weapon system used as a reconnaissance, surveillance, and target acquisition sensor. ITAS provides long-range anti-armor/precision assault fire capabilities to the Armys Infantry and Stryker Brigade Combat Teams (BCTs) as well as to the Marine Corps. ITAS is a major product upgrade that greatly reduces the number of components, minimizing logistics support and equipment requirements. Built-in diagnostics and improved interfaces enhance target engagement performance.

ITAS's second-generation forward-looking infrared sensors double the long-range surveillance of its predecessor, the M220 TOW system. It offers improved hit probability with aided target tracking, improved missile flight software algorithms, and an elevation brake to minimize launch transients. The ITAS includes an integrated day/night sight with eye-safe laser rangefinder, a position attitude determination subsystem, a fire-control subsystem, a lithium-ion battery power source, and a modified traversing unit. Soldiers can also detect and engage long-range targets with TOW missiles or, using the ITAS far-target location (FTL) enhancement, direct other fires to destroy the targets. The FTL enhancement consists of a position attitude determination subsystem (PADS) that provides the gunner with his own Global Positioning System (GPS) location and a 10-digit grid location to his target through the use of differential GPS. With the PAQ-4/PEQ-2 Laser Pointer, ITAS can designate .50 caliber or MK-19 grenade engagements. The ITAS can fire all versions of the TOW family of missiles.

The TOW 2B Aero and the TOW Bunker Buster have an extended maximum range to 4,500 meters. The TOW 2B Aero flies over the target (off-set above the gunners aim point) and uses a laser profilometer and magnetic sensor to detect and fire two downward-directed, explosively formed penetrator warheads into the target. TOW Bunker Buster, with its high-explosive blast-fragmentation warhead, is optimized for performance against urban structures, earthen bunkers, field fortifications, and light-skinned armor threats. ITAS operates from the High Mobility Multipurpose Wheeled Vehicle (HMMWV), the dismount tripod platform, and Stryker anti-tank guided missile (ATGM) vehicles.

SYSTEM INTERDEPENDENCIES

In this Publication
None

Other Major Interdependencies
The ITAS system is integrated on the M1121/1167 HMMWV and the Stryker Anti Tank Guided Missile (ATGM). The ITAS system provides the guidance for the TOW missile.

PROGRAM STATUS

- **Current:** ITAS has been fielded to 20 active and 16 reserve component Infantry BCTs and ten Stryker BCTs
- **Current:** The Marine Corps has begun fielding the ITAS to infantry and tank battalions to replace all Marine Corps M220A4 TOW 2 systems by 2012

PROJECTED ACTIVITIES

- **Continue:** ITAS total package fielding
- **FY13:** Complete fielding

ACQUISITION PHASE

| Technology Development | Engineering & Manufacturing Development | Production & Deployment | Operations & Support |

FOREIGN MILITARY SALES
NATO Maintenance and Supply Agency:
Canada
Direct Commercial Sale (DCS):
Portugal

CONTRACTORS
Raytheon (McKinney, TX)
Training Devices:
Intercoastal Electronics (Mesa, AZ)

Improvised Explosive Device Defeat/Protect Force (IEDD/PF)

INVESTMENT COMPONENT

Modernization

Recapitalization

Maintenance

MISSION
Provides both mounted and dismounted Soldiers with rapid and enduring capabilities to detect, defeat, and neutralize explosive hazards.

DESCRIPTION
The Improvised Explosive Device Defeat/Protect Force (IEDD/PF) Defeat product line is comprised of several highlighted systems:

The Autonomous Mine Detection System (AMDS) will provide the warfighter protection from explosive hazards through the use of robotic platforms and marking and neutralization systems.

The Entry Control Point (ECP) in a Box is a collection of non-lethal equipment that provides Soldiers protection against personal and vehicle borne IEDs. This is a coordinated effort with PdM Force Protection Systems (FPS).

The Jackal and Rhino are IED defeat systems integrated with Mine Resistant Ambush Protected (MRAP) vehicle platforms.

SYSTEM INTERDEPENDENCIES
In this Publication
None

Other Major Interdependencies
Mine Resistant Ambush Protected Vehicles (MRAP), Manual Transport Robotic System (MTRS)

PROGRAM STATUS
- **Current:** AMDS is in pre-Engineering and Manufacturing Development phase
- **Current:** The ECP in a Box is fielded in support of Overseas Contingency Operations (OCO)
- **Current:** The Jackal and Rhino were fielded in support of OCO; they now are potential candidates for Army Preposition Stock 5 (APS-5)

PROJECTED ACTIVITIES
- **4QFY13:** AMDS Milestone B

CE LAID AND BURIED THREATS DETECTION (SLBTD)

LIZATION PAYLOAD

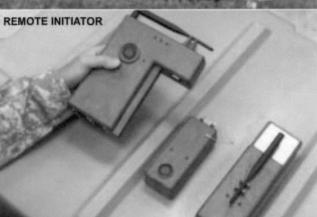
REMOTE INITIATOR

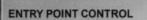
ENTRY POINT CONTROL

Improvised Explosive Device Defeat/Protect Force (IEDD/PF)

FOREIGN MILITARY SALES
None

CONTRACTORS
ECP:
Aardvark Technical (Azusa, CA)
Jackal:
Raytheon Technical Services
  (Indianapolis, IN)
Rhino:
Letterkenny Army Depot
  (Chambersburg, PA)

Installation Information Infrastructure Modernization Program (I3MP)

Modernization

Recapitalization

Maintenance

MISSION

Provides Soldiers with emerging Information Technology (IT) and infrastructure systems through lifecycle management, supporting Army enterprises and Joint networks.

DESCRIPTION

The Installation Information Infrastructure Management System (I3MP) modernizes installation infrastructure (voice, video, and data) connectivity by using a standard architecture and equipment from multiple vendors. I3MP supports network-centric operations and enterprise Unified Capabilities (UC).

The program also engineers, furnishes, installs, and tests integrated IT solutions for Strategic Command Centers. I3MP builds a converged UC ready infrastructure system that, connects the desktop to the Defense Information Systems Network (DISN) via an installation's campus area network. Based at Fort Belvoir, VA, I3MP modernizes installations in the Continental United States (CONUS), Europe, and Pacific.

SYSTEM INTERDEPENDENCIES

None

PROGRAM STATUS

- **FY12:** 39 contracting actions totaling over $225 million, in support of major modernization projects at various CONUS and Outside the Continental United States post, camps, stations

PROJECTED ACTIVITIES

- **FY13:** 30 modernization projects planned at various CONUS and Outside the Continental United States post, camps, and stations

Installation Information Infrastructure Modernization Program (I3MP)

FOREIGN MILITARY SALES
None

CONTRACTORS
The Infrastructure Modernization (IMOD) contract fulfills requirements of I3MP. IMOD is a group of 10 Indefinite Delivery/ Indefinite Quantity contracts which will support the upgrade and modernization of enterprise enabled voice and data networks worldwide.

AT&T Government Solutions
    (Bedminster, NJ)
General Dynamics Network Systems, Inc.
    (Arlington VA)
Science Applications International Corp.
    (McLean, VA)
Avaya Government Solutions (Fairfax, VA)
NextiraOne Federal LLC (Herndon, VA)

Instrumentable Multiple Integrated Laser Engagement System (I-MILES)

INVESTMENT COMPONENT

- Modernization
- Recapitalization
- Maintenance

MISSION
Provides force-on-force and force-on-target collective training at home stations and Combat Training Centers(CTCs).

DESCRIPTION
The Instrumentable Multiple Integrated Laser Engagement System (I-MILES) is the Army's primary live simulation system. I-MILES products include man-worn systems, combat vehicle systems, target systems, shoulder-launched systems, and controller devices. The system operates within a live, virtual, and constructive integrated architecture that supports Army and Joint exercises.

The I-MILES Combat Vehicle Tactical Engagement Simulation System (CV TESS) provides live training devices for armored vehicles with fire control systems including Bradley

Fighting Vehicles and Abrams Tanks. It interfaces and communicates with CTCs and home station instrumentation, providing casualty and battlefield damage assessments for after-action reporting. I-MILES CV TESS provides real-time casualty effects necessary for tactical engagement training in direct fire force-on-force and instrumented training scenarios.

The I-MILES Individual Weapons System (IWS) is a man-worn dismounted system, providing event data that can be downloaded for use in an after-action review and training assessment.

The Tactical Vehicle System (TVS) encompasses the Wireless Independent Target System (WITS) and replaces the previously fielded Independent Target System (ITS) and other Basic MILES currently fielded on non-turret military vehicles. TVS/WITS designs include Stryker variants, tactical wheeled vehicle configurations, and a separate configuration for tracked/oversized vehicles such as the M113 and Mine Resistant Ambush Protected (MRAP) vehicles.

The Shoulder Launched Munitions (SLM) system provides real-time casualty effects necessary for tactical engagement training in direct fire, force-on-force training scenarios, and instrumented scenarios. It replaces Basic MILES and provides better training fidelity for blue forces AT4 & Bunker Defeat Munition weapons and threat weapons using opposing force RPG7 visual modifications.

The Universal/Micro Controller Devices (UCD/MCD) are low-cost, lightweight devices used by observer controllers and maintenance personnel to initialize, set up, troubleshoot, reload, reset, resurrect, and manage participants during live force-on-force training exercises. These modular, self-contained devices interact and provide administrative control over all other MILES devices.

SYSTEM INTERDEPENDENCIES
None

PROGRAM STATUS
IWS:
- **Current:** Fielded approximately 14,000 IWS kits to the National

Training Center (NTC) and over 64,000 kits Army-wide
MXXI CVS:
- **Current:** Fielded over 400 systems to the NTC and Joint Readiness Training Center
SLM:
- **Current:** Fielded over 1,000 SLM kits to NTC and over 6,000 kits Army-wide
UCD/MCD:
- **Current:** Fielded over 14,000 CD kits Army-wide
TVS/WITS:
- **Current:** Fielded approximately 11,000 WITS kits to various home stations

PROJECTED ACTIVITIES
IWS:
- IWS testing completed and begin fielding
UCD/MCD:
- Complete basis of issue
CV TESS:
- CV TESS will complete testing
TVS:
- TVS will begin fielding

ACQUISITION PHASE

| Technology Development | Engineering & Manufacturing Development | Production & Deployment | Operations & Support |

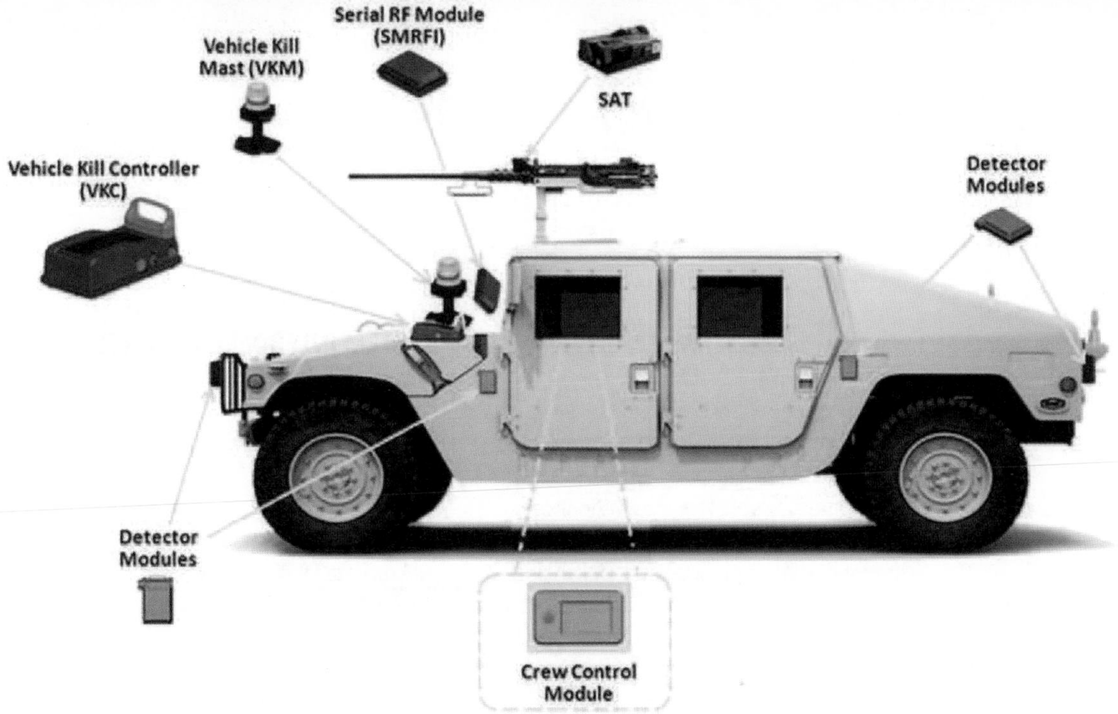
Serial RF Module
(SMRFI)
Vehicle Kill
Mast (VKM)
SAT
Vehicle Kill Controller
(VKC)
Detector
Modules
Detector
Modules
Crew Control
Module

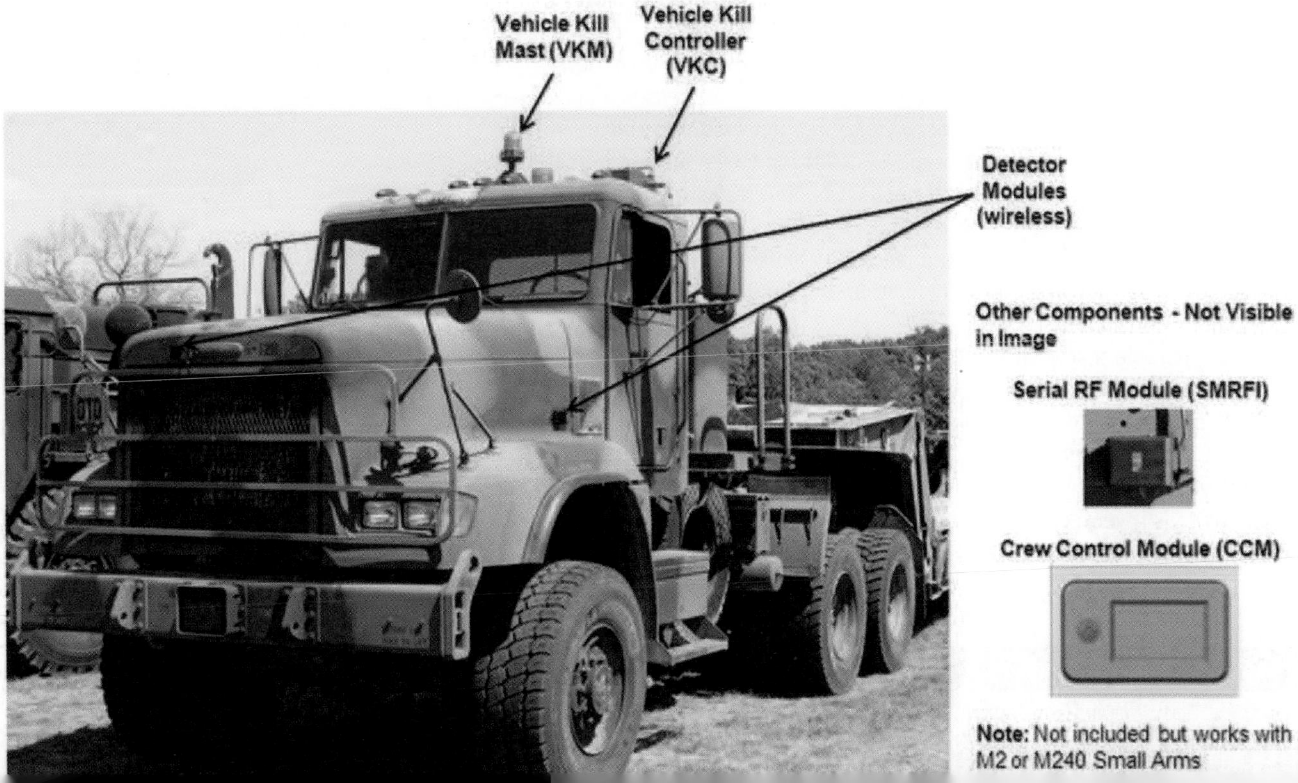
Vehicle Kill
Mast (VKM)
Vehicle Kill
Controller
(VKC)
Detector
Modules
(wireless)
Other Components - Not Visible
in Image
Serial RF Module (SMRFI)
Crew Control Module (CCM)
Note: Not included but works with
M2 or M240 Small Arms

Instrumentable Multiple Integrated Laser Engagement System (I-MILES)

FOREIGN MILITARY SALES
Armenia, Bulgaria, Croatia, Egypt, Georgia, Jordan, Latvia, Lithuania, Saudi Arabia, Romania, Ukraine

CONTRACTORS

IWS:
CUBIC Defense Sys. (San Diego, CA)

WITS:
Lockheed Martin (Orlando, FL)

TVS:
CUBIC Defense Systems (San Diego, CA)

MXXI CVS:
Lockheed Martin (Orlando, FL)

SLM:
Lockheed Martin (Orlando, FL)

CV TESS:
SAAB Training USA LOC (Orlando, FL)

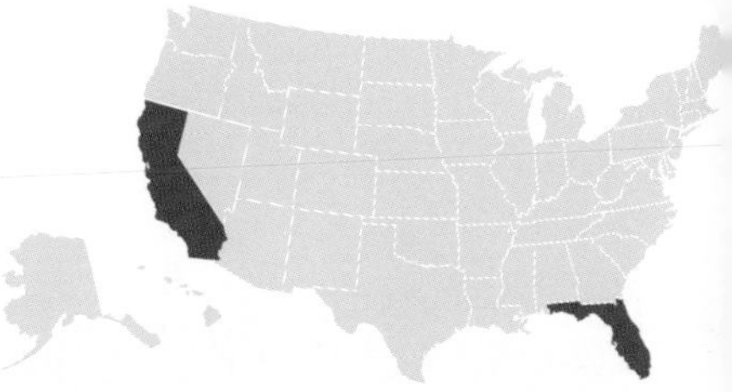

Integrated Family of Test Equipment (IFTE)

INVESTMENT COMPONENT

- Modernization
- Recapitalization
- Maintenance

MISSION

Develops, acquires, fields, and sustains automatic test equipment with the capability to troubleshoot, isolate, and diagnose faults, as well as verify the operational status of the weapon system.

DESCRIPTION

The Integrated Family of Test Equipment (IFTE) consists of interrelated, integrated, mobile, tactical, and man-portable systems. These rugged, compact, lightweight, general-purpose systems enable verification of the operational status of weapon systems, as well as fault isolation to the line-replaceable unit at all maintenance levels, both on and off the weapon system platform.

Base Shop Test Facility–Version 3 (BSTF(V)3):

The BSTF(V)3 is an off-platform automatic test system that tests electronic Line Replaceable Units (LRU) and Shop Replaceable Units (SRU) of ground and aviation systems.

Electro-Optics Test Facility (EOTF):

The EOTF tests the full range of Army electro-optical systems, including laser transmitters, receivers, spot trackers, forward-looking infrared systems, and television systems. It is fully mobile with VXI instrumentation, touch-screen operator interface, and an optical disk system for test program software and electronic technical manuals.

Next Generation Automatic Test System (NGATS):

The NGATS is the follow-on reconfigurable, rapidly deployable, expeditionary, interoperable tester and screener that supports Joint operations, reduces the logistics footprint, and replaces and consolidates obsolete, unsupportable automatic test equipment in the Army's inventory.

Maintenance Support Device– Version 3 (MSD-V3):

The latest generation MSD is a lightweight, rugged, compact, man-portable, general-purpose, at-platform automatic tester that has a docking station, detachable core tablet, and swivel and touch screen capabilities. It is used to verify the operational status of aviation, automotive, watercraft, electronic, and missile weapon systems and to isolate faulty components for immediate repair or replacement. MSD-V3 hosts Interactive Electronic Technical Manuals and the Digital Logbook, acts as a software uploader/ verifier to provide or restore mission software to weapon systems, and supports condition-based maintenance data collection and reporting. MSD-V3 supports more than 50 weapon systems and is used by more than 40 military occupational specialties.

SYSTEM INTERDEPENDENCIES

None

PROGRAM STATUS

- **1QFY12:** NGATS Limited Users Test 2
- **2QFY12:** NGATS delivery to Anniston Army Depot
- **4QFY12:** NGATS delivery to Red River Army Depot
- **Current:** MSD-V3 fielding
- **Current:** EOTF operations and support

PROJECTED ACTIVITIES

- **3QFY14:** MSD-V4 Contract Award
- **3QFY14:** NGATS first unit equipped

ACQUISITION PHASE

- Technology Development
- Engineering & Manufacturing Development
- Production & Deployment
- Operations & Support

ICE Test Adapter Set (bottom)   Digital Data Bus Set (top)

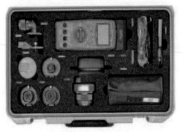

SWICE Kit   =   +

**Communicates via Wireless
Diagnostic Sensor (WDS)**

**Transmits Analog and Digital
Vehicle Health Data to MSD
for Analysis**

Notes:
* For use with analog vehicles
** For use with digital vehicles

MSD-V3
AAO: 40,059

**FOREIGN MILITARY SALES
MSD:**
Afghanistan, Australia, Bahrain, Chile, Djibouti, Egypt, Ethiopia, Germany, Israel, Iraq, Jordan, Korea, Kuwait, Lithuania, Macedonia, Morocco, Netherlands, Oman, Poland, Portugal, Saudi Arabia, Taiwan, Turkey, United Arab Emirates, Uzbekistan, Yemen

**CONTRACTORS
MSD-V3:**
Vision Technology Miltope Corp.
   (Hope Hull, AL)
BSTF(V)3:
Northrop Grumman (Rolling Meadows, IL)
EOTF:
Northrop Grumman (Rolling Meadows, IL)
NGATS:
Northrop Grumman (Rolling Meadows, IL)
DRS-TEM (Huntsville, AL)

Integrated Personnel and Pay System–Army (IPPS-A)

INVESTMENT COMPONENT

Modernization

Recapitalization

Maintenance

MISSION

Develops and delivers a single, integrated personnel and pay system that streamlines Army Human Resources, enhances the efficiency and accuracy of Army personnel and pay procedures, and support Soldiers and their families.

DESCRIPTION

The Integrated Personnel and Pay System–Army (IPPS-A) is a web-based, human resource (HR) system designed to provide integrated, multi-component personnel and pay capabilities across the Army.

SYSTEM INTERDEPENDENCIES

In this Publication

None

Other Major Interdependencies

IPPS-A will interface with more than 50 current HR and pay systems including:

- Digital Training Management System
- Army/American Council on Education Registry Transcript System
- Interactive Personnel Electronic Records Management System

PROGRAM STATUS

- **4QFY11:** Increment I acquisition strategy approved
- **2QFY12:** Increment I system integrator contract awarded
- **2QFY12:** Increment I acquisition program baseline approved
- **2QFY12:** Increment I development begins
- **4QFY12:** Product level testing begins
- **4QFY12:** Increment II system integrator request for proposal released to vendors

PROJECTED ACTIVITIES

- **Q2FY13:** Increment I Milestone C
- **Q2FY13:** Increment II Milestone B
- **Q2FY13:** Increment I full deployment decision

ACQUISITION PHASE

| Technology Development | Engineering & Manufacturing Development | Production & Deployment | Operations & Support |

UNITED STATES ARMY

Integrated Personnel and Pay System–Army (IPPS-A)

FOREIGN MILITARY SALES
None

CONTRACTORS
Prime:
EDC Consulting, LLC (Alexandria, VA)
Booz Allen Hamilton (Alexandria, VA)
Subcontractors:
IBM Global Business Services
  (Alexandria, VA)
Oracle (Alexandria, VA)
SRA International (Alexandria, VA)
ALTA IT Services (Alexandria, VA)
Maryn Consulting (Alexandria, VA)
Integrated Global Solutions
  (Alexandria, VA)
Susan L Berger, LLC (Alexandria, VA)
Joint Technology Solutions, Inc.
  (Alexandria, VA)
Trioh Consulting (Alexandria, VA)

Interceptor Body Armor (IBA)

Modernization

Recapitalization

Maintenance

MISSION

Increases warfighter lethality and mobility by optimizing Soldier protection while effectively managing all lifecycle aspects of personal protection equipment.

DESCRIPTION

The U.S. Army's Interceptor Body Armor (IBA) offers highly effective ballistic protection to Soldiers. The system includes two types of "soft armor": The Improved Outer Tactical Vest (IOTV), with mission tailored protective attachments, protects the neck, shoulders, groin, and lower back; and the Soldier Plate Carrier System (SPCS) offers decreased coverage area in order to increase Soldier mobility in various terrain conditions.

Two hard armor plate systems are available for use with the IOTV and SPCS: the Enhanced Small Arms Protective Inserts (ESAPI) and Enhanced Side Ballistic Inserts (ESBI); and the X Small Arms Protective Inserts (XSPI) and X Side Ballistic Inserts (XSBI). Both systems provide classified, multi-hit protection

against numerous stressing threats. The IOTV base vest (size medium) weighs 9.86 pounds. The IOTV with yoke and collar ensemble, groin, and lower back components weighs 15.09 pounds. All components provide robust fragmentation and 9mm protection. The total system with all components weighs 31.09 pounds (size medium). The IOTV is produced in 11 sizes to accommodate every Soldier in the Army.

ESAPI and XSAPI are produced in five sizes. ESBI and XSBI are produced in one size. Attachable throat, groin, shoulder, and lower back protectors increase fragmentation and 9mm protection. Webbing attachment loops on the vest accommodate Modular Lightweight Load-Carrying Equipment (MOLLE). A medical access panel on the IOTV allows for treatment to vital areas while a quick release mechanism on both systems allows for rapid doffing during emergency situations.

The complete IOTV system provides 1,085 square inches of fragmentation and 9mm handgun protection. The SPCS (size medium) provides 456 square inches of ballistic protection.

SYSTEM INTERDEPENDENCIES

None

PROGRAM STATUS

- **Current:** In sustainment, production, and being fielded
- **FY12:** Fielded 212,318 IOTV; 84,000 SPCS; 320,000 XSAPI; 320,000 XSBI

PROJECTED ACTIVITIES

- **2QFY13-1QFY15:** Continue fielding

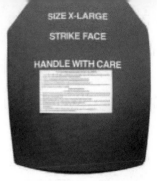

Interceptor Body Armor (IBA)

FOREIGN MILITARY SALES
None

CONTRACTORS
IOTV:
Carter Enterprises, LLC (Brooklyn, NY)
XSAPI, ESAPI:
Ceradyne, Inc. (Costa Mesa, CA)
IOTV, ESAPI, XSAPI:
BAE Systems (Phoenix, AZ)
SPCS:
KDH Defense Systems (Johnstown, PA)

Javelin

Modernization

Recapitalization

Maintenance

MISSION

Provides the dismounted Soldier a man-portable, fire-and-forget system that is highly lethal against targets ranging from main battle tanks to fleeting targets of opportunity found in current threat environments.

DESCRIPTION

The Close Combat Missile System Medium (CCMS-M) Javelin is highly effective against a variety of targets at extended ranges under day/night, battlefield obscurants, adverse weather, and multiple counter-measure conditions. The system's soft-launch feature permits firing from enclosures commonly found in complex urban terrain. Javelin's modular design allows the system to evolve to meet changing threats and requirements via both software and hardware upgrades. The system consists of a reusable command launch unit (CLU) with a built-in-test (BIT), and a modular missile encased in a disposable launch tube assembly. The CLU provides stand alone all-weather and day/night surveillance capability ideally suited for infantry operations in Afghanistan.

The Javelin missile and CLU together weigh 48.8 pounds. The system also includes training devices for tactical training and classroom training.

Javelin's imaging infrared fire-and-forget technology allows the gunner to fire and immediately take cover, to move to another fighting position, or to reload. The Javelin provides enhanced lethality through the use of a tandem warhead that will defeat all known armor threats. It is effective against both stationary and moving targets. This system also provides defensive capability against attacking/hovering helicopters. The performance improvements in current production Javelin Block I CLUs are: increased target identification range, increased surveillance time with new battery and software management of the on time, and external RS-170 interface for video output. The performance improvements in current production Javelin Block I missiles are: increased range, reduced weight, improved warhead lethality, and reduced time of flight. In current conflicts the CLU is being used as a stand-alone surveillance and target

acquisition asset. The Army is the lead for this Joint program with the Marine Corps.

SYSTEM INTERDEPENDENCIES

None

PROGRAM STATUS

- **Current:** Missile and CLU production
- **Current:** CLU total package fielding & CLU Retrofit
- **Current:** Javelin has been fielded to more than 95 percent of active duty units; fielding is underway to the National Guard

PROJECTED ACTIVITIES

- **Continue:** Multipurpose Warhead is planned to improve lethality against irregular/soft targets
- **Continue:** CLU total package fielding & CLU Retrofit
- **Continue:** Missile production
- **Complete** CLU production

Technology Development | Engineering & Manufacturing Development | Production & Deployment | Operations & Support

Javelin

FOREIGN MILITARY SALES
Australia, Czech Republic, France, Ireland, Jordan, Lithuania, New Zealand, Norway, Oman, Taiwan, United Arab Emirates, United Kingdom

CONTRACTORS
CLU:
Raytheon Missile Systems (Tucson, AZ)
DRS Technologies (Dallas, TX)
Raytheon (McKinney, TX; Dallas, TX; Garland, TX)
GEU:
Raytheon Missile Systems (Tucson, AZ)
Raytheon (McKinney, TX; DallasTX; Garland, TX)
ESAF/Seeker:
Lockheed Martin (Orlando, FL; Ocala, FL)
FPA:
DRS Technologies (Dallas, TX)
Propulsion Unit:
Aerojet (Camden, AR)
Missile Final Assembly:
Lockheed Martin (Troy, AL)
FTT, EPBST:
ECC International (Orlando, FL)
Batteries:
Acme Electric (Tempe, AZ)
Precursor Whd:
General Dynamics Ordnance and Tactical Systems (Camden, AR)
Containers:
Independent Pipe Products (Grand Prairie, TX)
Test Support:
Javelin Joint Venture (Huntsville, AL)

Joint Air-to-Ground Missile (JAGM)

INVESTMENT COMPONENT

Modernization

Recapitalization

Maintenance

MISSION

Provides a precision-guided, air-to-ground weapon for use by Joint service manned and unmanned aircraft to destroy stationary and moving high-value targets.

DESCRIPTION

The Joint Air-to-Ground Missile (JAGM) System will provide an air-to-surface, precision, standoff strike capability from the host platform. The adverse weather capable JAGM System will enable war-fighters to attack critical, high value, fixed and moving/fleeting targets day or night in battlefield limited visibility conditions from significant standoff ranges while remaining fully effective against a variety of countermeasures. The standoff capability of the JAGM System enables the weapon to place high value, heavily defended targets at risk while the aircrews remain outside the range of lethal point and area defenses. A terminal guidance capability enables the platform to launch the weapon and depart the launch area to enhance aircraft survivability. The precision accuracy enables the JAGM System to place point and moving/fleeting targets and target elements within a target complex at risk and reduces the probability of collateral damage.

The JAGM is an Army program with Joint requirements from the Navy and United Sates Marine Corps. The threshold platforms for JAGM are the U.S. Armys Apache (AH-64D) and the USMCs Super Cobra (AH-1Z).

Diameter: 7 inches
Weight: 115 pounds
Length: 70 inches
Range: 500-8,000 meters

SYSTEM INTERDEPENDENCIES

In this Publication
None

Other Major Interdependencies
Rotary-wing Launcher/Rack: M299

PROGRAM STATUS

- **4QFY12:** Extended Technology Development Phase contracts awarded

PROJECTED ACTIVITIES

- **1QFY13:** Delta Preliminary Design Review
- **1QFY14:** Critical Design Review (Guidance Section)
- **1QFY15:** Extended Technology Development Phase completed

ACQUISITION PHASE

Technology Development

Engineering & Manufacturing Development

Production & Deployment

Operations & Support

Joint Air-to-Ground Missile (JAGM)

FOREIGN MILITARY SALES
None

CONTRACTORS
Prime:
Raytheon (Tucson, AZ)
Lockheed Martin (Orlando, FL)

Joint Battle Command–Platform (JBC-P)

INVESTMENT COMPONENT

Modernization

Recapitalization

Maintenance

MISSION

Provides primary platform–level digital mission command for the Army, Marine Corps, and the Special Operations Forces (SOF) by distributing situational awareness (SA) data and mission command messages within/between platforms, dismounted leaders, and command centers/posts across celestial and terrestrial networks.

DESCRIPTION

Joint Battle Command–Platform (JBC-P) is the foundation for achieving information interoperability between Joint warfighting elements on current and future battlefields. As the next generation of the battle-tested Force XXI Battle Command Brigade-and-Below (FBCB2) technology, it will be the principal command and control system for the Army, Marine Corps, and Special Operations Forces units at the brigade-and-below level.

JBC-P provides users access to the tactical data necessary to gain and retain the tactical and operational initiative under all mission, enemy, terrain, troop, time, and civilian conditions despite an accelerated tempo. JBC-P contributes to the SA component of combat identification (CID) resulting in greater combat effectiveness and reduced fratricide. It consists of computer hardware and software integrated into tactical vehicles and aircraft, and is provided to dismounted forces.

JBC-P uses a product line approach to software development to reduce costs and promote a common architecture. Components include a core software module that provides common functionality required of all platforms, and tailored software modules with unique capabilities for dismounted, vehicle, logistics, aviation, and command post elements. It also provides one-way SA reporting called Beacons to track 100 percent of the vehicles in the operational environment or battlefield. JBC-P software is designed for use over the Blue Force Tracking II transceiver and associated satellite networks, as well as ground-based networks. Other key enhancements include a redesigned, more intuitive user interface with drag and drop icons, a faster map engine (includes touch-to-zoom), free draw, and group chat. It also provides an integrated Tactical Ground Reporting (TIGR) functionality.

SYSTEM INTERDEPENDENCIES

In this Publication

Advanced Field Artillery Tactical Data System (AFATDS), Battle Command Sustainment Support System (BCS3), Distributed Common Ground System Army (DCGS-A), Joint Tactical Radio System (JTRS), Movement Tracking System (MTS), Nett Warrior (NW), Warfighter Information Network Tactical (WIN-T) Increment 1, Warfighter Information Network Tactical (WIN-T) Increment 2, Warfighter Information Network Tactical (WIN-T) Increment 3

Other Major Interdependencies

Blue Force Tracking-Aviation (BFT-Avn), MCWS/CPOF, Joint Tactical COP Workstation (JTCW), Soldier Radio Waveform (SRW)

PROGRAM STATUS

- **1QFY12:** NIE 12.1 (SUE)
- **3QFY12:** NIE 12.2 (SUE)
- **4QFY12:** Milestone C Decision

PROJECTED ACTIVITIES

- **Continue:** JBC-P development and testing for Capability Sets (CS) 14 and 15
- **3QFY13:** NIE 13.2: Initial Operational Test & Evaluation
- **4QFY13:** Full-rate production decision
- **1QFY14:** Initial operating capability

ACQUISITION PHASE

Technology Development | Engineering & Manufacturing Development | Production & Deployment | Operations & Support

Joint Battle Command–Platform (JBC-P)

FOREIGN MILITARY SALES
None

CONTRACTORS
Software Development (Government Performing):
Software Engineering Directorate, AMRDEC (Huntsville, AL)
Program Support:
CACI (Aberdeen Proving Ground, MD)
Subject Matter Expert:
MITRE (Aberdeen Proving Ground, MD)
TIGR Production Engineering:
General Dynamics C4 Systems (Scottsdale, AZ)

Joint Biological Point Detection System (JBPDS)

INVESTMENT COMPONENT

Modernization

Recapitalization

Maintenance

BIDS

NBC RV

MISSION

Protects the Soldier by providing rapid and fully automated detection, identification, warning, and sample isolation of high-threat biological warfare agents.

DESCRIPTION

The Joint Biological Point Detection System (JBPDS) is the first automated, Joint biological warfare agent (BWA) detection system designed to meet the broad spectrum of operational requirements encountered by the services, across the entire spectrum of conflict.

It consists of a common biosuite that can be integrated onto a service platform, shipboard, or trailer-mounted to provide biological detection and identification to all service personnel. The JBPDS is portable and can support bare-base or semi-fixed sites. JBPDS will presumptively identify ten BWAs simultaneously. It will also collect a liquid sample for confirmatory analysis and identification. Technology refresh efforts will focus on reducing life cycle costs and obsolescence.

JBPDS can operate from a local controller on the front of each system, remotely, or as part of a network. JBPDS meets all environmental, vibration, and shock requirements of its intended platforms, as well as requirements for reliability, availability, and maintainability.

The JBPDS includes both military and commercial global positioning, meteorological, and network modem capabilities. The system will interface with the Joint Warning and Reporting Network (JWARN).

The JBPDS is currently fielded on the Stryker Nuclear, Biological, Chemical, Reconnaissance Vehicle (NBCRV) and M31A2 Biological Integrated Detection System.

SYSTEM INTERDEPENDENCIES
In this Publication
Nuclear Biological Chemical Reconnaissance Vehicle (NBCRV) Stryker, Force XXI Battle Command Brigade and Below (FBCB2), Single Channel Ground and Airborne Radio System (SINCGARS)

PROGRAM STATUS
- **1QFY12-4QFY12:** Continue unit fieldings
- **1QFY12-4QFY12:** Engineering Change Proposal developmental testing

PROJECTED ACTIVITIES
- **1QFY13-4QFY13:** Engineering change proposal First Article Testing
- **2QFY14:** Engineering Change Proposal production decision review
- **1QFY13-4QFY15:** Continue unit fieldings

ACQUISITION PHASE

Technology Development

Engineering & Manufacturing Development

Production & Deployment

Operations & Support

FOREIGN MILITARY SALES
None

CONTRACTORS
General Dynamics Armament and
Technical Products (Charlotte,NC)

BIO SUITE UNIT

BAWS INTAKE — COLLECTOR INTAKE

LCS COMPUTER AND DISPLAY   COLLECTOR   SWITCH, ROUTER, MEDIA CONVERTERS, & JCA

POWER PACK

BAWS W/ PURGE   IDENTIFIER   FLUID TRANSFER SYSTEM

Joint Biological Tactical Detection System (JBTDS)

INVESTMENT COMPONENT

Modernization

Recapitalization

Maintenance

MISSION

Provides a tactical, lightweight, battery-operated, biological warfare agent (BWA) system capable of detecting, warning, and presumptively identifying and collecting samples for follow-on confirmatory analysis.

DESCRIPTION

The Joint Biological Tactical Detection Systems (JBTDS) will be a lightweight, man-portable, battery-operated system that detects, warns, and provides presumptive identification and sample collection of BWA to provide near-real-time detection of biological attacks and hazards in the area of operation. It will have a local alarm and be networked to provide cooperative capability with reduced probability of false alarms. JBTDS will be employed organically at the battalion and lower levels by non-chemical, biological, radiological and nuclear personnel in tactical environments across multiple operational locations (e.g. forward operating bases, operationally engaged units, amphibious landing sites, air base operations, etc).

JBTDS will ultimately support force protection and maximize combat effectiveness by enhancing medical response decision making. When networked, JBTDS will augment existing biological detection systems to provide a theater-wide, seamless array capable of detection and warning.

SYSTEM INTERDEPENDENCIES

None

PROGRAM STATUS

- **1QFY12-4QFY12:** Continued Technology Development effort

PROJECTED ACTIVITIES

- **1QFY13:** Preliminary Design Review
- **3QFY13:** Milestone B decision
- **4QFY14:** Operational Assessment

ACQUISITION PHASE

Technology Development | Engineering & Manufacturing Development | Production & Deployment | Operations & Support

UNITED STATES ARMY

Joint Biological Tactical Detection System (JBTDS)

FOREIGN MILITARY SALES
None

CONTRACTORS
To be determined

NOTIONAL CONCEPT

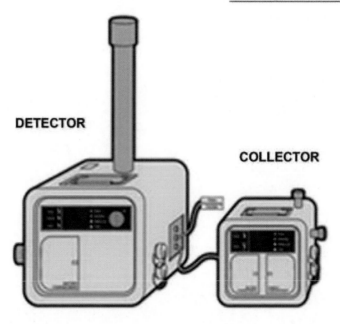

DETECTOR

COLLECTOR

MANUAL INDENTIFIER

Joint Chem/Bio Coverall for Combat Vehicle Crewman (JC3)

INVESTMENT COMPONENT

Modernization

Recapitalization

Maintenance

MISSION

Provides percutaneous protection against chemical and biological (CB) warfare agents to personnel who serve as crew members on armored vehicles.

DESCRIPTION

The Joint CB Coverall for Combat Vehicle Crewmen (JC3) is a lightweight, one-piece, flame-resistant (FR), CB protective coverall that resembles a standard Combat Vehicle Crewman (CVC) coverall. The JC3 may be worn over, or in lieu of, the CVC duty uniform in a CB environment. It resists ignition and provides thermal protection to allow emergency egress. The JC3 is not degraded by exposure to petroleum, oils, and lubricants present in the operational environment. The JC3 is compatible and provides complete percutaneous and respiratory protection when worn with current and developmental protective masks and mask accessories, headgear, gloves/ mittens, footwear, and other CVC ancillary equipment (e.g., Spall vest).

SYSTEM INTERDEPENDENCIES

In this Publication

Abrams Tank Upgrade (M1A2), Bradley Fighting Vehicle Systems Upgrade, Stryker Family of Vehicles

Other Major Interdependencies

Existing and co-developmental protective masks, appropriate mask accessories, protective headwear, hand-wear, footwear, and Army and Marine Corps armored vehicles

PROGRAM STATUS

- **1QFY12-4QFY12:** Continued production and fielding

PROJECTED ACTIVITIES

- **FY13-FY15:** Complete production and fielding

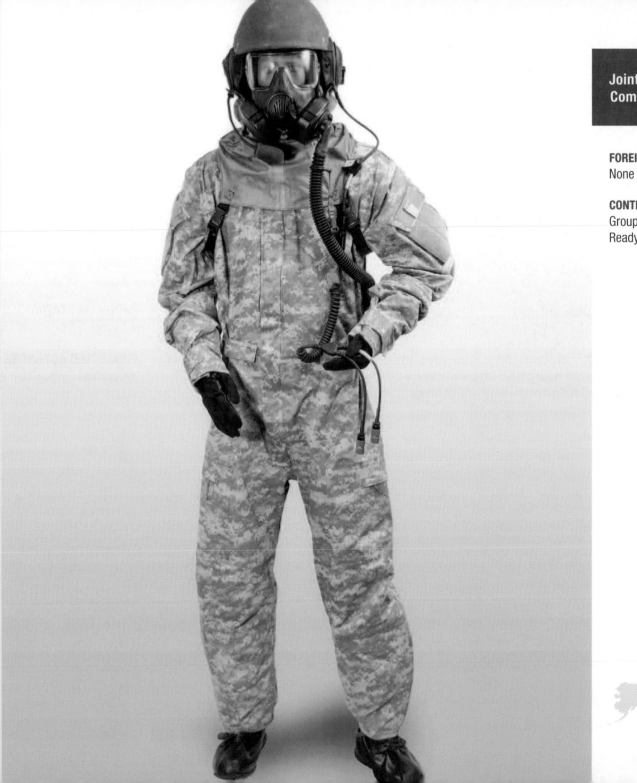

Joint Chem/Bio Coverall for Combat Vehicle Crewman (JC3)

FOREIGN MILITARY SALES
None

CONTRACTORS
Group Home Foundation, Inc. (Belfast, ME)
ReadyOne Industries (El Paso, TX)

Joint Chemical Agent Detector (JCAD) M4A1

INVESTMENT COMPONENT

Modernization

Recapitalization

Maintenance

MISSION
Protects U.S. forces by detecting, identifying, alerting, and reporting the presence of chemical warfare agents and toxic industrial chemical vapor.

DESCRIPTION
The Joint Chemical Agent Detector (JCAD) M4A1 is a pocket-size, rugged, handheld detector that automatically detects, identifies, and alarms to chemical warfare agents and toxic industrial chemical vapors.

The services can use the system on mobile platforms, at fixed sites, and on individuals designated to operate in a chemical threat area. The system can operate in a general chemical warfare environment, and can undergo conventional decontamination procedures by the warfighter.

The M4A1 JCAD commenced production in FY11. The M4A1 JCAD will reduce operation and sustainment costs, has an improved user interface, and is net ready.

The JCAD replaces the Automatic Chemical Agent Detector and Alarm (ACADA or M22), M90, and M8A1 systems. The JCAD may replace the Chemical Agent Monitor (CAM) and Improved Chemical Agent Monitor (ICAM).

Specific capabilities include:
- Instant feedback of hazard (mask only or full Mission-Oriented Protective Posture)
- Real-time detection of nerve, blister, and blood agents
- Stores up to 72 hours of detection data

The M4A1 will be net-ready through implementation of the common chemical, biological, radiological, and nuclear standard interface.

SYSTEM INTERDEPENDENCIES
In this Publication
None

Other Major Interdependencies
Modular Lightweight Load-carrying Equipment (MOLLE)

PROGRAM STATUS
- **Current:** Production and deployment

PROJECTED ACTIVITIES
- **Continue:** Production and deployment

ACQUISITION PHASE

| Technology Development | Engineering & Manufacturing Development | Production & Deployment | Operations & Support |

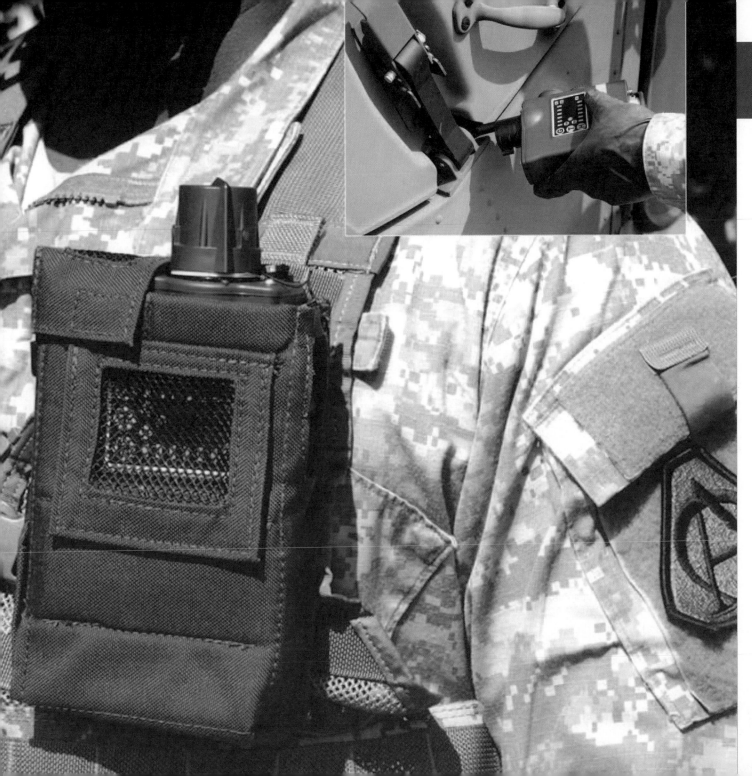

Joint Chemical Agent Detector (JCAD) M4A1

FOREIGN MILITARY SALES
None

CONTRACTORS
Smiths Detection, Inc. (Edgewood, MD)

Joint Effects Model (JEM)

INVESTMENT COMPONENT

Modernization

Recapitalization

Maintenance

MISSION

Provides enhanced operational and tactical-level situational awareness of the battlespace and provides near real-time hazard information before, during, and after an incident to influence and minimize effects on current operations.

DESCRIPTION

Joint Effects Model (JEM) is a web-based software program. It is the only accredited DoD computer-based tactical and operational hazard prediction model capable of providing common representation of chemical, biological, radiological, nuclear (CBRN) and toxic industrial chemicals/toxic industrial material hazard areas and effects. It may be used in two variants: as a standalone system, or as a resident application on host command, control, communications, computers, and intelligence systems. It is capable of modeling hazards in various scenarios, including counterforce, passive defense, accidents, incidents, high-altitude releases, urban environments, building interiors, and human performance degradation.

JEM supports planning to mitigate the effects of weapons of mass destruction and to provide rapid estimates of hazards and effects integrated into the Common Operational Picture.

The planned JEM Increment 2 will address the following targeted capabilities:

- Enhanced modeling to support biological and medical events
- STRATCOM RN requirements
- Urban Dispersion Modeling Improvements
- Modeling CBRN Defense effects of threat Missile Intercept
- Improved Toxic Industrial Material handling
- Improved agent fate predictive modeling
- Access to higher fidelity weather
- Common CBRN Modeling Interface (CCMI)
- Improved CBRN analyst support

JEM interfaces and communicates with the Joint Warning and Reporting Network (JWARN), associated weather systems, intelligence systems and various databases.

SYSTEM INTERDEPENDENCIES

In this Publication

Global Command and Control System Army (GCCS-A), Joint Warning and Reporting Network (JWARN)

Other Major Interdependencies

Global Command and Control System, Joint Tactical Common Operational Picture Workstation (JTCW)/Command and Control Personal Computer (C2PC), Meteorological Data Server

PROGRAM STATUS

- **1QFY12-4QFY12:** Continued Increment 1 deployment

PROJECTED ACTIVITIES

- **2QFY14:** Increment 2 Milestone B
- **4QFY14:** Increment 2 Milestone C

ACQUISITION PHASE

| Technology Development | Engineering & Manufacturing Development | Production & Deployment | Operations & Support |

Joint Effects Model (JEM)

FOREIGN MILITARY SALES
To be determined

CONTRACTORS
Northrop Grumman Mission Systems
(San Diego, CA)

Joint Effects Targeting System (JETS) Target Location Designation System (TLDS)

INVESTMENT COMPONENT

Modernization

Recapitalization

Maintenance

MISSION

Provides the dismounted Forward Observer and Joint Terminal Attack Controller the ability to acquire, locate, mark, and designate for precision Global Positioning System (GPS)-guided and laser-guided munitions, and provides connectivity to the Joint forces through fire and close air support digital planning/messaging devices.

DESCRIPTION

The Joint Effects Targeting System (JETS) is an Army-led, Joint interest program with the Air Force and Marine Corps to develop and field a one-man portable targeting system for forward observers and Joint Terminal Attack Controllers (JTACS).

This future system will answer the need for a lightweight, highly accurate targeting system that will allow target engagements with precision munitions (e.g., the Joint Direct Attack Munition, Excalibur, and laser-guided weapons) and provide crucial digital connectivity to request and control indirect fires and close air support from all Joint assets. The JETS' light weight will allow small units supported by Army forward observers or JTACs to have access to precision targeting in all operational environments.

The JETS consists of two major subsystems: the Target Location Designation System (TLDS) and the Target Effects Coordination System (TECS).

The TLDS will provide the dismounted observer and JTAC with a common enhanced lightweight hand-held capability to rapidly acquire, accurately locate, positively identify, and precisely designate targets. The TECS will interface with the TLDS and will provide a networked, automated communications capability to plan, coordinate, and deliver fire support, as well as provide terminal close air support guidance. Based on a strategy approved in FY11 by the Army Acquisition Executive and endorsed by the Joint Fire Support Executive Steering Committee (JFS ESC), the TECS requirement will be satisfied by continued development of existing service-specific forward entry systems which will comply with a Joint common minimum messaging set.

SYSTEM INTERDEPENDENCIES

In this Publication
None

Other Major Interdependencies
U.S. Army Portable Forward Entry Device, U.S. Air Force Tactical Air Control Party-Close Air Support System, and U.S. Marine Corps StrikeLink

PROGRAM STATUS

- **FY12:** Technology Development Activities for JETS TLDS for Milestone B
- **3QFY12:** Successful JETS TLDS pre-EMD review
- **4QFY12:** JETS TLDS Technology Demonstrator Early User Assessment
- **4QFY12:** JETS TLDS EMD Request for Proposal release

PROJECTED ACTIVITIES

- **FY13:** Milestone B Decision for JETS TLDS

ACQUISITION PHASE

| Technology Development | Engineering & Manufacturing Development | Production & Deployment | Operations & Support |

Joint Effects Targeting System (JETS) Target Location Designation System (TLDS)

FOREIGN MILITARY SALES
None

CONTRACTORS
BAE Systems (Nashua, NH)
Northrop Grumman Guidance and
    Electronics, Laser Systems (Apopka, FL)
DRS Technologies (Dallas, TX)
L-3 Warrior Systems (Londonderry, NH)

Joint Land Attack Cruise Missile Defense Elevated Netted Sensor System (JLENS)

INVESTMENT COMPONENT

Modernization

Recapitalization

Maintenance

MISSION

Provides a persistent surveillance and tracking capability for unmanned aerial vehicle and cruise missile defense. The system also provides fire control quality data to missile systems such as Army PATRIOT and Navy Aegis and fighter aircraft, allowing these systems to engage hostile threats from extended ranges.

DESCRIPTION

The Joint Land Attack Cruise Missile Defense Elevated Netted Sensor System (JLENS) orbit comprises two systems: a fire control radar system, and a wide-area surveillance radar system. Each system has a 74-meter tethered aerostat, a mobile mooring station, radar, communications payload, processing station, and associated ground support equipment. The JLENS mission is achieved by both the fire control radar and the surveillance radar systems operating as an "orbit," however, each system can operate autonomously and contribute to the JLENS mission.

JLENS uses its advanced sensor and networking technologies to provide 360-degree wide-area surveillance and tracking of cruise missiles and other aircraft. Operating as an orbit, the surveillance radar generates information that enables the fire control radar to readily search for, detect, and track low-altitude cruise missiles and other airborne threats. Once the fire control radar develops tracks, this information is provided to tactical data networks so other network participants can assess threat significance and assign systems to counter the threat. The fire control data supports extended engagement ranges by other network participants by providing high-quality track data on targets that may be terrain-masked from surface-based radar systems. JLENS information is distributed through the Joint service networks and contributes to the development of a single integrated air picture and the Army Integrated Air & Missile

Defense, communicating with Air & Missile Defense systems of systems. JLENS also performs as a multirole platform, enabling extended range communication and control linkages, communications relay, battlefield situational awareness, and can be configured to detect and track surface moving targets. JLENS can stay aloft up to 30 days, providing 24-hour radar coverage of the assigned areas. The radar systems can be transported by aircraft, railway, ship, or roadway.

SYSTEM INTERDEPENDENCIES
In this Publication
None

Other Major Interdependencies
The JLENS System is dependent on capabilities provided by Cooperative Engagement Capability (CEC), Multifunctional Information Distribution System (MIDS), and the Integrated Broadcast System (IBS). The JLENS program is interdependent with PATRIOT Advanced Capability-Three, Medium Extended Air Defense System, and Navy Integrated Fire Control–Counter Air (NIFC-CA).

PROGRAM STATUS
- **3QFY12:** Successful PAC-3 Integrated Fire Control Live Fire Test
- **4QFY12:** Naval Integrated Fire Control-Counter Air Demo
- **4QFY12:** Developmental Test 2

PROJECTED ACTIVITIES
- **1QFY13:** Limited user test
- **4QFY13:** Developmental Test 3
- **4QFY13:** Completes Engineering and Manufacturing Development Testing

ACQUISITION PHASE

| Technology Development | Engineering & Manufacturing Development | Production & Deployment | Operations & Support |

Joint Land Attack Cruise Missile Defense Elevated Netted Sensor System (JLENS)

FOREIGN MILITARY SALES
None

CONTRACTORS
Radar and Systems Engineering:
Raytheon (Andover, MA)
Surveillance Radar:
Raytheon (El Segundo, CA)
Platform:
TCOM (Columbia, MD; Elizabeth City, NC)
SETA Support:
Various SETA Contractors SETA
(Huntsville, AL)
Engineering and Technical Support:
E&TS Ktrs (Huntsville, AL)
Software:
Raytheon Solipsys (Fulton, MD)
Software Engineering:
Northrop Grumman (Huntsville, AL)

Joint Land Component Constructive Training Capability (JLCCTC)

INVESTMENT COMPONENT

Modernization

Recapitalization

Maintenance

MISSION
Provides unit commanders and their battle staffs the capability to train in an operationally relevant constructive simulation environment in Army Decisive Action operations.

DESCRIPTION
The Joint Land Component Constructive Training Capability (JLCCTC) supports Army Title X training worldwide at specific training facilities. JLCCTC supports Decisive Action to include offensive, defensive, stability, and civil support operations.

JLCCTC is a modeling and simulation software capability that provides the appropriate levels of modeling and simulation resolution as well as the fidelity needed to support both Army and Joint training requirements. JLCCTC is comprised of two separate federations, JLCCTC Multi-Resolution

Federation (MRF) and JLCCTC-Entity Resolution Federation (ERF). The MRF is a federated set of constructive simulation software that will support training of commanders and their staffs in maneuver, logistics, intelligence, air defense, and artillery.

The JLCCTC MRF-WARSIM trains Army commanders and their staffs in support of Command Post Exercises (CPXs), Warfighter Exercises (WFXs), and Mission Rehearsal Exercise (MRXs). JLCCTC provides the simulated environment in which computer-generated forces stimulate and respond to the Mission Command (MC) processes of the commanders and staffs. JLCCTC models will provide full training functionality for the Army and the Joint, intergovernmental, interagency and multinational (JIIM) spectrum. JLCCTC allows commanders and their staffs to train with their organizational real-world MC equipment.

SYSTEM INTERDEPENDENCIES
None

PROGRAM STATUS
MRF-W:
- **1QFY12-4QFY12:** 2ID Warpath II Exercise (South Korea)
- **1QFY12-4QFY12:** 2ID Full Spectrum Exercise (FSX) (South Korea)
- **1QFY12-4QFY12:** 56th BCT/36th ID - IBCT (Brownwood, TX)
- **1QFY12-4QFY12:** Intel School House Exercise (Fort Huachuca, AZ)
- **1QFY12-4QFY12:** 1st BDE/10th Mountain Div (Fort Drum, NY)
- **1QFY12-4QFY12:** 32nd IBCT Warfighter Exercise (WFX) (Fort McCoy, WI)
- **1QFY12-4QFY12:** III Corps Ramp-Up Exercise (Fort Hood, TX)
- **1QFY12-4QFY12:** III Corps WFX (Fort Hood, TX)
- **1QFY12-4QFY12:** 48th IBCT BDE WFX (Fort Stewart, GA)
- **1QFY12-4QFY12:** Fielded ERF V5.3 to numerous Army Sites
- **4QFY12:** Three additional BDE WFXs supported by MCTP utilizing MRF-W

PROJECTED ACTIVITIES
- **FY13:** 2ID Warpath Exercises (South Korea)
- **FY13:** Yama Sakura (YS) 63 Exercise (Japan)
- **FY13:** Ulchi-Freedom Guardian (UFG) 13 Exercise (South Korea)
- **FY13:** DIV WFX supported by MCTP
- **FY13:** Approximately twelve BDE WFXs supported by MCTP
- **FY13:** Numerous WIM Upper Enclave (UE) fielding's with ERF v5.3 sites
- **FY13:** Sustain ERF v5.3.x
- **FY14:** 2ID Warpath Exercises (South Korea)
- **FY14:** YS 65 Exercise (Japan)
- **FY14:** KR 14 Exercise (South Korea)
- **FY14:** UGF 14 Exercise (South Korea)
- **FY14:** Approximately twelve BDE WFXs supported by MCTP
- **FY14:** DIV/corps WFX supported by MCTP
- **FY14:** Sustain ERF v5.3.x
- **FY15:** 2ID Warpath Exercises (South Korea)
- **FY15:** YS 67 Exercise (Japan)
- **FY15:** KR 15 Exercise (South Korea)
- **FY15:** UGF 15 Exercise (South Korea)
- **FY15:** Approximately twelve BDE WFXs supported by MCTP
- **FY15:** DIC/Corps WFX supported by MCTP
- **FY15:** Sustain ERF v5.3.x

Joint Land Component Constructive Training Capability (JLCCTC)

FOREIGN MILITARY SALES
None

CONTRACTORS
Lockheed Martin Global Training and
    Logistics (Orlando, FL)
Tapestry Solutions Inc. (San Diego, CA)
Booz/Allen/Hamilton (Orlando, FL)

Joint Light Tactical Vehicle (JLTV)

Modernization

Recapitalization

Maintenance

MISSION

Provides a family of vehicles capable of performing multiple mission roles that will be designed to provide protected, sustained, networked mobility for personnel and payloads across the full range of military operations.

DESCRIPTION

The Joint Light Tactical Vehicle (JLTV) Family of Vehicles (FoVs) is an Army-led, Joint service program that will be capable of operating across a broad spectrum of terrain and weather conditions. The Joint services require enhanced performance, exceeding the existing High Mobility Multipurpose Wheeled Vehicle, supporting the Joint Functional Concepts of Battlespace Awareness, Force Application, and Focused Logistics. The JLTV FoVs consist of two variants: the Combat Tactical Vehicle (CTV) and the Combat Support Vehicle (CSV). The JLTV is transportable by a range of lift assets, including rotary-wing aircraft, to support operations across the range of military operations. Its maneuverability enables operations across the spectrum of terrain,

including urban areas, while providing inherent and supplemental armor against direct fire and improvised explosive device threats.

Payloads: CTV–3,500 pounds, CSV–5,100 pounds
Transportability: Internal–C-130, External–CH-47 at Curb Weight plus 2,000 pounds and CH-53, Sea–Height-restricted decks
Protection: Scalable armor to provide mission flexibility while protecting the force.
Mobility: Maneuverability to enable operations across the spectrum of terrain, including urban areas.
Networking: Connectivity for improved battlespace awareness and responsive, well-integrated command and control for embarked forces.
Sustainability: Reliable, maintainable, maximum commonality across mission role variants, onboard and exportable power, and reduced fuel consumption.

The JLTV FoV balances the "Iron Triangle" of payload, protection, and performance.

SYSTEM INTERDEPENDENCIES
None

PROGRAM STATUS
- **2QFY12:** Capability Development Document approved
- **2QFY12:** Engineering and Manufacturing Development (EMD) Request for Proposal released
- **4QFY12:** Milestone B, Enter EMD
- **4QFY12:** Award three full and open competition EMD contracts

PROJECTED ACTIVITIES
- **4QFY12-4QFY14:** Fabricate and Test 22-prototypes from each JLTV contractor

| Technology Development | Engineering & Manufacturing Development | Production & Deployment | Operations & Support |

UNITED STATES ARMY

Combat Support Vehicle (CSV)

Utility / Prime Mover

Shelter Carrier

Combat Tactical Vehicle (CTV)

General Purpose

Close Combat Weapons Carrier

Heavy Guns Carrier

FOREIGN MILITARY SALES
None

CONTRACTORS
AM General, LLC (South Bend, IN)
Lockheed Martin (Grand Praire, TX)
Oshkosh Corp. (Oshkosh, WI)

JLTV

Variant	CSV (2 Seat)	CTV (4 Seat)	
Base Vehicle Platform	UTL	GP	CCWC
Mission Package Configuration	Utl/PM/SC	GP	HGC	CCWC

wo Variants - Three Base Platforms - Multiple Mission Packages

Joint Personnel Identification Version 2 (JPIv2)

INVESTMENT COMPONENT

Modernization

Recapitalization

Maintenance

MISSION
Provides tactical biometrics collection capability configurable for multiple operational mission environments, enabling identity superiority.

DESCRIPTION
Joint Personnel Identification Version 2 (JPIv2) is an Army-led, Joint-interest program to develop a mobile system to provide multimodal (fingerprint, iris, palm and facial) biometric capability to collect, match, and store adversaries, host nation personnel, and third country nationals' biometric identities in support of a multi-range of military operational environments. This mobile system will be optimized for enrollments, analysis, referencing and producing/verifying credentials at fixed site locations, and a man portable system that is ruggedized, lightweight and supports dismounted operations. JPIv2 will be interoperable with and capable of synchronizing data with the DoD Automated Biometric Information System. The JPIv2 solution will support Biometrically-Enabled Intelligence and Forensics Enabling Capability solutions, as these capabilities emerge. This Joint, common solution is a continuation of the Navy's Personnel Identification Version 1 program.

SYSTEM INTERDEPENDENCIES
In this Publication
Biometric Enabling Capability (BEC)

PROGRAM STATUS
- **4QFY12:** Joint Staff Review of the JPIv2 Capability Development Document (CDD)

PROJECTED ACTIVITIES
- **1QFY13:** Approval of the JPIv2 CDD
- **2QFY13:** Complete pre-Engineering and Manufacturing Development (EMD) Review
- **3QFY13:** Milestone B for JPIv2
- **3QFY13:** Proceed to EMD Phase

ACQUISITION PHASE

| Technology Development | Engineering & Manufacturing Development | Production & Deployment | Operations & Support |

Joint Personnel Identification Version 2 (JPIv2)

FOREIGN MILITARY SALES
None

CONTRACTORS
CACI (Arlington, VA)
The Research Associates (TRA)
  (New York City, NY)
Technology Management Group (TMG)
  (King George, VA)

Joint Precision Airdrop System (JPADS)

INVESTMENT COMPONENT

Modernization

Recapitalization

Maintenance

MISSION

Provides the warfighter with precision airdrop capability, ensuring an accurate delivery of supplies to forward-operating forces, reducing vehicular convoys, and allowing aircraft to drop cargo at safer altitudes and off-set distances.

DESCRIPTION

The Joint Precision Airdrop System (JPADS) is a precision-guided airdrop system that provides rapid, precise, high-altitude delivery capabilities that do not rely on ground transportation. The system ensures accurate and timely delivery in support of operational missions, while providing aircraft with increased survivability. JPADS integrates a parachute decelerator, an autonomous guidance unit, and a load container or pallet to create a system that can accurately deliver critical supplies with great precision along a predetermined glide and flight path. The system is being developed in two weight classes: 2,000 pounds and 10,000 pounds. The guidance system uses military global positioning satellite data for precise navigation and interfaces with a wirelessly updatable mission planning module on board the aircraft to receive real-time weather data and compute multiple aerial release points. JPADS is being designed for aircraft to drop cargo from altitudes of up to 24,500 feet mean sea level. It will release cargo from a minimum off-set of eight kilometers from the intended point of impact, with an objective capability of 25 kilometers off-set. This off-set allows aircraft to stay out of range of many anti-aircraft systems. It also enables aircraft to drop systems from a single aerial release point and deliver them to multiple or single locations, thus reducing aircraft exposure time. Once on the ground, the precise placement of the loads greatly reduces the time needed to recover the load as well as minimizes exposure to ground forces.

SYSTEM INTERDEPENDENCIES

None

PROGRAM STATUS

- **1QFY12:** Completed and fielded product improvements to provide increased capabilities for the 2,000-pound variant in accordance with Joint Urgent Operational Needs Statement to include: accuracy improvements, adding terrain avoidance capability, and reducing the retrograde burden
- **1QFY12:** Completed Initial Operational Test & Evaluation of the 10,000-pound JPADS variant
- **3QFY12:** Completed high altitude follow-on tests of JPADS 10K for standard extraction from C-17 and C-130 aircraft
- **4QFY12:** Milestone C (Full-rate production and fielding decision) for 10,000-pound variant with subsequent award of remaining production contract option

PROJECTED ACTIVITIES

- **3QFY13:** Fielding begins for 10,000-pound variant

Airborne Guidance Unit (AGU)

Joint Precision Airdrop System (JPADS)

FOREIGN MILITARY SALES
None

CONTRACTORS
Airborne Systems North America (Pennsauken, NJ)
Draper Laboratories (Cambridge, MA)

Joint Service Aircrew Mask–Rotary Wing (JSAM RW) (MPU-5)

INVESTMENT COMPONENT

Modernization

Recapitalization

Maintenance

MISSION

Provides individual respiratory, ocular, and percutaneous protection for general purpose rotary wing aircrews against chemical/biological (CB) agents, toxic industrial materials, and radioactive particulate matter.

DESCRIPTION

The Joint Service Aircrew Mask–Rotary Wing (JSAM RW) is a part of the JSAM Family of Systems and will replace the currently fielded legacy non-Apache Aircrew Chemical-Biological Masks. The mask components will be optimized to minimize impact on the wearer's performance, maximize ability to interface with aircrew protective clothing, and provide improved field-of-view when compared to current aircrew protective masks. The JSAM provides core CB protective capabilities that are tailored to the RW airframes:
- Capable of being donned and doffed while in flight
- Flame retardant hood
- Continuous 16-hour CB protection
- Greater comfort and less physiological burden for aircrew

- Greater flexibility of use with man-mounted systems
- Compatible with appropriate aircraft life-support equipment
- No aircraft modifications required

It will be worn as part of the Air Warrior System ensemble in Force Protection (FP) conditions 1-3 and worn with the Aircrew Integrated Recovery Survival Armor Vest and Equipment (AIRSAVE) vest in FP 4 to increase an aviator's effectiveness in a CB environment. JSAM RW will be worn primarily by individuals on flying status whose primary duty is serving as a flight crew member.

The JSAM RW will integrate with aircraft subsystems of Aircrew Life Support Equipment, seating, portable aircrew systems, restraint systems, night vision goggles, sighting systems, communications systems, and aircraft mounted oxygen systems. Integration with aircraft and aircraft subsystems will follow applicable aircraft, safety, and airworthiness certification processes.

SYSTEM INTERDEPENDENCIES
In this Publication
Air Warrior (AW), Black Hawk/UH-60, CH-47F Chinook, Kiowa Warrior

PROGRAM STATUS
- **1QFY12-4QFY12:** Developmental testing

PROJECTED ACTIVITIES
- **1QFY14:** Milestone C Low-rate initial production decision
- **1QFY15-2QFY15:** Multi-Service Operational Test and Evaluation

Technology Development | Engineering & Manufacturing Development | Production & Deployment | Operations & Support

Joint Service Aircrew Mask–Rotary Wing (JSAM RW) (MPU-5)

FOREIGN MILITARY SALES
None

CONTRACTORS
Avox Systems (Lancaster, NY)

Joint Service General Purpose Mask (JSGPM) M-50/M-51

MISSION

Provides face, eye, and respiratory protection from battlefield concentrations of chemical and biological (CB) agents, toxins, toxic industrial materials, and radiological particulate matter.

DESCRIPTION

The Joint Service General Purpose Mask (JSGPM) is a lightweight, protective mask system incorporating state-of-the-art technology to protect U.S. Joint forces from actual or anticipated threats. There are two variants: The M-50 for ground and shipboard personnel and the M-51 for armored combat vehicle crewman. The JSGPM provides:

- Provide 24 hours of above-the-neck protection from CB agents and radioactive particles
- Head-eye-respiratory protection against chemical, biological, radiological, and nuclear (CBRN) threats (including toxic industrial materials)
- Flame resistant Hood for M-51 (Combat Vehicle Version)
- Improved CB protection

- Improved compatibility with current and emerging CB garments
- Improved reliability, improved comfort
- Enhanced field of view (greater than or equal to 80 percent)
- Lower breathing resistance (less than or equal to 30 millimeter of water)
- Reduced weight/bulk
- Improved drinking system design
- Improved mask carrier system

The JSGPM is interoperable with existing legacy and commercial radio systems, while ensuring future operation with the next generation of communications equipment. The mask system replaces the M40/M42 series of field protective masks for the Army and Marine Corps ground and combat vehicle operations, as well as the MCU-2/P series of protective masks for Air Force and Navy shore-based and shipboard applications. This mask is currently being fielded to all four Services – a first in the history of development and fielding.

SYSTEM INTERDEPENDENCIES
In this Publication
Abrams Tank Upgrade (M1A2), Bradley Fighting Vehicle Systems Upgrade, Joint Chem/Bio Coverall for Combat Vehicle Crewman (JC3), Stryker Family of Vehicles

Other Major Interdependencies
The JSGPM will interface with Joint service vehicles, weapons, communication systems, individual clothing and protective equipment, and CBRN personal protective equipment.

PROGRAM STATUS
- **1QFY12-4QFY12:** Continued production and fielding to the Air Force, Marines, and Navy warfighters

PROJECTED ACTIVITIES
- **3QFY13:** U.S. Navy Initial Operational Capability
- **2QFY14:** U.S. Army Initial Operational Capability

Joint Service General Purpose Mask (JSGPM) M-50/M-51

FOREIGN MILITARY SALES
None

CONTRACTORS
Avon Protection Systems (Cadillac, MI)

Joint Service Transportable Small Scale Decontaminating Apparatus (JSTSS DA) M26

INVESTMENT COMPONENT

Modernization

Recapitalization

Maintenance

MISSION
Provides the capability to conduct operations and support thorough decontamination operations.

DESCRIPTION
The Joint Service Transportable Small Scale Decontaminating Apparatus (JSTSS DA) M26 will enable warfighters to conduct operations and support through decontamination of non-sensitive military materiel and limited facility decontamination at logistics bases, airfields (and critical airfield assets), naval ships, ports, key command and control centers, as well as fixed facilities that have been exposed to chemical, biological, radiological, and nuclear (CBRN) warfare agents/contamination and toxic industrial materials. The system may also support other hazard abatement missions as necessary. The M26 is supported with one accessory kit and one water blivet per-system.

The M26 is transportable by a non-dedicated platform [e.g., High Mobility Multipurpose Wheeled Vehicle (HMMWV)/Trailer, Family of Medium Tactical Vehicles (FMTV)/Trailer] off-road over any terrain.

The M26 will decontaminate Chemical Warfare Agents (Nerve–G, Nerve–V_ Blister H) on tactical vehicles and crew served weapons below detection levels of M8 detector paper within five minutes contact time after an attack. The M26 will have a reliability of greater than or equal to 0.89.

SYSTEM INTERDEPENDENCIES
In this Publication
Family of Medium Tactical Vehicles (FMTV), High Mobility Multipurpose Wheeled Vehicle (HMMWV) Family of Vehicles

Other Major Interdependencies
All individual protective equipment, decontaminants, and detectors

PROGRAM STATUS
- **1QFY12-4QFY12:** Continued production and fielding

PROJECTED ACTIVITIES
- **4QFY13:** Complete fielding to achieve Full Operational Capability (FOC) for all Services

ACQUISITION PHASE

Technology Development

Engineering & Manufacturing Development

Production & Deployment

Operations & Support

Joint Service Transportable Small Scale Decontaminating Apparatus (JSTSS DA) M26

FOREIGN MILITARY SALES
None

CONTRACTORS
DRS Technologies (Florence, KY)

Joint Tactical Ground Station (JTAGS)

Modernization

Recapitalization

Maintenance

MISSION

Disseminates early-warning, alerting, and cueing information of ballistic missile attack and other infrared events to theater combatant commanders by using real-time, direct down-linked satellite data.

DESCRIPTION

Joint Tactical Ground Station (JTAGS) are forward-deployed, echelon-above-corps, transportable systems designed to receive, process, and disseminate direct down-linked infrared data from space-based sensors. Ongoing product improvement efforts will integrate JTAGS with the next-generation Space Based Infrared System (SBIRS) satellites. SBIRS sensors will significantly improve theater missile warning parameters. Expected improvements include higher quality cueing of active defense systems, decreased missile launch search area, faster initial report times, and improved impact ellipse prediction.

JTAGS processes satellite data and disseminates ballistic missile warning or special event messages

to warfighters in support of regional combatant commanders over multiple theater communication systems. Five JTAGS are deployed worldwide as part of the U.S. Strategic Command's Tactical Event System. The Army Space and Missile Defense Command Soldiers operate JTAGS, providing 24/7/365 support to theater operations.

SYSTEM INTERDEPENDENCIES

In this Publication
None

Other Major Interdependencies
U.S. Air Force's ACAT I, SBIRS satellite program

PROGRAM STATUS

- **1QFY12-3QFY12:** Complete fielding of JTAGS block upgrades including: commercial antenna systems, and information assurance
- **2QFY12-4QFY12:** Fielding of the Initial SBIRS Geosynchronous Orbit (GEO) satellite capability
- **3QFY12:** Begin new contract for support of Pre-Planned Product Improvement program; Includes full GEO satellite integration and de-shelter
- **4QFY12:** Initial SBIRS GEO certification for operational use

PROJECTED ACTIVITIES

- **1QFY12-4QFY13:** Software support, contractor logistics support, and depot operations continue

Technology Development | Engineering & Manufacturing Development | Production & Deployment | Operations & Support

Joint Tactical Ground Station (JTAGS)

FOREIGN MILITARY SALES
None

CONTRACTORS
Develop, Deploy, Sustain (CLS):
Northrop Grumman Electronic Systems
(Colorado Springs, CO)
SETA support:
BAE Systems (Huntsville, AL)

Joint Tactical Radio System Ground Mobile Radios (JTRS GMR)

INVESTMENT COMPONENT

Modernization

Recapitalization

Maintenance

MISSION

Develops, demonstrates, certifies, fields, and sustains an affordable, multi-channel networking radio system that meets DoD ground vehicle digitization and tactical communication requirements.

DESCRIPTION

Joint Tactical Radio System Ground Mobile Radios (JTRS GMR) are a key enabler of the DoD and Army Transformation and will provide critical communications capabilities across the full spectrum of Joint operations.

Through software reconfiguration, JTRS GMR can emulate current force radios and operate new Internet Protocol-based networking waveforms offering increased data throughput by utilizing self-forming, self-healing, and managed communication networks. The GMR route and retransmit functionality links various waveforms in different frequency bands to form one internetwork. GMR can scale from one to four channels supporting multiple security levels, and effectively use the frequency spectrum within the two megahertz to two gigahertz frequency range. The radios are Software Communications Architecture compliant with increased bandwidth through future waveforms. GMR are interoperable with more than four legacy radio systems and the JTRS family of radios.

SYSTEM INTERDEPENDENCIES

In this Publication
None

Other Major Interdependencies
Enhanced Position Locating Reporting System (EPLRS), High Frequency (HF), Satellite Communications (SATCOM), Soldier Radio Waveform (SRW), Ultra-High Frequency (UHF), Wideband Networking Waveform (WNW)

PROGRAM STATUS

- **1QFY12:** Under Secretary of Defense for Acquisition Technology and Logistics cancelled GMR and directed smart closeout of the GMR Engineering, Manufacturing, and Development (EMD) contract, and use of artifacts for a follow-on new start program
- **2QFY12:** National Security Agency Certified GMR
- **2QFY12:** GMR EMD contract closed out
- **4QFY12:** Initiated new start program referred to as Mid-Tier Networking Vehicular Radio (MNVR)

PROJECTED ACTIVITIES

To be determined

ACQUISITION PHASE

Technology Development

Engineering & Manufacturing Development

Production & Deployment

Operations & Support

Joint Tactical Radio System Ground Mobile Radios (JTRS GMR)

FOREIGN MILITARY SALES
None

CONTRACTORS*
Prime:
Boeing (Huntington Beach, CA)
Hardware:
BAE Systems (Wayne, NJ)
Rockwell Collins (Cedar Rapids, IA)
Northrop Grumman (Carson, CA)

*Contractors listed above are through 2QFY12.

Joint Tactical Radio System Handheld, Manpack, Small Form Fit (JTRS HMS)

INVESTMENT COMPONENT

Modernization

Recapitalization

Maintenance

MISSION

Provides the warfighter with a software re-programmable, networkable, multi-mode system-of-systems capable of simultaneous voice, data, and video communications, and meets the radio requirements for Soldiers and small platforms, such as missiles and ground sensors.

DESCRIPTION

The Joint Tactical Radio System (JTRS) Handheld, Manpack, and Small Form Fit (HMS) is a materiel solution meeting the requirements of the DoD Chief Information Officer for a Software Communications Architecture (SCA) compliant hardware system hosting SCA-compliant software waveforms (applications). HMS is an Acquisition Category ID program that encompasses specific requirements to support Special Operations Command, Army, Marine Corps, Air Force, and Navy communication needs.

SYSTEM INTERDEPENDENCIES

In this Publication
None

Other Major Interdependencies
SRW, HF, UHF SATCOM, MUOS

PROGRAM STATUS

- **1QFY12:** RR Initial Operational Test & Evaluation (IOT&E)
- **3QFY12:** MP IOT&E
- **4QFY12:** RR Initial Operational Capability (IOC)

PROJECTED ACTIVITIES

- **4QFY13:** MP IOC
- **1QFY14:** MP Full-rate production decision
- **1QFY14:** RR Full-rate production decision

ACQUISITION PHASE

| Technology Development | Engineering & Manufacturing Development | Production & Deployment | Operations & Support |

Joint Tactical Radio System Handheld, Manpack, Small Form Fit (JTRS HMS)

FOREIGN MILITARY SALES
None

CONTRACTORS
MP, 2 CH HH, SFFs -A, -B, -K,
AN/PRC-154 (Rifleman Radio):
General Dynamics (Scottsdale, AZ)
2 CH HH, SFF-B, AN/PRC-154
(Rifleman Radio):
Thales (Clarksburg, MD)
SFF-A, -D, -K:
BAE Systems (Wayne, NJ)
MP, SFF-D:
Rockwell Collins (Cedar Rapids, IA)
PM Support:
Science Applications International Corp.
    (SAIC) (San Diego, CA)

Joint Tactical Radio System Network Enterprise Domain (JTRS NED)

INVESTMENT COMPONENT

Modernization

Recapitalization

Maintenance

MISSION

Develops portable, interoperable, mobile ad-hoc networking waveforms/applications, providing the combatant commanders with the ability to command, control, and communicate with their forces via secure voice, video, and data media forms during military operations.

DESCRIPTION

The Joint Tactical Radio System Network Enterprise Domain (JTRS NED) is responsible for the development, sustainment, and enhancement of JTRS interoperable networking and legacy software waveforms. NED's product line consists of: 14 legacy waveforms (Bowman VHF, COBRA, EPLRS, Have Quick II, HF SSB/ALE, HF 5066, Link-16, SINCGARS, UHF DAMA SATCOM 181/182/183/184, UHF LOS, VHF LOS); three mobile ad-hoc networking waveforms (Wideband Networking Waveform [WNW], Soldier Radio Waveform [SRW], and Mobile User Objective System [MUOS]–Red Side Processing); and Network Enterprise Services (NES) including the JTRS WNW Network Manager (JWNM), SRW Network Manager (SRWNM), JTRS Enterprise Network Manager (JENM), and Enterprise Network Services (ENS).

JTRS NED manages the development of software waveforms targeted to operate on platforms such as the Mid-Tier Networking Vehicular Radio (MNVR), the Handheld, Manpack, and Small Form Fit (HMS) radios, the Airborne and Maritime/Fixed Station (AMF) radios, and the Multifunctional Information Distribution System (MIDS) radios. The JTRS NED software development and sustainment efforts leverage commercial technology and employ open-system architecture to better ensure interoperability and portability of each waveform. JTRS NED develops networking waveforms to support wireless networking with Global Information Grid connectivity for deployed warfighters at the tactical edge. In addition, NED provides network management and network services software for the planning, execution, configuration, and monitoring of the JTRS radios and networks, including route and retransmit services between networking and legacy waveforms.

SYSTEM INTERDEPENDENCIES

In this Publication
None

Other Major Interdependencies
Enhanced Position Location and Reporting System (EPLRS), MUOS, Link-16

PROGRAM STATUS

- **1QFY12:** SRW facilitated the successful completion of the Rifleman Radio Initial Operational Test & Evaluation (IOT&E)
- **1QFY12:** SRWNM evaluated as Operationally Suitable & Effective in Rifleman Radio IOT&E
- **3QFY12:** SRW operating in either the lower UHF band or L Band, supported both dismounted and mounted platoon operations during Network Integration Evaluation (NIE) 12.2, while JENM 1.2 loaded and managed over 1100+ SRW capable radios in 42 subnets
- **3QFY12:** Supporting both 173rd Airborne Brigade Combat Team and the 2nd battalion of 75th Ranger Regiment fielding of JTRS
- **4QFY12:** Complete MUOS Formal Qualification Test (FQT)

PROJECTED ACTIVITIES

- **1QFY13:** Support the continued deployment of JTRS throughout the Army and other service components
- **1QFY13:** Complete JENM FQT
- **1QFY13:** The JTRS NED becomes the Joint Tactical Networking Center (JTNC) Joint Tactical Networking (JTN) Program Management Office
- **1QFY15:** OH-58F KW Limited User Testing
- **1QFY15:** OH-58F KW Limited User Testing

ACQUISITION PHASE

Technology Development

Engineering & Manufacturing Development

Production & Deployment

Operations & Support

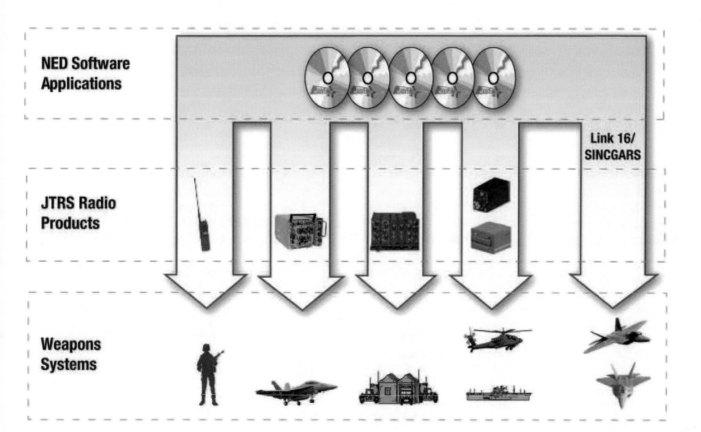

NED Software Applications

JTRS Radio Products

Link 16/ SINCGARS

Weapons Systems

Joint Tactical Radio System Network Enterprise Domain (JTRS NED)

FOREIGN MILITARY SALES
None

CONTRACTORS
MUOS:
Lockheed Martin (Sunnyvale, CA)
SRW, SRWNM, ENS Phase 1 (SoftINC):
ITT Corp. (Fort Wayne, IN)
PM Support:
SRA (Fairfax, VA)
JWNM, WNW, JENM:
Boeing (Huntington Beach, CA)
ENS Phase 1 Tactical Data Controller (TDC):
Rockwell Collins (Cedar Rapids, IA)

Joint Warning and Reporting Network (JWARN)

INVESTMENT COMPONENT

Modernization

Recapitalization

Maintenance

MISSION
Provides the Joint Forces with a capability to report, analyze, and disseminate detection, identification, location, and warning information to accelerate the warfighter's response to a chemical, biological, radiological, or nuclear (CBRN) attack.

DESCRIPTION
The Joint Warning and Reporting Network (JWARN) is a computer-based application that networks CBRN sensors directly with Joint and service command and control systems to collect, analyze, identify, locate, and report information on CBRN threats, and to disseminate that information to decision-makers for battlespace situational awareness. JWARN has a software and hardware component: JWARN Mission Application Software (JMAS) and JWARN Component Interface Device (JCIDS).

JMAS is the software component and resides on the Joint and service Command, Control, Communications, Computers, Intelligence, and Surveillance Reconnaissance (C4ISR) systems. It will generate warning and dewarning information to affected forces via nuclear, biological, and chemical (NBC) reports. It reduces the time from incident observation to warning to within two minutes (NBC-1 and -4), enhances warfighters' situational awareness throughout the area of operations, and supports battle management tasks. The application allows operator selection of automatic, delayed, or on-command dissemination of all NBC Reports (1-6) and single, validated and accredited implementation of Allied Tactical Publication-45 with the capability to generate and display CBRN/Toxic Industrial Material hazard plots. JWARN automates the recording and archiving of exposure data for effective force protection. JWARN interfaces with the Joint Effects Model to provide detailed, high fidelity hazard prediction plume overlays.

JCIDS is the hardware component that will provide interface between the CBRN sensors and relays warnings to C4ISR systems via advanced wired or wireless networks.

JWARN Increment 2 will provide a single, accredited, net-centric capability to detect, identify, communicate, correlate, analyze, and display results on the Common Operational Picture.

SYSTEM INTERDEPENDENCIES
In this Publication
Global Command and Control System–Army (GCCS-A), Joint Effects Model (JEM)

Other Major Interdependencies
Global Command and Control System–Joint (GCCS-J), Joint Tactical Common Operational Picture (COP) Workstation (JTCW)/Command and Control Personal Computer (C2PC)

PROGRAM STATUS
- **1QFY12-4QFY12:** Continued JMAS Increment 1 deployment to Army as part of Capability Set (CS) 13-14
- **1QFY12-4QFY12:** Modernization and updates to JMAS Increment 1
- **2QFY12:** JWARN Increment 2 Materiel Development Decision

PROJECTED ACTIVITIES
- **3QFY13:** JCIDS Increment 2 Milestone A decision
- **1QFY13-3QFY14:** Continue JMAS Increment 1 deployment as part of CS 13-14
- **4QFY14-4QFY15:** Continue JMAS Increment 1 deployment as part of CS 15-16
- **1QFY15:** JMAS Increment 2 Milestone B decision

ACQUISITION PHASE

| Technology Development | Engineering & Manufacturing Development | Production & Deployment | Operations & Support |

UNITED STATES ARMY

Joint Warning and Reporting Network (JWARN)

FOREIGN MILITARY SALES
None

CONTRACTORS
Northrop Grumman Information
    Technology (Orlando, FL)

Kiowa Warrior

INVESTMENT COMPONENT

Modernization

Recapitalization

Maintenance

MISSION
Performs aerial reconnaissance and security in support of ground maneuver forces.

DESCRIPTION
The Kiowa Warrior is a single-engine, two-man, lightly armed reconnaissance helicopter with advanced avionics, navigation, communication, weapons, and cockpit integration systems. Its mast-mounted sight houses a thermal imaging system, low-light television, and a laser rangefinder/designator permitting target acquisition and engagement at standoff ranges and in adverse weather. Sensor imagery from compatible Unmanned Aerial Systems and manned aircraft can be received and relayed to other aircraft or ground stations. The navigation system can convey precise target locations to other aircraft or artillery via its advanced digital communications system. Modifications to address safety, obsolescence, and weight will keep the aircraft viable beyond FY25. The Cockpit and Sensor Upgrade Program is the Program of Record that will convert the OH-58D into the OH-58F

Kiowa Warrior; addressing additional capabilities, safety enhancements, weight reduction and obsolescence issues. Additionally, the Kiowa Warrior Wartime Replacement Aircraft program addresses attrition with delivery of new aircraft.

SYSTEM INTERDEPENDENCIES
In this Publication
2.75 Inch Rocket Systems (Hydra), HELLFIRE Family of Missiles, Air Warrior (AW), Aviation Combined Arms Tactical Trainer (AVCATT), Shadow Tactical Unmanned Aerial Vehicle (TUAV), Global Command and Control System–Army (GCCS-A), Distributed Common Ground System–Army (DCGS-A), Single Channel Ground and Airborne Radio System (SINCGARS)

Other Major Interdependencies
M3P .50 Caliber Machine Gun, Common Sensor Payload, ARC-231, BFT, CXP (APX-118), CXP ((APX-123), IDM, AMPS

PROGRAM STATUS
- **1QFY12:** OH-58D Kiowa Warrior Fielding Single Channel Full Authority Digital Electronic Control improvements
- **2QFY12:** OH-58F Kiowa Warrior Cockpit and Sensor Upgrade Program Critical Design Review (CDR)
- **3QFY12:** First Wartime Replacement Aircraft delivery from Corpus Christi Army Depot
- **4QFY12:** OH-58F Kiowa Warrior complete prototype build of first OH-58F aircraft

PROJECTED ACTIVITIES
- **3QFY13:** OH-58F prototype aircraft first flight
- **1QFY15:** OH-58F Kiowa Warrior Limited User Testing

Kiowa Warrior

FOREIGN MILITARY SALES
None

CONTRACTORS
Airframe:
Bell Helicopter Textron (Fort Worth, TX)
Sensor (OH-58D):
DRS Optronics Inc. (Palm Bay, FL)
Sensor (OH-58F):
Raytheon (McKinney, TX)
Engine:
Rolls Royce Corp. (Indianapolis, IN)
Mission Computer:
Honeywell (Albuquerque, NM)
Cockpit Displays:
Elbit Systems of America (Fort Worth, TX)

Korea Transformation, Yongsan Relocation Plan, Land Partnership Plan (KT/YRP/LPP)

INVESTMENT COMPONENT

Modernization

Recapitalization

Maintenance

MISSION

Develops, engineers and delivers the command, control, communications, computers and intelligence (C4I) systems and services and the migration of existing systems and services throughout Korea as part of a seamless transition to United States Army Garrison – Camp Humphreys (USAG-H), with no impact to the warfighter's ability to conduct operations.

DESCRIPTION

Product Director (PD) Korea Transformation, Yongsan Relocation Plan, Land Partnership Plan (KT/YRP/LPP) facilitates the relocation of U.S. Forces from Seoul and other outlying areas to USAG-H. USAG-H is projected to grow from approximately 9,000 personnel to more than 29,000 personnel, and will be adding 25 million square feet of facilities by 2016.

PD KT/YRP/LPP will provide and sustain classified and unclassified C4I capabilities from the end user building to the Wide Area Network, to include telephone, data, audio-visual

information, security, and control systems for United States Forces Korea (USFK).

PD KT/YRP/LPP will be closely synchronized with the transition of wartime Operational Control from the United States to the Republic of Korea (ROK).

SYSTEM INTERDEPENDENCIES

In this Publication
None

Other Major Interdependencies
The U.S. and ROK jointly share costs and responsibilities for the program. The Far East District Corp of Engineers controls the Master Schedule. Program execution is influenced by bilateral, international agreements and relationships.

PROGRAM STATUS

- **1QFY12:** Establishment of KT/YRP/LPP Identify Gap Coverage. Develop and define Acquisition Plan and Systems Engineering Plan. Increment 1: Request For Proposal (RFP) release

- **2QFY12:** Award Increment 1. Initiate Increment 2: Contracting Activities
- **3QFY12:** Initiate Increment 3: Contracting Activities. Execute Increment 1: Engineering and Installation

PROJECTED ACTIVITIES

- **1QFY13:** Increment 2: RFP release. Award Increment 2
- **2QFY13:** Increment 2: Engineering and Installation
- **3QFY13:** Increment 3: RFP release
- **1QFY14:** Award Increment 3
- **2QFY14:** Increment 3: Engineering and Installation

Korea Transformation, Yongsan Relocation Plan, Land Partnership Plan (KT/YRP/LPP)

FOREIGN MILITARY SALES
None

CONTRACTORS
Approximately three individual delivery orders will be awarded; contractors are to be determined pending contract awards.

Lakota/UH-72A

INVESTMENT COMPONENT

Modernization

Recapitalization

Maintenance

MISSION

Provides a flexible response to Homeland Security requirements such as search and rescue operations, reconnaissance and surveillance, and medical evacuation (MEDEVAC) missions.

DESCRIPTION

The UH-72A LAKOTA Light Utility Helicopter (LUH) will conduct general support utility helicopter missions and execute tasks as part of an integrated effort with other Joint Services, government agencies, and non-governmental organizations. The LUH is to be deployed only to non-combat, non-hostile environments. The UH-72A is a variant of the American Eurocopter U.S.-produced EC-145.

The UH-72A is a twin-engine, single-main-rotor commercial helicopter. It has seating for two pilots and up to six passengers or two NATO standard litters. Two Turbomeca Arriel 1E2 engines, combined with an advanced four-blade rotor system, provide lift and speed in a wide range of operating conditions. The LUH can be configured with two NATO standard litters, passenger seating for a medical attendant and a crew chief.

The UH-72A is equipped with modern communication and navigation avionics. It includes a 3-axis autopilot and single pilot Instrument Flight Rules capability. The cockpit is compatible with night vision devices. In addition to the MEDEVAC configuration, the UH-72A is also being fielded in a VIP, ARNG Security & Support (S&S) and Combined Training Center (CTC) configurations.

The United States Navy Test Pilot School ordered five UH-72A aircraft in 2008. These were fielded in early FY10 and support experimental pilot training at the school.

In 2011 The Battalion Mission Equipment Package (MEP) and the CTC MEP were added to the UH-72A fleet. The S&S MEP provides the National Guard to conduct Homeland Security, patrol and counter drug missions. One hundred UH-72A will be equipped with the MEP and fielded across the Continental United States to include Puerto Rico and Hawaii.

The CTC MEP provides the ability to conduct Opposing Force and Observor/Controller missions to support training at the National Training Center, Joint Readiness Training Center and the Joint Multinational Readiness Center. Forty aircraft will be retrofitted with the MEP.

SYSTEM INTERDEPENDENCIES

In this Publication
None

Other Major Interdependencies
ARC-231, C-5 (RERP), C-17, Civil Communications GATM, OH-58A/C, UH-1, Sealift, USCG Communications, VHF/UHF Communications

PROGRAM STATUS

- **1QFY12:** 271 aircraft are on contract with 49 to be delivered
- **2QFY12:** First production delivery of S&S MEP aircraft
- **4QFY12:** Complete production of 235 aircraft

PROJECTED ACTIVITIES

- **1QFY13:** 34 Aircraft on contract
- **4QFY13:** Complete CTC MEP retrofits

Lakota/UH-72A

FOREIGN MILITARY SALES
None

CONTRACTORS
Airframe:
American Eurocopter (Columbus, MS; Grand Prairie, TX)
CLS:
Helicopter Support, Inc. (Trumbull, CT)
American Eurocopter (Grand Prairie, TX)
Training:
American Eurocopter (Grand Prairie, TX)
CFSR:
American Eurocopter (Grand Prairie, TX)
Program Management:
EADS North America (Huntsville, AL; Arlington, VA)
Helicopter Support, Inc. (Huntsville, AL; Grand Prairie, TX)

Light Capability Rough Terrain Forklift (LCRTF)

INVESTMENT COMPONENT

Modernization

Recapitalization

Maintenance

MISSION

Provides a mobile, variable reach, rough terrain forklift capable of handling all classes of supply.

DESCRIPTION

The Light Capability Rough Terrain Forklift (LCRTF) is a C-130 and CH-47 sling load transportable, 5,000 pound capacity, variable reach rough terrain forklift with fork tine oscillation and side-shift cab controls. The LCRTF enters, stuffs and un-stuffs Army International Organization for Standardization (ISO) containers, and the extendable boom fork carriage also un-stuffs pallets from ISO containers on trucks. The LCRTF is a significant improvement over the existing 4,000 pound capacity forklift fleet (enclosed air conditioned cab, moveable tines, TIER III engine and improved helicopter lift).

SYSTEM INTERDEPENDENCIES

None

PROGRAM STATUS

- **Current:** The LCRTF is currently going through testing at Aberdeen Proving Ground

PROJECTED ACTIVITIES

- **FY13:** Complete testing and logistics development
- **FY14-FY17:** 602 systems are projected to be fielded

ACQUISITION PHASE

Technology Development

Engineering & Manufacturing Development

Production & Deployment

Operations & Support

UNITED STATES ARMY

Light Capability Rough Terrain Forklift (LCRTF)

FOREIGN MILITARY SALES
None

CONTRACTORS
Kalmar RT Center, LLC (Cibolo, TX)

Lightweight 155mm Howitzer System (LW155)

INVESTMENT COMPONENT

Modernization

Recapitalization

Maintenance

MISSION
Provides direct, reinforcing, and general artillery fire support to maneuver forces.

DESCRIPTION
The Lightweight 155mm Howitzer (M777A2) will replace all M198 155mm Howitzers in operation with the Army and Marine Corps. The extensive use of titanium in all its major structures makes it 7,000 pounds lighter than its predecessor, the M198, with no sacrifice in range, stability, accuracy, or durability, and it can be dropped by parachute. The M777A2's independent suspension, smaller footprint, and lower profile increase strategic deployability and tactical mobility. The system uses numerous improvements to enhance reliability and accuracy, and significantly increase system survivability. The M777A2 is Jointly managed; the Marine Corps led the development of the Howitzer and the Army led the development of Towed Artillery Digitization, the digital fire control system for the M777A2.

Software upgrades incorporating the Enhanced Portable Inductive Artillery Fuze Setter and the Excalibur Platform Integration Kit hardware give the M777A2 the capability to program and fire the Excalibur precision-guided munition.

Specifications for the M777A2 Excalibur-compatible howitzer are:

Weight: 10,000 pounds
Emplace: Less than three minutes
Displace: Two to three minutes
Maximum range: 30 kilometers (rocket assisted round)
Rate-of-fire: Four rounds per minute maximum; two rounds per minute sustained
Ground mobility: Family of Medium Tactical Vehicles, Medium Tactical Vehicle Replacement, five-ton trucks
Air mobility: CH-53D/E; CH-47D; MV-22; two per C-130; six per C-17; 12 per C-5
155mm compatibility: All fielded and developmental NATO munitions
Digital and optical fire control: Self-locating and pointing; digital and voice communications; self-contained power supply

SYSTEM INTERDEPENDENCIES
In this Publication
None

Other Major Interdependencies
Army Software Blocking, Defense Advanced Global Positioning System Receiver

PROGRAM STATUS
- **4QFY12-4QFY12:** Continued Full-rate production

PROJECTED ACTIVITIES
- **2QFY13-4QFY13:** Continue Full-rate production. Continue Army and Marine Corps New Equipment Training and Fieldings. Support FMS partners (Canada and Australia)
- **3QFY13:** Performance Based Life Cycle Support Contract Award (Army and USMC)

ACQUISITION PHASE

Technology Development | Engineering & Manufacturing Development | Production & Deployment | Operations & Support

Lightweight 155mm Howitzer System (LW155)

FOREIGN MILITARY SALES
Australia and Canada

CONTRACTORS
Prime:
BAE Systems (Hattiesburg, MS)
Barrow-in-Furness (United Kingdom)
Cannon Assembly (GFE):
Watervliet Arsenal (Watervliet, NY)
Titanium Castings:
Precision CastParts Corp. (Portland, OR)
Body:
Triumph Structures (Chatsworth, CA)
Castings:
Howmet Castings (Whitehall, MI)
Optical Fire Control (GFE):
Seiler Instruments (St. Louis, MO)
Digital Fire Control:
General Dynamics Canada (Ottawa, Canada)

Lightweight Counter Mortar Radar (LCMR)

INVESTMENT COMPONENT

Modernization

Recapitalization

Maintenance

MISSION

Identifies indirect fire threats by providing the ability to rapidly locate rockets, artillery, and mortar firing positions automatically by detecting and tracking the shell and backtracking to the weapon position. Provides observed fires (for friendly fires), will provide accurate "did hit" data of friendly fires, and will detect and template hostile locations.

DESCRIPTION

The AN/TPQ-50 Lightweight Counter Mortar Radar (LCMR) is a man-portable and High Mobility Multipurpose Wheeled Vehicle 1152A mountable lightweight radar system used to locate rocket, artillery, and mortar Points of Origin (POO) and Points of Impact (POI) out to a range of 10 kilometers. The radar accomplishes this by detecting and tracking the projectile then extrapolating the POO and POI to within 50 meters Circular Error Probability. The AN/TPQ-50 has a continuous 360 degree surveillance using an electronically scanned antenna. The radar can be rapidly deployed by two Soldiers. The AN/TPQ-50 sends a warning message to indicate an incoming round and is a critical sensor to the Counter Rocket Artillery and Mortar (C-RAM) system of systems construct. The radar also is digitally interoperable with the Advanced Field Artillery Tactical Data System and Forward Area Air Defense Command and Control.

SYSTEM INTERDEPENDENCIES
In this Publication
Counter-Rocket, Artillery and Mortar (C-RAM)

PROGRAM STATUS
- **FY12:** Continued Low-rate initial production
- **3QFY12:** Initial Operational Test and Evaluation

PROJECTED ACTIVITIES
- **1QFY13:** Full Materiel Release
- **2QFY13:** Full-rate production decision
- **3QFY14:** Follow-on Test and Evaluation

ACQUISITION PHASE

Technology Development	Engineering & Manufacturing Development	Production & Deployment	Operations & Support

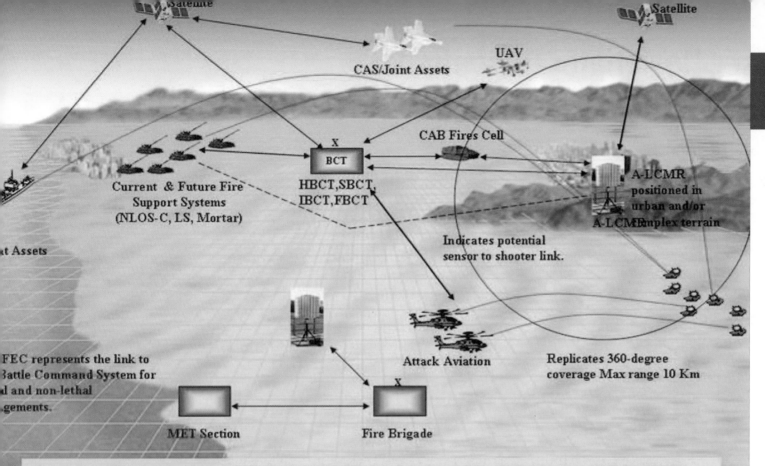

Satellite

CAS/Joint Assets

UAV

CAB Fires Cell

X
BCT

HBCT,SBCT,
IBCT,FBCT

Satellite

A-LCMR
positioned in
urban and/or
A-LCMR complex terrain

Current & Future Fire
Support Systems
(NLOS-C, LS, Mortar)

Indicates potential
sensor to shooter link.

t Assets

Attack Aviation

Replicates 360-degree
coverage Max range 10 Km

FEC represents the link to
Battle Command System for
d and non-lethal
gements.

MET Section

X

Fire Brigade

The A-LCMR is a lightweight, 360-degree, rapidly man-transportable, digitally connected, day/night mortar, cannon and rocket locating radar that uses computer-controlled signal processing of the radar signal data to perform target detection, verification, and tracking of enemy and friendly mortars. POO and POI are determined and provided to C2 nodes for engagement of Threat firing platform and early warning for force protection.

Lightweight Counter Mortar Radar (LCMR)

FOREIGN MILITARY SALES
None

CONTRACTORS
SRCTec (North Syracuse, NY)
Syracuse Research Corp. (SRC) (Syracuse, NY; Statewide, MD)
Yuma Proving Ground (Yuma, AZ)
Tobyhanna AD (Tobyhanna, PA)

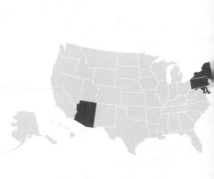

WEAPON SYSTEMS 2014

Lightweight Laser Designator Rangefinder (LLDR) AN/PED-1 & AN/PED-1A

MISSION

Provides the dismounted Fire Support Teams, Combat Observation and Lasing Teams, and Scouts with a precision target location and laser designation system that allows them to call for fire using precision, near-precision, and area munitions.

DESCRIPTION

The AN/PED-1 Lightweight Laser Designator Rangefinder (LLDR) is a crew-served man-portable, modular target locator and laser designation system. The primary components are the Target Locator Module (TLM) and the Laser Designator Module (LDM).

The TLM incorporates a thermal imager, day camera, laser designator spot imaging electronic display, eye-safe laser rangefinder, digital magnetic compass, Selective Availability/Anti-Spoofing Module Global Positioning System (SAASM GPS), and digital export capability. The original LLDR 1 operates on one BA-5699 battery, but it can also use a Single Channel Ground and Airborne Radio System (SINCGARS) battery when laser

designation is not required. A new compact laser designator is being fielded with the LLDR2, which requires less power and operates on one common SINCGARS battery (BA-5390 or BA-5590).

To provide a precision targeting capability to the dismounted Soldier, Product Manager Soldier Precision Targeting Devices has developed the LLDR 2H (AN/PED-1A), which integrates a celestial navigation system with the digital magnetic compass in the TLM to provide highly accurate target coordinates to allow the Soldier to call for fire with precision GPS guided munitions. A Modification of in Service Equipment program will retrofit fielded LLDR 1 and 2 systems with the LLDR 2H precision targeting capability beginning in FY13.

The TLM can be used as a stand-alone device or in conjunction with the LDM. At night and in obscured battlefield conditions, the operator can recognize vehicle-sized targets at more than three kilometers. During day operations, targets can be recognized at more than seven kilometers. The LDM emits coded

laser pulses compatible with DoD and NATO laser-guided munitions. Targets can be designated at ranges greater than five kilometers.

Weight: (total system) 35 pounds (LLDR 1), less than 30 pounds (LLDR 2), and less than 32 pounds (LLDR 2H) for a 24-hour mission

SYSTEM INTERDEPENDENCIES
In this Publication
None

Other Major Interdependencies
SAASM

PROGRAM STATUS
- **3QFY12:** First LLDR 2H retrofit contract awarded
- **4QFY12:** First production deliveries of LLDR 2H

PROJECTED ACTIVITIES
- **FY13:** Follow-on developmental test for LLDR 2H
- **FY13:** LLDR 2H materiel release and intial fielding
- **FY13:** LLDR 2H multiple-year retrofit contract award

Lightweight Laser Designator Rangefinder (LLDR) AN/PED-1 & AN/PED-1A

FOREIGN MILITARY SALES
None

CONTRACTORS
Northrop Grumman Guidance and Electronics, Laser Systems (Apopka, FL)

Line Haul Tractor

MISSION

Supports combat service and support units with transportation of bulk petroleum products, containerized cargo, general cargo, and bulk water.

DESCRIPTION

The M915A5 Truck Tractor is a 6x4 semi-tractor used to perform the Line Haul mission. The M915A5 is a block upgrade of the M915A3 system, incorporating enhanced suspension and power train components. This block upgrade allows the M915A5 to readily accept armor packages without reducing mission capability.

Gross vehicle weight rating: 120,000 pounds
Unarmored Gross vehicle weight: 26,500 pounds
Armored Gross vehicle weight: 33,500 pounds
Fifth-wheel capacity: six-inch, 30,000 pounds
Diagnosis: Electronic

Brake system: Anti-lock brake system (ABS)
Towing speed: 65 miles per hour with full payload
Engine: Detroit Diesel S60 (500 horse power, 1650 pound-foot torque, DDEC IV engine controller)
Transmission: Allison HD4500SP (six-speed automatic)

The M915A5 truck is equipped with a two-passenger cab and has an updated power distribution module, upgraded wiring harnesses, and a Roll Stability Control system. Auxiliary power connections have been added to supply emerging systems and added command, control, communications, computers, and intelligence communication systems. A pair of 60-gallon fuel tanks increase fuel capacity by 20 gallons to extend driving range. The cab is ten inches wider and extends 34 inches behind the driver and passenger seats. The vehicle has an improved ABS and an updated collision warning system.

The M915A3 Line Haul Tractor is the Army's key line haul distribution platform. It is a 6x4 tractor with a two-inch kingpin and 105,000-pound gross combination weight capacity.

Gross vehicle weight: 52,000 pounds
Fifth-wheel capacity: two-inch, 30,000 pounds
Diagnosis: Electronic
Brake system: ABS
Towing speed: 65 miles per hour with full payload
Engine: Detroit Diesel S60 (430 horse power, 1,450 pound-foot torque, DDEC IV engine controller)
Transmission: Allison HD5460P (six-speed automatic) with power take-off
The M916A3 Light Equipment Transport (LET) is a 6x6 tractor with 68,000-pound gross vehicle weight tractor with 3-1/2-inch, 40,000-pound capacity, 45,000-pound winch for recovery and transport and compensator fifth wheel. It has an electronic diesel engine, automatic electronic transmission, ABS, and is capable of operating at speeds up to 60 miles per hour on flat terrain. This Non-Developmental Item vehicle is used primarily to transport the M870 40-ton low-bed semi-trailer.

The M917A2 and M917A2 Truck Chassis, 75,000 gross vehicle weight rating, 8x6 (for 20-ton dump truck), 12-cubic yard dump truck vehicles are authorized in Corps units, primarily the construction and combat support companies and the combat heavy battalions. It has an electronic diesel engine, automatic electronic transmission, ABS, and is capable of operating at speeds up to 55 miles per hour on flat terrain.

SYSTEM INTERDEPENDENCIES

In this Publication
None

Other Major Interdependencies
M872, 34-ton flatbed semi-trailer; M1062A1, 7,500-gallon semi-trailer; M967/M969, 5,000-gallon semi-trailer

PROGRAM STATUS

- Production completed of all M915A5s currently on contract, 220 produced
- Fielding completed to all CONUS based units

PROJECTED ACTIVITIES

- Continue divestment of older M915 variants (A0, A1, A2, A4)
- Development of M915A3, M915A5, & M916A3 C-Kit armor
- Initiation and development of Heavy Dump Truck (HDT) program

Line Haul Tractor

FOREIGN MILITARY SALES
Afghanistan

CONTRACTORS
Prime:
Daimler Trucks North America
    LLC/Freightliner (Portland, OR;
    Cleveland, NC)
Engine:
Detroit Diesel (Detroit, MI)
ABS Brakes:
Meritor (Troy, MI)
Dump body:
Casteel Manufacturing (San Antonio, TX)

Load Handling System Compatible Water Tank Rack (Hippo)

INVESTMENT COMPONENT

Modernization

Recapitalization

Maintenance

MISSION

Enhances and expedites the delivery of bulk potable water into the division and brigade areas, providing the Army with the capability to receive, store, and distribute potable water to units deployed throughout the battlefield.

DESCRIPTION

The Load Handling System (LHS) Compatible Water Tank Rack (Hippo) represents the latest in bulk water distribution systems technology. It replaces the 3,000 and 5,000 Semi-trailer Mounted Fabric Tanks. The Hippo consists of a 2,000-gallon potable water tank in an International Organization for Standardization frame with an integrated pump, engine, alternator, filling stand, and 70-foot hose reel with bulk suction and discharge hoses. It has the capacity to pump 125 gallons of water per minute.

The Hippo is fully functional, mounted or dismounted, and is air transportable and ground transportable when full, partially full, or empty. It is Heavy Expanded Mobility Tactical Truck, Palletized Load System (PLS)

and PLS-Trailer-compatible, and designed to operate in cold weather environments and can prevent water from freezing at -25 degrees Fahrenheit. The Hippo can be moved, set up, and established rapidly using minimal assets and personnel. No site preparation by engineer assets is required, and its modular configuration supports Expeditionary Joint Forces Operations.

SYSTEM INTERDEPENDENCIES

None

PROGRAM STATUS

- Production and fielding of Hippos
- Competitive contract award selection
- **Continue:** RESET program

PROJECTED ACTIVITIES

- **Continue:** production and fielding of Hippos
- **Continue:** RESET program

Technology Development   Engineering & Manufacturing Development   Production & Deployment   Operations & Support

Load Handling System Compatible Water Tank Rack (Hippo)

FOREIGN MILITARY SALES
None

CONTRACTORS
Mil-Mar Century, Inc. (Miamisburg, OH)

Longbow Apache (AH-64D) (LBA)

INVESTMENT COMPONENT

Modernization

Recapitalization

Maintenance

MISSION

Conducts armed reconnaissance, close combat, mobile strike, and vertical maneuver missions when required, in day, night, obscured battlefield, and adverse weather conditions.

DESCRIPTION

The AH-64D Longbow Apache (LBA) is the Army's only heavy attack helicopter for both the current and Future Force. Capable of destroying armor, personnel, and materiel targets in obscured battlefield conditions the Longbow Apache is a 2-engine, 4-bladed, tandem-seat attack helicopter with 30mm cannon, Hydra 70 2.75-inch rockets, laser, and Radio Frequency (RF) HELLFIRE missiles. It upgrades 634 Apaches into AH-64D Longbow Block III configuration with procurement of 259 Fire Control Radars (FCRs). There will also be 56 new Block III aircraft built to meet force requirements.

The fleet includes the A model Apache and D model Longbow, with the A model fleet mostly consumed by the Longbow remanufacturing program. The last A model was inducted into the remanufacture line in July 2012. The remanufacture line uses the A model and incorporates a millimeter-wave FCR, fire-and-forget radar-guided HELLFIRE missiles, and other cockpit management and digitization enhancements.

The Longbow is undergoing recapitalization modifications such as upgraded forward-looking infrared technology, non-line-of-sight communications, video transmission/ reception, and maintenance cost reductions. Longbow supports Brigade Combat Teams across the full spectrum of warfare. Apache is fielded to Active Army, National Guard and Army Reserve attack battalions, armed reconnaissance battalions, and cavalry units as defined in the Army Modernization Plan.

The Longbow Apache Block III (AB3) meets all the requirements for Army and Joint interoperability goals for the future and will add significant combat capability while addressing obsolescence issues, ensuring the aircraft remains a viable combat multiplier beyond 2035.

The Block III modernized Longbows will be designed and equipped with an open systems architecture to incorporate the latest communications, navigation, sensor, and weapon systems.

Combat mission speed: Longbow 145 knots (max speed); AB3 164 knots (max speed)
Combat range: 260 nautical miles
Combat endurance: 2.5 hours
Maximum gross weight: 20,260 pounds
Ordnance: 16 HELLFIRE missiles, 76 2.75-inch rockets, and 1,200 30mm chain gun rounds
Crew: Two (pilot and copilot gunner)

SYSTEM INTERDEPENDENCIES
In this Publication
Aviation Combined Arms Tactical Trainer (AVCATT), Aircraft System (UAS), HELLFIRE Family of Missiles, 2.75 Inch Rocket Systems (Hydra), Air Warrior (AW), Joint Tactical Radio System Airborne and Maritime/Fixed Station (JTRS AMF), Laser HELLFIRE, Shadow Tactical Unmanned Aerial Vehicle (TUAV)

Other Major Interdependencies
TCDL, Link 16, JSTARS, AWACS, GPS, AMPS, DCGS-A, GCCS-A, BFT, Have Quick, SATCOM, SINCGARS, Land Warrior, M-1 Tank, M-2 Bradley, Stryker, Fire Support, UH-60, CH-47, A2C2S, OH-58D, ERMP UAS

PROGRAM STATUS
- **2QFY12:** Block III system successfully completed Initial Operational Test & Evaluation
- **3QFY12:** Full-rate production decision
- **Current:** Upgrade Block I and II Longbow to Block III configuration with eventual acquisition objective of 634 remanufacture airframes and 56 new build airframes for a total of 690 Block III Longbows

PROJECTED ACTIVITIES
- **3QFY13:** Initial operating capability

Technology Development | Engineering & Manufacturing Development | Production & Deployment | Operations & Support

Longbow Apache (AH-64D) (LBA)

FOREIGN MILITARY SALES
Egypt, Greece, Israel, Kuwait, Netherlands, Saudi Arabia, Singapore, United Arab Emirates
Direct commercial sales:
Greece, Japan, United Kingdom

CONTRACTORS
Airframe: Boeing (Mesa, AZ)
MTADS: Lockheed Martin (Orlando, FL)
REU: Lockheed Martin (Orlando, FL)
Northrop Grumman (Linthicum, MD)
APU: Honeywell (Phoenix, AZ)
Technical: Aviation and Missile Solutions, LLC (Huntsville, AL)
FCR: Longbow LLC (Orlando, FL)
Radar: Northrop Grumman (Linthicum, MD)
Logistics: AEPCO (Huntsville, AL)
TADS/PNVS: Lockheed Martin (Goodyear, AZ)
Programmatics: DynCorp (Fort Worth, TX)
EGI: Honeywell (Clearwater, FL)
LRUs: Smiths (Clearwater, FL)
IPAS: Honeywell (Tempe, AZ)

M109 Family of Vehicles (FOV) (Paladin/FAASV, PIM SPH/CAT)

INVESTMENT COMPONENT

Modernization

Recapitalization

Maintenance

MISSION

Provides the primary indirect fire support for Heavy Brigade Combat Teams (HBCTs), armored and mechanized infantry divisions as well as an armored ammunition resupply vehicle.

DESCRIPTION

The M109A6 (Paladin) 155mm Howitzer is the most technologically advanced self-propelled cannon system in the Army. The Field Artillery Ammunition Support Vehicle (FAASV) provides an armored ammunition resupply vehicle in support of the Paladin.

The Paladin Integrated Management (PIM) program addresses concerns about obsolescence, space, weight, and power and ensures sustainment of the Paladin and FAASV fleet through 2050. The PIM, scheduled to be fielded in January 2017, consists of the Self-Propelled Howitzer (SPH)

and the Carrier, Ammunition, Tracked (CAT). PIM provides indirect fire support for full spectrum operations and can support HBCTs, infantry and Stryker BCTs. PIM improves: Survivability with Increased Force Protection and Survivability to enable "off-Forward Operating Base" operations; responsive fires by firing within 45 seconds of a complete stop with on-board communications, remote travel lock, automated cannon slew capability, and pivot steer technology; increased commonality and reliability by using Bradley powertrain, track, and suspension components; electric drive technology with electric elevation/traverse drives gun system and an electric rammer; and growth potential with future requirements and technology insertions that provide growth for future network requirements and increased power with a 600 volt system.

Other specifications:

Crew: Paladin and SPH = four; FAASV and CAT = four
Combat loaded weight: Paladin =

34.25 tons; FAASV = 29.26 tons, SPH = 39 tons, CAT = 36 tons
Paladin on-board ammo: 39 rounds, SPH: 42
FAASV on-board ammo: 95 rounds, CAT: 95
Rates of fire: Four rounds per minute for first three minutes maximum; one round per minute sustained
Maximum range: high-explosive/Rocket Assisted Projectile, 22/30 kilometers
Cruising range: Paladin and SPH = 180 miles; FAASV and CAT = 180 miles
Fire Support Network: Paladin Digital Fire Control System software supports Fire Support Network

SYSTEM INTERDEPENDENCIES

Advanced Field Artillery Tactical Data System (AFATDS), Excalibur (XM982), Force XXI Battle Command Brigade and Below (FBCB2), Artillery Ammunition, Precision Guidance Kit (PGK)

PROGRAM STATUS

- **2QFY12:** Capability Production Document JROC Approved
- **2QFY12:** Defense Acquisition Board approval for award of Comprehensive Contract Modification
- **2QFY12:** Completed Phase I Government Developmental Test (DT)
- **2QFY12:** PIM Acquisition Program Baseline approved
- **3QFY12:** Delta Critical Design Review
- **3QFY12:** Integrated Baseline Review
- **3QFY12:** Program Support Review
- **3QFY12:** Logistics Demonstration began
- **4QFY12-3QFY14:** Phase II Government DT

PROJECTED ACTIVITIES

- **3QFY11-3QFY14:** Developmental testing
- **3QFY13:** Milestone C

M109 Family of Vehicles (FOV) (Paladin/FAASV, PIM SPH/CAT)

FOREIGN MILITARY SALES
None

CONTRACTORS
PIM Development:
BAE Systems (York, PA)
PIM SW Support/FATB/Matrix Support:
Armaments R&D Center (Picatinny Arsenal, NJ)
Program Management Support:
Tank-Automotive and Armaments Command (TACOM) (Warren, MI)
Testing:
Yuma Test Center/Proving Ground (Yuma, AZ)
Aberdeen Test Center/Proving Ground (Aberdeen, MD)

M1200 Armored Knight

INVESTMENT COMPONENT

Modernization

Recapitalization

Maintenance

MISSION
Assists Heavy and Infantry Brigade Combat Teams (HBCTs and IBCTs) in performing terrain surveillance, target acquisition and location, and fire support for combat observation lasing team missions.

DESCRIPTION
The M1200 Armored Knight provides precision-strike capability by locating and designating targets for both ground- and air-delivered laser-guided ordnance and conventional munitions. It replaces the M707 Knight High Mobility Multipurpose Wheeled Vehicle (HMMWV) base and M981 fire support team vehicles used by combat observation lasing teams (COLTs) in both HBCTs and IBCTs. It operates as an integral part of the brigade reconnaissance element, providing COLT and fire support mission planning and execution.

The M1200 Armored Knight is built on an M1117 Armored Security Vehicle (ASV) chassis and includes an armored cupola capable of 360-degree rotation and an integrated Mission Equipment Package (MEP). The MEP includes: a cupola-mounted Fire Support Sensor System (FS3), Targeting Station Control Panel, Mission Processor Unit (MPU), Inertial Navigation Unit (INU), Defense Advanced Global Positioning System Receiver (DAGR), Power Distribution Unit, Rugged Handheld Computer (RHC2), 3 Channel Ground to Air Radio Systems (SINCGARS), Force XXI Battle Command Brigade and Below (FBCB2) or Blue Force Tracker (BFT), Driver's Display Unit (DDU), and Vehicle Intercom System (VIS).

The mission equipment package includes: Fire Support Sensor System (FS3) mounted sensor, Targeting Station Control Panel, Mission Processor Unit, Inertial Navigation Unit, Defense Advanced Global Positioning System Receiver (DAGR), Power Distribution Unit, Rugged Hand-Held Computer Unit (RHC) Forward Observer Software (FOS).

Other M1200 Armored Knight specifications:

Crew: Three COLT members
Combat loaded weight: Approximately 15 tons
Maximum speed: 63 miles per hour
Cruising range: 440 miles
Target location accuracy: less than 20 meters circular error probable

SYSTEM INTERDEPENDENCIES
In this Publication
Advanced Field Artillery Tactical Data System (AFATDS), Force XXI Battle Command Brigade and Below (FBCB2)

Other Major Interdependencies
Fire Support Sensor System (FS3), Forward Observer Software (FOS)

PROGRAM STATUS
- **FY12:** Cumulative total of 414 M1200 Armored Knight vehicle systems produced out of 465 vehicle systems procured out of Army Acquisition Objective (AAO) of 465 systems

PROJECTED ACTIVITIES
- **FY13:** Last 51 M1200 Armored Knights produced to meet AAO of 465
- **FY14:** M1200 system sustainment begins

M1200 Armored Knight

FOREIGN MILITARY SALES
None

CONTRACTORS
M1117 ASV Hull:
Textron Marine & Land Systems
(New Orleans, LA)
Precision Targeting Systems
Production/Vehicle Integration:
DRS Sustainment Systems, Inc.
(St. Louis, MO; West Plains, MO)
FS3 Sensor:
Raytheon (McKinney, TX)
Inertial Navigation Unit:
Honeywell (Clearwater, FL)
Common Display Unit:
DRS Tactical Systems (Melbourne, FL)

Medical Communications for Combat Casualty Care (MC4)

Modernization

Recapitalization

Maintenance

MISSION
Integrates, fields, and supports a comprehensive medical information system, enabling lifelong electronic medical records, streamlined medical logistics and enhanced situational awareness for Army operational forces.

DESCRIPTION
Medical Communications for Combat Casualty Care (MC4) is a ruggedized system-of-systems containing medical software packages fielded to operational medical forces worldwide, providing the tools to digitally record and transfer critical medical data from the foxhole to medical treatment facilities worldwide. MC4 helps ensure service members have secure, accessible, lifelong electronic medical records, resulting in better informed health care providers and easier access to Veteran's Affairs medical benefits.

SYSTEM INTERDEPENDENCIES
In this Publication
None

Other Major Interdependencies
MC4 relies on software developers such as the Defense Health Information Management System to provide global software databases to store data generated by the MC4 system, providing medical situational awareness for operational commanders and patient record visibility to medical staffs worldwide.

PROGRAM STATUS
- **Current-4QFY12:** Fielding Theater Medical Information Program (TMIP) Block 2 Release 1 Service Pack 1 worldwide

PROJECTED ACTIVITIES
- **1QFY14:** TMIP Increment 2 Release 2 (I2R2) software Full Deployment Decision Review
- **2QFY14:** Begin Fielding TMIP I2R2 software

Medical Communications for Combat Casualty Care (MC4)

FOREIGN MILITARY SALES
None

CONTRACTORS
System Integration Support:
L-3 Communications (Reston, VA)
Fielding, Training, and System Administration Support:
General Dynamics Information Technology (Fairfax, VA)
Program Management and Support Services:
Booz Allen Hamilton (Herndon, VA)

Medical Simulation Training Center (MSTC)

Modernization

Recapitalization

Maintenance

MISSION

Conduct sustainment and enhanced medical training for Combat Medics and Combat Lifesavers in support of Unified Land Operations.

DESCRIPTION

A Medical Simulation Training Center (MSTC) is a regional Army training asset that provides standardized medical training to both medical and non-medical personnel in both a classroom environment and under simulated battlefield conditions. MSTCs sustain and enhance the skills of Soldier Medics in Tactical Combat Casualty Care and validate annual Medical Education and Demonstration of Individual Competence. In addition, non-medical personnel are trained in the latest Combat Lifesaver Program of Instruction (POI). All training conducted in the MSTC is managed and overseen by the U.S. Army Medical Department Center and School.

MSTCs provide centralized locations where Soldier Medics can learn the latest battlefield techniques and train on the newest equipment. The MSTC turn-key solution affords commanders the latitude and ease to provide readiness and sustainment training throughout the year. In addition, the POIs have been developed in modules so that the training can be tailored to meet mission/pre-deployment requirements, or specific skill-level rehearsals.

MSTCs provide Active Duty, Army Reserve, and Army National Guard Soldiers with realistic medical training with anatomically correct and weighted bleed-breathe mannequins along with medical part-task trainers.

SYSTEM INTERDEPENDENCIES

None

PROGRAM STATUS

- **3QFY12:** Award of the Instructor Support System/Turn Key Operation contract
- **4QFY12:** Award of the Virtual Patient Simulator (VPS) contract

PROJECTED ACTIVITIES

- **1QFY13-4QFY13:** Develop, procure and field the VPS mannequin, the Medical Training Evaluation System, and the Medical Training Command and Control system
- **4QFY13-FY19:** Continue to field MSTCs to achieve the Full Operational Capability of 34 sites

Medical Simulation Training Center (MSTC)

FOREIGN MILITARY SALES
None

CONTRACTORS
Computer Science Corp. (Orlando, FL)
PULAU Corp. (Orlando, FL)
Firehouse Medical Inc (Anaheim, CA)
KForce Government Solutions (Fairfax, VA)
Genesis Concepts and Consultants LLC
   (San Antonio, TX)
Medical Training Consultants, Inc
   (Lakewood, WA)
Bound Tree Medical LLC (Dublin, OH)
EMS Safety Services, Inc
   (San Clemente, CA)
Moulage Sciences and Training, LLC
   (Downers Grove, IL)
SKEDCO, Inc (Tualatin, OR)

Medium Caliber Ammunition (MCA)

INVESTMENT COMPONENT

Modernization

Recapitalization

Maintenance

MISSION

Provides overwhelming lethality in medium caliber ammunition and point- and area-target engagement via medium handheld and crew-served weapons.

DESCRIPTION

Medium caliber ammunition includes 20mm, 25mm, 30mm, and 40mm armor-piercing, high-explosive, smoke, illumination, training, and anti-personnel cartridges with the capability to defeat light armor, materiel, and personnel targets. The 20mm cartridge is a multi-purpose tracer, used in the Counter-Rocket, Artillery, and Mortar (C-RAM) weapon system. The 25mm target practice (TP), high-explosive incendiary and armor-piercing cartridges are fired from the M242 Bushmaster Cannon for the Bradley Fighting Vehicle. The 30mm TP and high-explosive dual-purpose (HEDP) cartridges are used in the Apache helicopter's M230 Chain Gun. A variety of 40mm TP, HEDP, and specialty cartridges are designed for use in the M203 Grenade Launcher, M320 Grenade Launcher, and the MK19 Grenade Machine Gun.

SYSTEM INTERDEPENDENCIES

In this Publication
None

Other Major Interdependencies
Medium calibur ammunition is dependent upon the weapons platforms currently in use.

PROGRAM STATUS

- **Current:** In production

PROJECTED ACTIVITIES

- **FY13:** Multiple year family acquisition for 20mm, 25mm and 30mm ammunition
- **FY14:** Multiple year family acquisition for 40mm ammunition
- **FY14:** Initiation of Development Program for new 40mm TP day/night/thermal training ammunition

Medium Caliber Ammunition (MCA)

20mm
M940 PGU27 PGU28 PGU30

25mm
M791 M792 M910 M793 PGU23 PGU25 M919

30mm x 113
M788 M789

30mm x 173
MK266 MK268 MK310 PGU14 PGU15 PGU-13B

40mm L60
PGU-9 M81

High Velocity
M430A1 M918A1 M385A1

40mm Grenades
M781 M433 M583A1 M585 M661 M662 M992

Low Velocity

Medium Extended Air Defense System (MEADS)

INVESTMENT COMPONENT

Modernization

Recapitalization

Maintenance

MISSION

Defends maneuver forces and critical assets against the theater ballistic missile, cruise missile, and air-breathing threats in contingency and mature theaters.

DESCRIPTION

The Medium Extended Air Defense System (MEADS) provides a robust, 360-degree defense using the PATRIOT Advanced Capablity-Three(PAC-3) hit-to-kill missile segment enhancement (MSE) against the full spectrum of theater ballistic missiles, anti-radiation missiles, cruise missiles, unmanned aerial vehicles, tactical air-to-surface missiles, and rotary- and fixed-wing threats. MEADS will also provide defense against multiple and simultaneous attacks by short-range ballistic missiles, cruise missiles, and other air-breathing threats. MEADS can be immediately deployed by air for early entry operations. MEADS also has the mobility to displace rapidly and protect maneuver force assets during offensive operations. Netted, distributed, open architecture, and modular components are utilized in the MEADS to increase survivability and flexibility of use in a number of operational configurations. The PAC-3 MSE improves upon the current missile configuration ranges/altitudes and improves performance against evolving threats.

The MEADS weapon system will use its netted and distributed architecture to ensure Joint and allied interoperability, and to enable a seamless interface to the next generation of battle management command, control, communications, computers, and intelligence (BMC4I). The system's improved sensor components and its ability to link other airborne and ground-based sensors facilitate the employment of its battle elements.

The MEADS weapon system's objective battle management tactical operations center (TOC) will provide the basis for the future common air and missile defense (AMD) TOC, leveraging modular battle elements and a distributed and open architecture to facilitate continuous exchange of information to support a more effective AMD system-of-systems.

SYSTEM INTERDEPENDENCIES
None

PROGRAM STATUS
- **1QFY12:** Successful MEADS System Launcher Missile Characterization Test
- **1QFY12:** Multifunction Fire Control Radar (MFCR)delivered for integration/testing at Pratica di Mare, Italy
- **2QFY12:** Low Frequency Sensor Outdoor Demonstration

PROJECTED ACTIVITIES
- **Current:** Remaining activities to implement a "Demonstration of Capabilities" through 2013 with the remaining Memorandum of Understanding funds to provide a meaningful capability for Germany and Italy and a possible future option for the U.S.
- **Current:** A new and detailed program/schedule for design and development was developed by NATO Medium, Extended Air Defense System Management Agency for the Board of Directors review and National Armament Director approval, and the associated contract amendment was signed in October 2011
- **4QFY12:** MFCR#1 delivery for integration/testing
- **1QFY13:** Flight Test–1 (FT-1, Air-Breathing Threat Intercept)
- **4QFY13:** Sensor Characterization Test
- **4QFY13:** Flight Test–2 (FT-2, Tactical Ballistic Missile Intercept)
- **2QFY14:** Development Program Completion

ACQUISITION PHASE

| Technology Development | Engineering & Manufacturing Development | Production & Deployment | Operations & Support |

UNITED STATES ARMY

Medium Extended Air Defense System (MEADS)

FOREIGN MILITARY SALES
None

CONTRACTORS
D&D Contract:
MEADS, Intl. (Syracuse, NY; Orlando, FL; Huntsville, AL)
Lockheed Martin (Grand Prairie, TX)
PM/SYS:
Government (Statewide, AL)
MSE:
Lockheed Martin (Grand Prairie, TX)
Security/Exciter:
Lockheed Martin (Syracuse, NY)
SETA:
Intuitive Research and Technology (Huntsville, AL)

Meteorological Measuring Set-Profiler (MMS-P)/Computer Meteorological Data-Profiler (CMD-P)

INVESTMENT COMPONENT

Modernization

Recapitalization

Maintenance

MISSION
Provides on-demand, real-time meteorological data over an extended battlespace.

DESCRIPTION
The AN/TMQ-52 Meteorological Measuring Set-Profiler (MMS-P) uses a suite of meteorological sensors, meteorological data from satellites, and an advanced mesoscale atmospheric model to provide highly accurate meteorological data for indirect fire artillery forces. The system uses common hardware, software, and operating systems; is housed in a command post platform shelter; and is transported on an M1152A High Mobility Multipurpose Wheeled Vehicle.

The mesoscale atmospheric model receives large-scale atmospheric data from the Air Force Weather Agency and other meteorological sensors, and produces a vertical profile of wind speed and direction, temperature, relative humidity, cloud base height, type precipitation, and horizontal visibility in the target area, all of which are necessary for precise targeting and terminal guidance of various munitions. Profiler transmits this data to indirect fire direction centers for use in developing the firing solution. The current Profiler provides meteorological coverage throughout a 60-kilometer radius. For the first time, Army field artillery systems can apply meteorological data along the trajectory from the firing platform to the target area.

The Profiler Block III or Computer Meteorological Data–Profiler (CMD-P) AN/GMK-2 System is the next evolutionary block of the Profiler system and is designed to reduce the logistical footprint to a laptop configuration located in the Tactical Operations Center (TOC), thus eliminating the Standard Integrated Command Post Shelter/Command Post Platform, support vehicle, and crew. The CMD-P software on the laptop will port MMS-P software that presently runs on three operating systems (OS) and three separate computing processors onto one OS and processor. Additionally, the local ground sensor will be removed to further reduce the logistical footprint. The system interface with the Advanced Field Artillery Tactical Data System will change from the Single Channel and Airborne Radio Systems to a Local Area Network connection in the TOC. The CMD-P will no longer require a dedicated Global Broadcast Service (GBS) receiver suite (AN/TSR-8) but instead will rely on the TOC GBS. The system software will be capable of providing Field Artillery Computer MET (METCM) and Gridded MET (METGM) messages on demand with or without an operator in-the-loop while extending coverage up to 500 kilometers. CMD-P completed Development Testing in FY11 and Operational Testing in FY12. Fielding is planned to begin in FY13. The CMD-P will reduce the system footprint and result in a significant Operations and Support cost avoidance for the Army as it replaces the MMS-P.

SYSTEM INTERDEPENDENCIES
In this Publication
None

Other Major Interdependencies
Navy Operational Global Atmospheric Prediction System, Global Broadcast System

PROGRAM STATUS
- **1QFY12-4QFY12:** Completed development and testing of CMD-Profiler Block III system, i.e. DT, OT, FQT
- **1QFY12-4QFY12:** Received PEO IEW&S approval to initiate developmental effort for next generation CMD-Profiler System
- **1QFY12-4QFY12:** Continued fielding of the MMS–Profiler Block I GBS Modification Work Order (MWO) (86 systems fielded to date)

PROJECTED ACTIVITIES
- **2QFY12-2QFY14:** Complete fielding of last systems and GBS MWO to Army units
- **2QFY12-2QFY14:** Procurement and fielding of the CMD-P starting in FY13

ACQUISITION PHASE

Technology Development	Engineering & Manufacturing Development	Production & Deployment	Operations & Support

Meteorological Measuring Set-Profiler (MMS-P)/Computer Meteorological Data-Profiler (CMD-P)

FOREIGN MILITARY SALES
None

CONTRACTORS
MMS-P – Block I:
Smiths Detection, Inc. (Edgewood, MD)
Pennsylvania State University (University Park, PA)
CMD-P – Block III:
Prime:
Mantech Sensor Technologies, Inc. (Red Bank, NJ)
Sub:
CGI Federal (Lawton, OK)

Mine Protection Vehicle Family (MPVF), Area Mine Clearing System (AMCS), Interrogation Arm

INVESTMENT COMPONENT

Modernization

Recapitalization

Maintenance

MISSION
Provides blast-protected platforms capable of locating, interrogating, and classifying suspected explosive hazards, including improvised explosive devices (IEDs).

DESCRIPTION
The Mine Protection Vehicle Family (MPVF) consists of the Medium Mine Protected Vehicle (MMPV), the Vehicle Mounted Mine Detection (VMMD) system, and the Mine Protected Clearance Vehicle (MPCV). All are blast protected as each of the systems in the MPVF has a blast-deflecting, V-shaped hull, and each conducts a specific mission.

The MMPV command and control vehicle is adaptable to a wide range of security and force protection activities. The MMPV Type I (Panther) will support Explosive Ordnance Disposal (EOD) Companies as the rapid response vehicle. The Panther will also support Chemical Biological Response Teams. The MMPV Type II (RG-31) will support Engineer Units in route and area clearance operations.

The VMMD is a vehicle-mounted mine-detection and lane-proofing system capable of finding and marking metallic explosive hazards, including metallic-encased IEDs. It consists of two mine detection "Husky" vehicles, and a set of three mine detonation trailers used for proofing. The system is designed to be quickly repairable in the field after a mine blast.

The MPCV is capable of interrogating and classifying suspected explosive hazards, including IEDs. The MPCV has an articulating arm with a digging/lifting attachment and camera to remotely interrogate a suspected explosive hazard and allow the crew to confirm, deny, and/or classify the explosive hazard. It also transports Soldiers allowing them to dismount to mark and/or neutralize explosive hazards.

The Area Mine Clearing System (AMCS) is a vehicle designed to clear anti-tank and anti-personnel mines by means of a rotating flail. The AMCS is a manually operated, self-powered vehicle with the capability to adjust flailing depth.

The Interrogation Arm (IA) is a mechanical counter IED asset that provides Soldiers with the capability for standoff detection and interrogation of suspected IEDs utilizing a probing/digging tool to expose objects and a visible camera to identify targets. The IA system can be utilized on the MMPV Type II (RG-31) and Husky platforms.

SYSTEM INTERDEPENDENCIES
None

PROGRAM STATUS
MMPV Type I
- **1QFY12-4QFY12:** Vehicle testing

MPCV
- **1QFY12:** First Unit Equipped
- **1QFY12-4QFY12:** Fielding

VMMD
- **1QFY12:** First Unit Equipped
- **1QFY12-4QFY12:** Fielding

AMCS
- **2QFY12:** Transfer of system from PdM Countermine/EOD
- **2QFY12-4QFY12:** Complete overhaul of logistics effort rescheduled; provisioning conferences held
- **4QFY12:** Operator's Technical Manual Validation/Verification; Acquisition Program Re-Baseline

Interrogation Arm
- **2QFY12:** Transfer of system from PdM Countermine/EOD

PROJECTED ACTIVITIES
MMPV Type I
- **2QFY13:** Full-rate production for recapitalization of RG33L+ vehicles into Panther-like vehicles, Letterkenny Army Depot (LEAD)
- **2QFY14:** Full Materiel Release/Type Classification–Standard; First Unit Equipped

PROJECTED ACTIVITIES (Cont.)

MMPV Type II
- **2Q-3QFY13:** Pilot Build
- **3QFY13-QFY14:** Production Verification Testing
- **3Q-4QFY14:** Operational Test
- **4QFY14-FY18:** Reset/ recapitalization

MPCV
- **1QFY13-1QFY15:** Recapitalization of assets at LEAD; complete fielding to units

VMMD
- **1QFY13-1QFY15:** Recapitalization of assets at LEAD; complete fielding to units

AMCS
- **1QFY13-1QFY14:** Continue logistics development (Technical Manuals, provisioning) to include Logistics Demonstration
- **3QFY13:** Operational Test
- **3QFY14:** Full Materiel Release/Type Classification-Standard; First Unit Equipped

Interrogation Arm
- **3QFY13-3QFY14:** Program of Record testing of IA mounted on the MMPV Type II (RG-31)

Mine Protection Vehicle Family (MPVF), Area Mine Clearing System (AMCS), Interrogation Arm

FOREIGN MILITARY SALES
MPCV:
United Kingdom
VMMD:
Canada, Kenya, Australia
IA:
Australia

CONTRACTORS
MMPV Type I (Panther):
BAE Systems (York, PA)
MMPV Type II (RG-31):
General Dynamics Land Systems-Canada (Ontario, Canada)
MPCV:
General Dynamics Land Systems-Force Protection Industries (Ladson, SC)
VMMD:
Critical Solutions International, Inc. (Dallas, TX)
AMCS:
Hydrema (Støvring, Denmark)
IA:
FASCAN, International, Inc. (Sparrows Point, MD)

Mine Resistant Ambush Protected Vehicles (MRAP), Army

INVESTMENT COMPONENT

Modernization

Recapitalization

Maintenance

MISSION

Provides tactical mobility for warfighters with multi-mission platforms capable of mitigating the effects of improvised explosive devices (IEDs), underbody mines, and small arms fire threats.

DESCRIPTION

The Joint Mine Resistant Ambush Protected (MRAP) Vehicle Program (JMVP) is a multi-service program currently supporting the Army, Navy, Marine Corps, Air Force, and the U.S. Special Operations Command. The program procures, tests, integrates, fields, and supports highly survivable vehicles that provide protection from IEDs and other threats. These four- to six-wheeled vehicles are configured with government-furnished equipment to meet unique warfighting requirements. Vehicle combat weights (fully loaded without add-on armor) range from approximately 34,000 to 60,000 pounds, with payloads ranging from 1,000 to 18,000 pounds. Key components (e.g., transmissions, engines) vary between vehicles and manufacturers, but generally consist of common commercial and military parts.

Three categories of vehicles support the following Army missions:

- **Category I (CAT I):** Carries four to six passengers. Designed to provide increased mobility and reliability in rough terrain.
- **CAT II:** Multimission operations (such as convoy lead, troop transport, and ambulance). Carries 10 passengers.
- **MRAP All Terrain Vehicle (M-ATV):** Carries four Soldiers plus a gunner. Supports small-unit combat operations in complex and highly restricted rural, mountainous, and urban terrains. The M-ATV provides better overall mobility characteristics than the original CAT I, II, and III MRAP vehicles yet retains the same survivability threshold.

In addition, CAT III vehicles support Mine/IED clearance operations and explosive ordnance disposal (EOD). CAT III vehicles carry six passengers, plus specialized equipment to support EOD operations. The Force Protection Industries Buffalo is the only CAT III variant and is the largest MRAP vehicle. The Army does not own CAT III MRAPs.

SYSTEM INTERDEPENDENCIES

In this Publication

None

Other Major Interdependencies

MRAP vehicles are equipped with multiple Government Furnished Equipment items, including communications equipment, mine and IED counter-measure equipment, in addition to weapons and crew protection systems.

PROGRAM STATUS

- **1QFY12-4QFY12:** Completed production and fielding of MRAP vehicles to Army, Marine Corps, Air Force, Navy, U.S. Special Operations, and foreign military sales customers
- **1QFY12-4QFY12:** Continued modernization of MRAP vehicles returning from theater in preparation for transition to enduring force requirements

PROJECTED ACTIVITIES

- **2QFY13-2QFY15:** Continue support of MRAP vehicles fielded in response to urgent theater requirements
- **2QFY13-2QFY15:** Continue modernization of MRAP vehicles returning from theater in preparation for transition to enduring force requirements
- **1QFY14:** Complete transition of program management functions from Navy to Army for all variants except Cougar

ACQUISITION PHASE

Technology Development | Engineering & Manufacturing Development | Production & Deployment | **Operations & Support**

Mine Resistant Ambush Protected Vehicles (MRAP), Army

FOREIGN MILITARY SALES
Canada, France, Italy, United Kingdom

CONTRACTORS
BAE Systems Land & Armaments, Ground Systems Division (York, PA)
BAE-TVS (Sealy, TX)
General Dynamics Land Systems, Canada (Ontario, Canada)
Navistar Defense (Warrenville, IL)
Oshkosh Corp. (Oshkosh, WI)

Modular Fuel System (MFS)

INVESTMENT COMPONENT

Modernization

Recapitalization

Maintenance

MISSION
Provides the ability to rapidly establish fuel distribution and storage capability at any location regardless of materiel handling equipment availability.

DESCRIPTION
The Modular Fuel System (MFS), formerly known as the Load Handling System Modular Fuel Farm (LMFF), is composed of 14 tank rack modules (TRM) and two each of the pump and filtration modules, commonly known as pump rack modules (PRMs). The TRM can be used with the MFS PRMs, the Heavy Expanded Mobility Tactical Truck (HEMTT) Tankers, or as a stand-alone system. TRM, when used with the HEMTT Tanker, doubles the HEMTT Tanker capacity. The MFS is transported by the HEMTT–Load Handling System (HEMTT-LHS) and the Palletized Load System (PLS).

The TRM is air-transportable with fuel and includes a baffled, 2,500-gallon-capacity fuel storage tank that can provide unfiltered, limited retail capability through gravity feed or the 25-gallon-per-minute (gpm) electric pump. The TRM also includes hose assemblies, refueling nozzles, fire extinguishers, grounding rods, a NATO slave cable, and a fuel spill control kit.

TRM full retail capability is being developed and will include replacing the existing electric pump with a continuous operating electric 20 gpm pump, a filtration system, and a flow meter for fuel accountability. The projected date for the TRM retail capability to be fielded is 4QFY12. The PRM includes a self priming 600 gpm diesel engine-driven centrifugal pump, filter separator, valves, fittings, hoses, refueling nozzles, aviation fuel test kit, fire extinguishers, grounding rods, flowmeter, and NATO Connectors. The PRM has an evacuation capability that allows the hoses in the system to be purged of fuel prior to recovery and is capable of refueling both ground vehicles and aircraft. The MFS is capable of receiving, storing, filtering, and issuing all kerosene-based fuels.

SYSTEM INTERDEPENDENCIES
In this Publication
HEMTT Palletized Load System (PLS), LHS (for transportation)

PROGRAM STATUS
- **4QFY12:** TRM Full-rate production approval
- **4QFY12:** TRM Retail Engineering Change Proposal contract award

PROJECTED ACTIVITIES
- **2QFY13:** Technical Manual Verification
- **2QFY13:** Complete Logistics Demonstration
- **3QFY13:** Type Classification/Full Materiel Release
- **3QFY13:** First Unit Equipped

ACQUISITION PHASE

| Technology Development | Engineering & Manufacturing Development | Production & Deployment | Operations & Support |

Modular Fuel System (MFS)

FOREIGN MILITARY SALES
None

CONTRACTORS
DRS Sustainment Systems, Inc.
 (St. Louis, MO)
E.D. Etnyre and Co. (Oregon, IL)

Mortar Systems

INVESTMENT COMPONENT

Modernization

Recapitalization

Maintenance

MISSION
Provides enhanced lethality, accuracy, responsiveness, and crew survivability while reducing the logistics footprint.

DESCRIPTION
The Mortar Fire Control System (MFCS)-equipped mortar systems provide organic, indirect fire support to the maneuver unit commander.

All three Army variants of 120mm mortar systems, have been qualified and are being equipped with MFCS. All of the mortar systems fire a full family of ammunition including high-explosive, infrared and visible light illumination, smoke, and training. The M120 120mm Towed Mortar System is transported by the M1101 trailer and is emplaced and displaced using the M326 Mortar Stowage Kit (MSK). The mounted variants are the M121 120mm mortar, used on the M1064A3 Mortar Carrier (M113 variant), and the 120mm Recoiling Mortar System, used on the M1129 Stryker Mortar Carrier.

Lightweight variants of the M252 81mm Mortar System and M224 60mm Mortar System have been qualified and are in production/fielding. Both systems provide high-rate-of-fire capability and are man-portable.

The M95/M96 MFCS-M, used on the M1064A3 and M1129, and the M150/M151 MFCS-D, used with the M120, combine a fire control computer with an inertial navigation and pointing system, allowing crews to fire in under a minute, improving lethality, accuracy and crew survivability.

The M32 Lightweight Handheld Mortar Ballistic Computer (LHMBC) has a tactical modem and embedded Global Positioning System, allowing mortar crews to send and receive digital call-for-fire messages, calculate ballistic solutions, and navigate.

The XM395 Accelerated Precision Mortar Initiative (APMI) achieved an Urgent Materiel Release (UMR) in March 2011 to field to eight Infantry Brigade Combat Teams (IBCTs) in Operation Enduring Freedom based on an Operational Need Statement (ONS). UMR for an additional Stryker BCT (SBCT) was approved in

May 2012. The Army fully funded the ONS requirement of 5,480 cartridges and associated fuze setting systems.

In 2011, Training and Doctrine Command and Army Test and Evaluation Command completed an assessment, with a recommendation for this capability to proceed to the Capabilities Development for Rapid Transition process. Any follow-on program of record will include full and open competition.

SYSTEM INTERDEPENDENCIES
In this Publication
Advanced Field Artillery Tactical Data System (AFATDS)

Other Major Interdependencies
M95/M96 MFCS-M and M150/M151 MFCS-D

PROGRAM STATUS
- **2QFY12-4QFY12:** MFCS-M fielded to one Stryker Brigade Combat Team (SBCT), one Heavy Brigade Combat Team (HBCT) Reset, and one SBCT Reset
- **1QFY12-4QFY12:** MFCS-D fielded to ten IBCTs and one Infantry Battalion
- **1QFY12-4QFY12:** LHMBC fielded to two SBCTs, one SFG, and four IBCT Resets

- **1QFY12-4QFY12:** 454 Mortar Weapon Systems (Lightweight 60mm, 81mm, 120mm) fielded to numerous IBCTs, HBCTs, SBCTs and FSGs
- **1QFY12-4QFY12:** MSK fielded to ten IBCTs and one Infantry Battalion
- **1QFY12-4QFY12:** Continued production of 60mm, 81mm, and 120mm mortar weapon systems
- **1QFY12-4QFY12:** Continued production of MSKs, and MFCS-D
- **4QFY12:** Complete production of LHMBCs
- **1QFY12-4QFY12:** Successfully conducted New Equipment Training for APMI to eight IBCTs, one Ranger Battalion, and 1 SBCT in OEF

PROJECTED ACTIVITIES
- **1QFY13-4QFY13:** Continue production of 81mm mortar systems
- **1QFY13-4QFY13:** Continue fielding of the 60mm Lightweight Mortar
- **2QFY13:** Complete production of 60mm lightweight mortar systems
- **3QFY13:** Initial Fielding of the 81mm Lightweight Mortar (M252A1)
- **1QFY13-4QFY13:** Continue production and fielding of MFCS-D
- **3QFY13:** Complete production of MSK

Mortar Fire Control System (MFCS-Mounted)

Electronics Rack w/ FCC, Portable Universal Battery Supply, DAGR, and Enhanced Power Distribution Assembly

Motar Fire Control System (MFCS-Dismounted)

Lightweight Handheld Mortar Ballistic Computer (LHMBC)

Mounted Weapon System Production

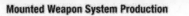

M120 120mm Mortar   M121 120mm Battalion Mortar System

M120 Towed Mortar on M326 Quick Stow Mount

60mm Lightweight Mortar System   81mm Lightweight Mortar System

M1064A3 Motar Carrier

Mortar Systems

FOREIGN MILITARY SALES
Afghanistan, Australia

CONTRACTORS
60mm and 81mm Mortar Bipod Production:
MaTech (Salisbury, MD)
60mm and 81mm Baseplate Production:
AMT (Fairfield, NJ)
MFCS-D and MFCS-M production, fielding, and installation:
Elbit Systems of America (Fort Worth, TX)
M32 LHMBC (R-PDA):
General Dynamics C4 Systems, Inc. (Taunton, MA)
120mm, 81mm, and 60mm cannons, 120mm baseplates:
Watervliet Arsenal (Watervliet, NY)

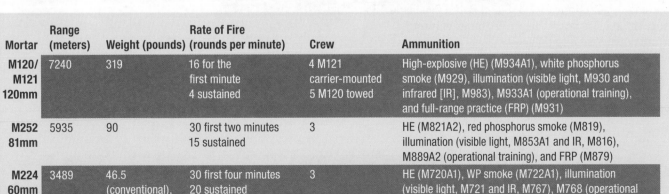

Mortar	Range (meters)	Weight (pounds)	Rate of Fire (rounds per minute)	Crew	Ammunition
M120/ M121 120mm	7240	319	16 for the first minute 4 sustained	4 M121 carrier-mounted 5 M120 towed	High-explosive (HE) (M934A1), white phosphorus smoke (M929), illumination (visible light, M930 and infrared [IR], M983), M933A1 (operational training), and full-range practice (FRP) (M931)
M252 81mm	5935	90	30 first two minutes 15 sustained	3	HE (M821A2), red phosphorus smoke (M819), illumination (visible light, M853A1 and IR, M816), M889A2 (operational training), and FRP (M879)
M224 60mm	3489	46.5 (conventional), 18.0 (handheld)	30 first four minutes 20 sustained	3	HE (M720A1), WP smoke (M722A1), illumination (visible light, M721 and IR, M767), M768 (operational training), and FRP (M769)

Movement Tracking System (MTS)

INVESTMENT COMPONENT

Modernization

Recapitalization

Maintenance

MISSION

Tracks the location of vehicles and logistics assets, communicates with vehicle operators, and redirects missions on a worldwide, near real-time basis during peacetime operations, operations other than war, and war.

DESCRIPTION

The Movement Tracking System (MTS) brings tactical logistics into the digitized battlefield of the 21st century. The system provides the technology necessary to communicate with and track Tactical Wheeled Vehicles and other select Combat Support (CS)/Combat Service Support (CSS) assets and cargo in near real-time, enabling safe and timely completion of distribution missions.

MTS is a Non-developmental Item integrated system consisting of a vehicle-mounted mobile unit and a control station. It is used to support missions through the full spectrum of military operations. Through the use of Global Positioning System (GPS), Radio Frequency Identification (RFID), and non-line-of-sight communications and mapping technologies, MTS provides the means for logistics commanders, transportation movement control, and CS/CSS operations sections to exercise assured positive control of assets anywhere in the world through the use of positioning and commercial satellites. Communications between MTS-equipped platforms and their control stations is conducted via text and pre-formatted messages and utilizes commercial satellites that enable units to send and receive traffic over the horizon, anytime, anywhere.

MTS plays a vital role in battlefield distribution operations. It helps to ensure that commanders and logisticians have the right information at the right time. It provides near real-time data for In-Transit Visibility (ITV) and velocity management of logistics and other Army Combat Support assets, from the sustaining base to the theater of operations. MTS facilitates the rapid movement of supplies through a streamlined distribution system, bypassing routine warehouse/storage functions from the source to the combatant.

Common user logistics transport vehicles and CS/CSS units in the Active and Reserve Components and National Guard will be fitted with MTS systems according to the Army Acquisition Objective for system distribution. When employed within the distribution system, MTS improves the effectiveness and efficiency of limited distribution assets, provides the ability to identify and re-route supplies to higher priority needs, avoids identified hazards, and informs operators of unit location changes.

MTS as a program of record in 2011 has been merged into the Force XXI Battke Command Brigade and Below/Joint Battle Command – Platform programs of record.

SYSTEM INTERDEPENDENCIES

In this Publication
None

Other Major Interdependencies
PM Joint Automatic Identification Technology (PM J-AIT) In-Transit Visibility (ITV), PD Battle Command Sustainment and Support System (BCS3)

PROGRAM STATUS

- **2QFY12:** Field Joint Capabilities Release – Logistics software (previously called MTS-ES)

PROJECTED ACTIVITIES

- **1QFY14:** Transition to Joint Battle Command – Platform (Logistics) [JBC-P (Log)] software and Blue Force Tracking-2 transceiver

ACQUISITION PHASE

| Technology Development | Engineering & Manufacturing Development | Production & Deployment | Operations & Support |

Movement Tracking System (MTS)

FOREIGN MILITARY SALES
None

CONTRACTORS
System Hardware (Military Ruggedized Tablets):
DRS Technologies (Melbourne, FL)
System Hardware (transceivers):
Comtech Mobile Datacom Corp. (CMDC) (Germantown, MD)
Field Service Support:
Engineering Solutions and Products, Inc. (ESP) (Eatontown, NJ)
Software:
v5.16 - Comtech Mobile Datacom Corp. (CMDC) (Germantown, MD)
Software:
JCR-Log - Northrop Grumman (Redondo Beach, CA)

MQ-1C Gray Eagle Unmanned Aircraft System (UAS)

INVESTMENT COMPONENT

Modernization

Recapitalization

Maintenance

MISSION

Provides combatant commanders a real-time responsive capability to conduct long-dwell, persistent stare, wide-area reconnaissance, surveillance, target acquisition, communications relay, and attack missions.

DESCRIPTION

The MQ-1C Gray Eagle Unmanned Aircraft System (UAS) addresses the need for a long-endurance, armed (up to four HELLFIRE missiles), UAS that offers greater range, altitude, and payload flexibility.

The Gray Eagle UAS is powered by a heavy fuel engine (HFE) for higher performance, better fuel efficiency, common fuel on the battlefield, and a longer lifetime.

Its specifications include the following:

Length: 28 feet
Wingspan: 56 feet
Gross take-off weight: 3,600 pounds
Maximum speed: 150 knots
Ceiling: 25,000 feet
Range: 2,500 nautical miles via satellite communications (SATCOM)
Endurance: 27+ hours

The Gray Eagle UAS is fielded in platoon sets, consisting of four unmanned aircraft, two Universal Ground Control Stations (UGCS), two Ground Data Terminals (GDT), one Portable Ground Control Station (PGCS), one Portable Ground Data Terminal (PGDT), one Satellite Ground Data Terminal (SGDT), an Automated Take-off and Landing System (ATLS), Light Medium Tactical Vehicles (LMTV), and other ground-support equipment, operated and maintained by a company of 128 Soldiers within the Combat Aviation Brigade.

SYSTEM INTERDEPENDENCIES

In this Publication
None

Other Major Interdependencies
PM–Robotic Unmanned Sensors (PM-RUS) provides the electro-optical/infrared (EO/IR) and SAR/GMTI payloads; PM–Joint Attack Munition Systems (PM-JAMS) provides HELLFIRE missiles; PM–Warfighter Information Network–Terrestrial (PM-WIN-T) provides communications relay payload

PROGRAM STATUS

- **Current:** Low-rate initial production

PROJECTED ACTIVITIES

- **4QFY12:** Initial operational test and evaluation
- **3QFY13:** Full-rate production decision

ACQUISITION PHASE

Technology Development | Engineering & Manufacturing Development | Production & Deployment | Operations & Support

UNITED STATES ARMY

MQ-1C Gray Eagle Unmanned Aircraft System (UAS)

FOREIGN MILITARY SALES
None

CONTRACTORS
Aircraft:
General Atomics, Aeronautical Systems
Inc. (San Diego, CA)
Ground Control Station:
AAI Corp. (Hunt Valley, MD)
Tactical Common Data Link:
L-3 Communications (Salt Lake City, UT)

Multiple Launch Rocket System (MLRS) M270A1

MISSION

Provides coalition ground forces with highly lethal, responsive, and precise long-range rocket and missile fires that defeat point and area targets in both urban/complex and open terrain with minimal collateral damage, via a highly mobile, responsive multiple launch system.

DESCRIPTION

The combat-proven Multiple Launch Rocket System (MLRS) M270A1 is a mechanized artillery weapon system that provides the combat commander with round-the-clock, all-weather, lethal, close- and long-range precision rocket and missile fire support for Joint forces, early-entry expeditionary forces, contingency forces, and modular fire brigades supporting Brigade Combat Teams.

The M270A1 is an upgraded version of the M270 launcher. The program entailed the concurrent incorporation of the Improved Fire Control System (IFCS) and the Improved Launcher Mechanical System (ILMS) on a rebuilt M993 Carrier (derivative of the Bradley Fighting Vehicle). With the IFCS, the M270A1 can fire future munitions and the ILMS reduces system load and reload times. The M270A1 provides responsive, highly accurate, and extremely lethal surface-to-surface, close- to long-range rocket and missile fires from 15 kilometers to a depth of 300 kilometers. It carries and fires either two launch pods containing six MLRS rockets each, or two Army Tactical Missiles, and is capable of firing all current and future MLRS family of rockets and missiles. It operates with the same MLRS command, control, and communications structure and has the same size crew as the M142 High Mobility Artillery Rocket System (HIMARS).

SYSTEM INTERDEPENDENCIES

In this Publication
None

Other Major Interdependencies
M993 Bradley derivative chassis

PROGRAM STATUS

- **2QFY12:** Continue induction of M270A1 launchers into overhaul program at Red River Army Depot (RRAD)
- **3QFY12:** Awarded Fire Control System-Update (FCS-U) Contract to Lockheed Martin
- **3QFY12:** Awarded Improved Armor Cab (IAC) Contract to Lockheed Martin/BAE
- **4QFY12:** M270 retired from Army inventory

PROJECTED ACTIVITIES

- **Continue:** M270A1 launcher overhaul program
- **Continue:** Providing sustainment and support activities for MLRS strategic partners and foreign military sales customers
- **Continue:** Applying Blue Force Tracker (BFT), Driver Vision Enhancement (DVE), Long Range Communications (LRC) modifications
- **Continue:** Sustainment, development, and test of both the FCS-U and IAC programs

Multiple Launch Rocket System (MLRS) M270A1

FOREIGN MILITARY SALES
M270 and M270A1:
Bahrain, Denmark, Egypt, Finland, France, Germany, Greece, Israel, Italy, Japan, Korea
M270 and M270B1:
Norway, Turkey, United Kingdom

CONTRACTORS
Prime and Launcher:
Lockheed Martin (Dallas, TX; Camden, AR)
Chassis:
BAE Systems (York, PA)
Improved Weapons Interface Unit:
Harris Corp. (Melbourne, FL)
Position Navigation Unit:
L-3 Communications Space & Navigation (Budd Lake, NJ)

NAVSTAR Global Positioning System (GPS)

INVESTMENT COMPONENT

Modernization

Recapitalization

Maintenance

MISSION
Provides real-time positioning, navigation, and timing data to tactical and strategic organizations.

DESCRIPTION
The NAVSTAR Global Positioning System (GPS) is a space-based, Joint-service program led by the U.S. Air Force, which distributes positioning, navigation, and timing (PNT) data to tactical and strategic organizations. The GPS has three segments: a space segment (nominally 24 satellites), a ground control segment, and a user equipment segment. User equipment consists of receivers configured for handheld, ground, aircraft, and watercraft applications. Military GPS receivers use the Precise Positioning Service (PPS) signal to gain enhanced accuracy and signal protection not available to commercial equipment. GPS receivers in the Army today are: the Defense Advanced GPS Receiver (DAGR), with more than 168,000 as handheld receivers and 128,000 distributed for platform installations for a total of nearly 300,000 DAGRs fielded; and the Precision Lightweight GPS Receiver (PLGR), with more than

40,000 in handheld, installed, and integrated applications. In addition, GPS user equipment includes a Ground-Based GPS Receiver Applications Module (GB-GRAM). Over 95,000 GB-GRAMs have been procured and provide embedded PPS capability to a variety of weapon systems. The Army represents more than 80 percent of the requirement for user equipment.

DAGR
Size: 6.37 x 3.4 x 1.56 inches
Weight: one pound; fits in a two-clip carrying case that attaches to load-bearing equipment
Frequency: Dual (L1/L2)
Battery life: 19 hours (four AA batteries)
Security: Selective availability anti-spoofing module
Satellites: All-in-view

GB-GRAM
Size: 0.6 x 2.45 x 3.4 inches
Weight: 3.5 ounces
Frequency: Dual (L1/L2)
Security: Selective availability anti-spoofing module
Satellites: All-in-view

SYSTEM INTERDEPENDENCIES
Blue Force Tracking, mobile ballistic computers, laser rangefinders, movement tracking systems, and several unmanned aerial vehicle systems

PROGRAM STATUS
- **1QFY12-4QFY12:** Continue DAGR fielding for Army components
- **1QFY12-4QFY12:** DAGR designated as an ACAT II program

PROJECTED ACTIVITIES
- **4QFY13:** Continue DAGR fielding; DAGR entering sustainment
- **1QFY14:** Introduction of DAGR Selective Availability Anti-Spoofing Module (SAASM) version 3.7 and GB-GRAM SAASM version 3.7
- **Continue:** Materiel Solution Analysis Phase for Tactical Assured GPS Regional (TAGR) for GPS augmentation
- **Continue:** Military GPS User Equipment (MGUE) development

Technology Development | Engineering & Manufacturing Development | Production & Deployment | Operations & Support

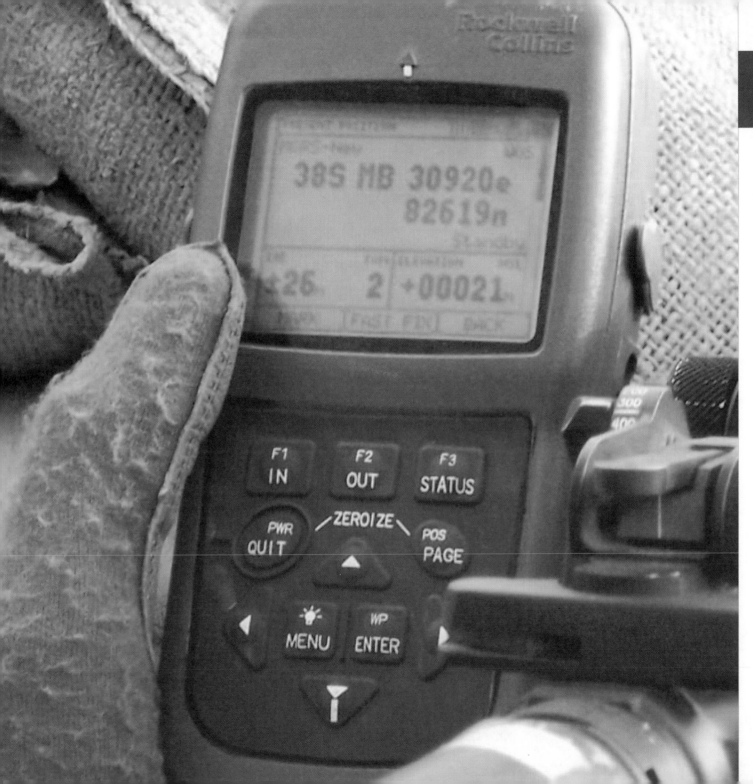

NAVSTAR Global Positioning System (GPS)

FOREIGN MILITARY SALES
PPS-capable GPS receivers have been sold to 41 authorized countries

CONTRACTORS
DAGR/GB-GRAM Acquisition and PLGR Support:
Rockwell Collins (Cedar Rapids, IA)

Nett Warrior (NW)

MISSION
Provides overmatch operational capabilities to all ground combat leaders and small unit operations.

DESCRIPTION
The Nett Warrior (NW) is an integrated dismounted leader situational awareness (SA) system for use during combat operations. The system provides unparalleled SA to the dismounted leader, allowing for faster and more accurate decisions in the tactical fight. With advanced navigation, SA, and information sharing capabilities, leaders are able to avoid fratricide and are more effective and more lethal in the execution of their combat missions.

The NW program focuses on the development of the SA system, which has the ability to graphically display the location of an individual leader's location on a digital geo-referenced map image. Additional Soldier and leader locations are also displayed on the smart device digital display. NW is connected through a secure radio that sends and receives information from one NW to another, thus connecting the dismounted leader to the network. These radios also connect the equipped leader to higher echelon data and information products to assist in decision-making and situational understanding. Soldier position location information will be added to the network via interoperability with the Army's Joint Tactical Radio System capability. All of this will allow the leader to easily see, understand, and interact in the method that best suits the user and the particular mission.

NW will employ a system-of-systems approach, optimizing and integrating capabilities while reducing the Soldier's combat load and logistical footprint.

SYSTEM INTERDEPENDENCIES
In this Publication
Joint Battle Command Product Line, Core Soldier System equipment, Joint Tactical Radio System

PROGRAM STATUS
- **1QFY12:** Army approved Capability Development Document in lieu of Capability Production Document with de-scoped requirements
- **2QFY12:** Defense Acquisition Executive delegated Milestone Decision Authority for NW to the Army Acquisition Executive (AAE)
- **3QFY12:** AAE signed Milestone C Acquisition Decision Memorandum for entry into Production & Deployment Phase and Low Rate Intial Production

PROJECTED ACTIVITIES
- **1QFY13:** Intial Operational Test & Evaluation
- **2QFY13:** First Unit Equipped
- **3QFY13:** Full-rate production decision
- **4QFY13-4QFY15:** Production

Nett Warrior (NW)

FOREIGN MILITARY SALES
None

CONTRACTORS
Various vendors provide components;
Government is prime integrator.

Night Vision Thermal Systems–Thermal Weapon Sight (TWS)

INVESTMENT COMPONENT

Modernization

Recapitalization

Maintenance

MISSION

Enables the Soldier to detect and engage targets, day or night, in all weather and visibility-obscured conditions.

DESCRIPTION

TheNight Vision Thermal Systems–Thermal Weapon Sight (TWS) family is a group of advanced infrared devices which can be both weapon-mounted or used in an observation mode. The TWS gives Soldiers with individual and crew served weapons the capability to see deep into the battlefield, increase surveillance and target acquisition range, and penetrate obscurants, day or night. The TWS systems use uncooled, forward-looking infrared technology and provide a standard video output for training, image transfer, or remote viewing. TWS systems are lightweight and mountable to a weapon rail. They operate to the maximum effective range of the weapon.

The TWS family comprises three variants, each of which is silent, lightweight, compact, durable, and battery-powered.

They include:
- **AN/PAS-13(V)1 Light Weapon Thermal Sight (LWTS)** for the M16 and M4 series rifles and carbines, as well as the M136 Light Anti-Armor Weapon:

Weight: 1.9 pounds
Field of view: 9 degrees/18 degrees (narrow/wide)
Operational time: 7 hours
Power: four lithium AA batteries

- **AN/PAS-13(V)2 Medium Weapon Thermal Sight (MWTS)** for the M249 Squad Automatic Weapon and M240B series medium machine guns:

Weight: 2.8 pounds
Field of view: 6 degrees/18 degrees (narrow/wide)
Operational time: 7 hours
Power: six lithium AA batteries

- **AN/PAS-13(V)3 Heavy Weapon Thermal Sight (HWTS)** for the squad leader's weapon M16 and M4 series rifles and carbines, M24 and M107 sniper rifles, and M2 HB and MK19 machine guns:

Weight: 3.9 pounds
Field of view: 3 degrees/9 degrees (narrow/wide)
Operational time: 7 hours
Power: six lithium AA batteries

SYSTEM INTERDEPENDENCIES

None

PROGRAM STATUS

- **FY12:** Awarded production contracts for TWS with 17 micron technology; systems delivered in FY13 will be lighter and consume less power

PROJECTED ACTIVITIES

- **FY13:** Continue to support and field in accordance with Headquarters Department of the Army guidance

ACQUISITION PHASE

| Technology Development | Engineering & Manufacturing Development | Production & Deployment | Operations & Support |

Night Vision Thermal Systems–
Thermal Weapon Sight (TWS)

FOREIGN MILITARY SALES
Thailand, Czech Republic, Sweden

CONTRACTORS
BAE Systems (Lexington, MA; Manchester,
  NH; Austin, TX; Manassas, VA)
DRS Optronics (Dallas, TX; Melbourne, FL)
Raytheon (Dallas, TX; Goleta, CA;
  McKinney, TX)

Non-Intrusive Inspection Systems (NIIS)

INVESTMENT COMPONENT

Modernization

Recapitalization

Maintenance

MISSION
Protects U.S. forces and critical warfighting materiel by inspecting cars, trucks, or cargo containers for the presence of explosives, weapons, drugs, or other contraband with nuclear (gamma rays) and X-ray technology.

DESCRIPTION
The Non-Intrusive Inspection Systems (NIIS) program is a layered force protection system which includes a range of approaches to accomplish the mission. These include: embedded commercial off-the-shelf (COTS) products, security personnel who are trained to maintain situational awareness, military working dogs, under-vehicle scanning mirrors, and handheld or desktop trace explosive detectors.

NIIS has a variety of distinct products that are added to the U.S. Army commander's "tool box." Differing characteristics include mobile; rail-mounted, but re-locatable; and fixed-site.

The primary active systems include the following:
- The **Mobile Vehicle and Cargo Inspection System (MVACIS)** is a truck-mounted system with a nuclear source that can penetrate approximately 6.5 inches of steel.
- The **Re-locatable Vehicle and Cargo Inspection System (RVACIS)** is a rail-mounted system with the same nuclear source as the MVACIS. Its versatility allows it to be used on either static locations, or deployed on rails within 24 hours to locations on a prepared rail system.
- The **Militarized Mobile VACIS (MMVACIS)** uses the same gamma source as the other VACIS products, but is mounted on a High Mobility Multipurpose Wheeled Vehicle.

- The **Z-Backscatter Van (ZBV)** is a van-mounted system that utilizes backscatter X-ray technology. It penetrates only approximately one-quarter inch of steel and is used in static locations where room is limited.
- The **Backscatter Vehicle Mounted Trailer** is a mobile inspection system for vehicles and cargo that uses the same backscatter X-ray technology as the ZBV. The BVMT trailer contains the X-Ray source and backscatter detectors and the forward scatter trailer contains the forward scatter detectors.
- **Personnel Scanners** utilizes backscatter X-ray technology to non-intrusively scan people for the presence of explosives, weapons or other contraband and are American National Standards Institute compliant. Depending on the model, these systems can scan between 140 to 240 people per-hour.
- The **T-10 Trailer** is a high-energy gantry vehicle and cargo scanner which uses a one Mega volts Liner Accelerator that penetrates up to four inches of steel while scanning.

SYSTEM INTERDEPENDENCIES
None

PROGRAM STATUS
- **4QFY12:** Fielded 14 Personnel Scanners

PROJECTED ACTIVITIES
- **1QFY13-4QFY13:** Replace obsolete systems which have reached their useful life and do not provide stand-off capabilities.

Technology Development | Engineering & Manufacturing Development | **Production & Deployment** | Operations & Support

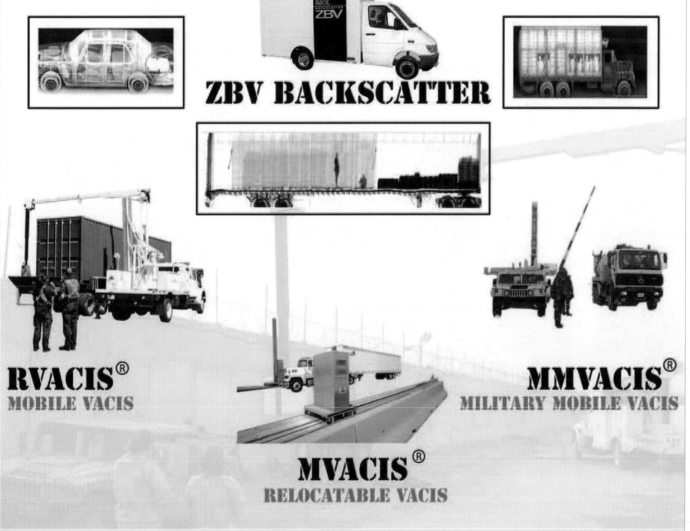

ZBV BACKSCATTER

RVACIS®
MOBILE VACIS

MVACIS®
RELOCATABLE VACIS

MMVACIS®
MILITARY MOBILE VACIS

Non-Intrusive Inspection Systems (NIIS)

FOREIGN MILITARY SALES
None

CONTRACTORS
American Science & Engineering, Inc.
(Billerica, MA)
Rapiscan Systems (Torrance, CA)
Science Applications International Corp.
(SAIC) (San Diego, CA)

Nuclear Biological Chemical Reconnaissance Vehicle (NBCRV) –Stryker Sensor Suites

INVESTMENT COMPONENT

Modernization

Recapitalization

Maintenance

MISSION

Performs nuclear, biological, and chemical (NBC) reconnaissance and locates, identifies, marks, samples, and reports NBC contamination on the battlefield.

DESCRIPTION

The Nuclear Biological Chemical Reconnaissance Vehicle (NBCRV)–Stryker is the chemical, biological, radiological, and nuclear (CBRN) reconnaissance configuration of the Infantry Carrier Vehicle in the Stryker Brigade Combat Teams (SBCTs), Heavy Brigade Combat Teams (HBCTs), and Chemical Companies.

The NBCRV-Stryker Sensor Suite consists of a dedicated system of CBRN detection, warning, and biological-sampling equipment. The NBCRV detects chemical, radiological, and biological contamination in its immediate environment through a variety of mechanisms which include: 1.) Chemical Biological Mass Spectrometer (CBMS), 2.) Automatic Chemical Agent Detector Alarm (ACADA), 3.) AN/VDR-2 Radiac

Detector, 4.) AN/UDR-13 Radiac Detector, 5.) Joint Biological Point Detection System (JBPDS), and 5.) at a distance, through the Joint Service Lightweight Standoff Chemical Agent Detector (JSLSCAD).

It automatically integrates contamination information from detectors with input from onboard navigation and meteorological systems and transmits digital NBC warning messages through the vehicle's command and control equipment to warn follow-on forces. NBCRV can collect samples for follow-on analysis.

SYSTEM INTERDEPENDENCIES
In this Publication
None

Other Major Interdependencies
ACADA, AN/UDR-13 Radiac Detector, CBMS, Chemical Vapor Sampler System (CVSS), JSLSCAD, Nuclear Biological Chemical Sensor Processing Group (NBCSPG)

PROGRAM STATUS
- **Current:** Full-rate production approved Dec 2011
- **Current:** Fielding to SBCT, select Active component HBCT and Chemical Companies

PROJECTED ACTIVITIES
- Fielding to select Active component and Reserve component HBCTs and Chemical Companies

Nuclear Biological Chemical Reconnaissance Vehicle (NBCRV)– Stryker Sensor Suites

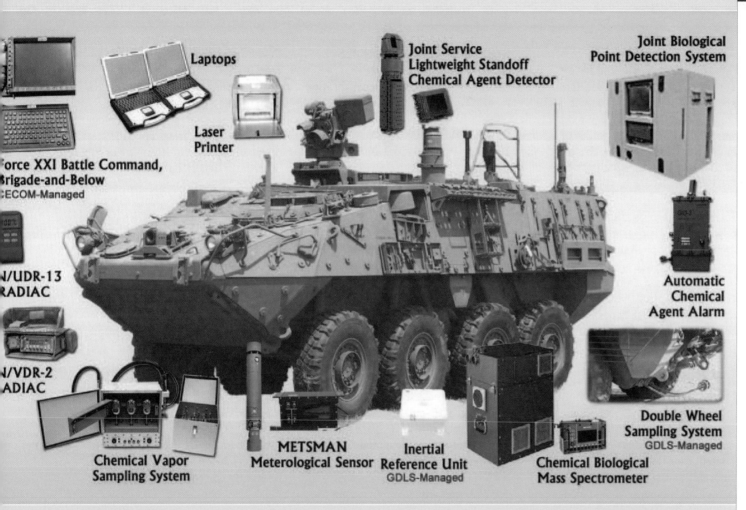

Laptops

Joint Service Lightweight Standoff Chemical Agent Detector

Joint Biological Point Detection System

Laser Printer

Force XXI Battle Command, Brigade-and-Below
CECOM-Managed

N/UDR-13 RADIAC

N/VDR-2 ADIAC

Automatic Chemical Agent Alarm

Chemical Vapor Sampling System

METSMAN Meterological Sensor

Inertial Reference Unit
GDLS-Managed

Chemical Biological Mass Spectrometer

Double Wheel Sampling System
GDLS-Managed

FOREIGN MILITARY SALES
None

CONTRACTORS
Prime Vehicle:
General Dynamics Land Systems
   (Sterling Heights, MI)
Sensor Software Integrator:
CACI Technologies (Manassas, VA)

One Semi-Automated Force (OneSAF)

INVESTMENT COMPONENT

Modernization

Recapitalization

Maintenance

MISSION
Provides simulation software that supports constructive and virtual training, computer-generated forces, and mission rehearsal designed for brigade-and-below, combat, and non-combat operations.

DESCRIPTION
One Semi-Automated Forces (OneSAF) is a next generation, entity-level simulation that supports both computer-generated forces and Semi-Automated Forces (SAF) applications. It is currently being integrated by the Synthetic Environment Core program as the replacement SAF for virtual trainers such as Aviation Combined Arms Tactical Trainer (AVCATT) and Close Combat Tactical Trainer (CCTT). OneSAF will serve as the basis for subsequent modernization activities for simulators across the Army. OneSAF was designed to represent the modular and Future Force and provides variables across the spectrum of military operations in the contemporary operating environment. OneSAF is unique in its ability to model unit behaviors from fire team to company level for all units—both combat and non-combat operations.

OneSAF represents a full range of capabilities in support of simulation applications applied to: advanced concepts and requirements; research, development, and acquisition; and training, exercise, and military operations. OneSAF is designed to meet the constructive training challenges presented by transformation. OneSAF displays a high fidelity environmental representation. Interoperability support is present for industry standards such as Distributed Interactive Simulation, High Level Architecture, Military Scenario Development Language, Joint Consultation Command and Control Information Exchange Data Model, and Army Battle Command System devices.

SYSTEM INTERDEPENDENCIES
In this Publication
OneSAF provides required capabilities for SE Core.

PROGRAM STATUS
Baseline International Deliveries:
- **FY12:** V5.1 Provided virtual product (CCTT/AVCATT) capability set
- **FY12:** PM Apache Longbow Crew Trainer (LCT) (Taiwan)
- **FY12:** PM Cargo Transportable Flight Proficiency Simulator (TFPS) (United Arab Emirates/Australia)

Co Developer Handover Integrations:
- **FY12:** U.S. Marine Corps. Combined Arms Command & Control Training Upgrade System (CACCTUS)
- **FY12:** AVCATT Handover & Merge with CCTT to establish OneSAF Virtual extension
- **FY12:** Battle Lab Simulated Environment(BLSE)/Space and Missile Defense Command (SMDC) capability sets
- **FY12:** IC WARGAMING capability set
- **FY12:** Australian Army capability set

New Capabilities:
- **FY12:** TRADOC Project Office (TPO) V6 Requirements and Integration Board requirements/CS 11-12/ARES
- **FY12:** 75th DIV (Engineering Module, Battle Book, Problem Tracking Reports)/Operational Environment Scenario Generation Tool
- **FY12:** TPO/Training, Exercise, and Military Operations/Army Modeling and Simulation Office PTRs
- **FY12:** Night Vision Goggle Tool Kit/After Action Review System components
- **FY12:** Bumble Bee product

Training:
- **FY12:** CGSC ALE 381/2 Operators courses/ 2 Developers courses

PROJECTED ACTIVITIES
- **FY13:** V6.0
- **FY14:** V7.0
- **FY15:** V8.0

Each release also includes the International release.

ACQUISITION PHASE

| Technology Development | Engineering & Manufacturing Development | Production & Deployment | Operations & Support |

ONESAF

One Semi-Automated Force (OneSAF)

FOREIGN MILITARY SALES
Australia, Bahrain, Canada, Czech Republic, Egypt, South Korea, New Zealand, United Kingdom

CONTRACTORS
Integration and Interoperability Support (I2S)
Prime:
Cole Engineering Services Inc. (CESI)
(Orlando, FL)
Subcontractors:
Lockheed Martin (Orlando, FL)
Dynamics Research Corp.
(Orlando, FL)
Camber Corp. (Orlando, FL)
Accenture (Orlando, FL)
CAE Inc. (Orlando, FL)
Production:
Prime:
Science Applications International Corp. (SAIC)
(Orlando, FL)
Subcontractors:
Lockheed Martin (Orlando, FL)
Northrop Grumman (Orlando, FL)
Dynamics Research Corp.
(Orlando, FL)
Aegis (Orlando, FL)
StackFrame (Sanford, FL)

Palletized Load System (PLS) and PLS Extended Service Program (ESP)

INVESTMENT COMPONENT

Modernization

Recapitalization

Maintenance

MISSION

Supports combat units by performing cross-country movement of configured loads of ammunition and other classes of supply loaded on flat racks or in containers.

DESCRIPTION

The base Palletized Load System (PLS) consists of the PLS truck, the PLS trailer, and demountable flat racks. The PLS is a 10x10, ten-wheel-drive, multi-drive truck with 16.5-ton capacity. It provides the timely delivery of unit equipment, a high tonnage of ammunition, and International Organization for Standardization (ISO) containers/shelters. It also carries all classes of supplies to units and weapon systems as far forward in the maneuver battalion area as the tactical situation allows. The PLS truck is equipped with an integral onboard load handling system that provides self-loading and unloading capability of flat racks, container roll-in/out platforms and 20-foot ISO containers.

There are two PLS truck variants: 1.) the basic PLS truck (M1075), and 2.) the PLS truck with materiel handling crane (M1074). The system also includes: the optional PLS trailer (M1076), a truck-mounted container handling unit for transporting 20-foot ISO containers; the M3/M3A1 container roll-in/out platform; and the M1/M1077A1 flat racks. The PLS trailer also has a matching payload capacity to the PLS truck of 16.5 tons.

The new PLSA1 truck model was fielded in 2011. It incorporates: independent front suspension, a new Caterpillar C-15 engine, the Allison 4500SP 6-speed transmission, J-1939 data-bus, and a cab that is common with the Heavy Expanded Mobility Tactical Truck A4 long-term armor strategy compliant cab.

The PLS Extended Service Program (ESP) is a recapitalization (RECAP) program that converts high-mileage base PLS trucks to zero miles/zero hours and to the current A1 production truck configurations. The base PLS trucks are disassembled and rebuilt with improved technology such as

an electronically controlled engine, electronic transmission, air ride seats, four-point seatbelts, bolt-together wheels, increased corrosion protection, enhanced electrical package, and independent front suspension on the PLSA1 truck.

SYSTEM INTERDEPENDENCIES

None

PROGRAM STATUS

- **Current:** To date, fielded approximately 6,000 PLS trucks and 13,000 PLS trailers

PROJECTED ACTIVITIES

- **FY10:** PLSA1 type classification/ materiel release
- **FY11:** PLSA1 first unit equipped
- **FY12:** PLSA0 RECAP begins
- **FY13:** PLSA1 New production ends

ACQUISITION PHASE

Technology Development | Engineering & Manufacturing Development | Production & Deployment | Operations & Support

Palletized Load System (PLS) and PLS Extended Service Program (ESP)

FOREIGN MILITARY SALES
Turkey, Israel, Jordan

CONTRACTORS
Prime:
Oshkosh Corp. (Oshkosh, WI)
Engine:
Detroit Diesel (Emporia, KS; Redford, MI)
Caterpillar C-15 (Peoria, IL)
Transmission:
Allison Transmission (Indianapolis, IN)
Tires:
Michelin (Greenville, SC)

	PLS	PLSA1
ENGINE	DDC 8V92 - 500 horsepower	CAT C-15 - 600 hp @ 2100 RPM
TRANSMISSION	Allison CLT-755 - 5-Speed	Allison HD 4500 - 6-Speed
TRANSFER CASE	Oshkosh 55,000 - 2-Speed	New Oshkosh - 2-Speed
AXLES FRONT: TANDEM	Rockwell SVI 5MR/Planetary Hub	Oshkosh TAK-4 with AxleTech carrier with differential lock and planetary wheelends
AXLES: REAR TRIDEM	Rockwell SVI 5MR	AxleTech carrier with differential lock and planetary wheel ends
SUSPENSION - AXLES #1 & #2	Hendrickson RT-340 - Walking Beam	Oshkosh TAK 4 Steel Spring
SUSPENSION - AXLE #3	Hendrickson-Turner Air Ride	Hendrickson-Turner Air Ride
SUSPENSION - AXLES #4 & #5	Hendrickson RT-400 - Walking Beam	Hendrickson RT-400 - Walking Beam
WHEEL ENDS	Rockwell	Rockwell
CONTROL ARMS	N/A	Standard MTVR on Front Tandem
STEERING GEARS - FRONT	492 Master/M110 Slave	M110 Master/M110 Slave
STEERING GEARS - REAR	492	M110
FRAME RAILS	14-inch	14-inch
CAB	PLS	LTAS Compliant Common Cab
RADIATOR	Roof Mount	Side Mount
MUFFLER	PLS	New
AIR CLEANER	United Air	United Air
LHS	Multilift MK V	Multilift MK V
CRANE	Grove	Grove
TIRES	Michelin 16.00 R20 XZLT	Michelin 16.00 R20 XZLT
SPARE TIRE	1 - Side Mounted	1 - Roof Mounted
CTI	CM Automotive	Dana
AIR COMPRESSOR	1400 Bendix	922 Bendix
STARTER	Prestolite	Prestolite
ALTERNATOR	12/24V	260 Amp Niehoff

PATRIOT Advanced Capability–3 (PAC-3)

INVESTMENT COMPONENT

Modernization

Recapitalization

Maintenance

MISSION
Protects ground forces and critical assets at all echelons from advanced aircraft, cruise missiles, and tactical ballistic missiles.

DESCRIPTION
The Phased Array Tracking to Intercept of Target (PATRIOT) Advanced Capability-Three (PAC-3) program is an air-defense, guided missile system with long-range, medium- to high-altitude, all-weather capabilities. It is designed to counter tactical ballistic missiles (TBMs), cruise missiles, and advanced aircraft. The combat element of the PATRIOT missile system is the fire unit, which consists of the following: a phased array radar set (RS); an engagement control station (ECS); a battery command post; an electric power plant; an antenna mast group; a communications relay group; and launching stations (LS) with missiles.

The RS provides the tactical functions of airspace surveillance, target detection, identification, classification, tracking, missile guidance, and engagement support. The ECS provides command and control. Depending upon configuration, the LS provides the platform for PAC-2 or PAC-3 missiles, which are sealed in canisters that serve as shipping containers and launch tubes.

The PAC-3 primary mission is to engage TBMs, and advanced cruise missile and aircraft threats. The PAC-3 missile uses hit-to-kill technology for greater lethality against TBMs armed with weapons of mass destruction. The PAC-3 system upgrades have provided improvements that increase performance against evolving threats, meet user requirements, and enhance Joint interoperability. PATRIOT's fast-reaction capability, high firepower, ability to track numerous targets simultaneously, and ability to operate in a severe environment make it the U.S. Army's premier air defense system. The PAC-3 Missile Segment Enhancement (MSE), currently in development, is planned to be used with the PAC-3 system and is the interceptor for the Medium Extended Air Defense System.

SYSTEM INTERDEPENDENCIES
In this Publication
None

Other Major Interdependencies
ABMOC, AEGIS, AMDTF, AOC, AWACS, CRC, HAWKEYE, P/M CAP, RIVET-JOINT, SHORAD, TACC, TAOC, THAAD

PROGRAM STATUS
- **2QFY11:** Post Deployment Build-6.5 (PDB-6.5) fielded

PROJECTED ACTIVITIES
- **3QFY11-3QFY13:** Missile Segment Enhancement (MSE) Developmental and Operational Testing
- **1QFY13:** Post Deployment Build-7 (PDB-7) fielding

ACQUISITION PHASE

| Technology Development | Engineering & Manufacturing Development | Production & Deployment | Operations & Support |

UNITED STATES ARMY

PATRIOT Advanced Capability–3 (PAC-3)

FOREIGN MILITARY SALES
Germany, Japan, Netherlands, Taiwan, United Arab Emirates

CONTRACTORS
Missile Program Management Team: Lockheed Martin (Dallas, TX)
Seeker Program Management Team: Boeing (Anaheim, CA)
Mods: Raytheon (Tewksbury, MA; Long Beach, CA)
Seeker Manufacturing/RFDL: Lockheed Martin Missiles and Fire Control (Chelmsford, MA)
ELES: Lockheed Martin (Lufkin, TX)
System Integration: Raytheon–El Paso (El Paso, TX); Raytheon (Huntsville, AL); Raytheon–Norfolk (Norfolk, VA); Raytheon–Burlington (Burlington, MA)
Missile Assembly: Lockheed Martin (Camden, AR)
Integration/GSE: Raytheon (Andover, MA)
Seeker: Boeing (Huntsville, AL)
Seeker Assembly: Boeing (El Paso, TX)
SRM/ACM: Aerojet (Camden, AR)
SETA: Intuitive Research and Technology (Huntsville, AL)

Precision Guidance Kit (PGK)

Modernization

Recapitalization

Maintenance

M795   M549A1

MISSION
Improves the accuracy of conventional 155mm high-explosive (HE) projectiles in the inventory.

DESCRIPTION
Precision Guidance Kit (PGK) technology is state-of-the-art and provides a first-of-its-kind capability. PGK contains a Global Positioning System (GPS) guidance kit with fuzing functions and an integrated GPS receiver to correct the inherent errors associated with ballistic firing solutions, reducing the number of artillery projectiles required to attack targets. The increase in efficiency that PGK's "near-precision" capability provides allows operational commanders to engage assigned targets and rapidly achieve desired effects while minimizing collateral damage.

The PGK program is following an incremental program approach. Increment 1, the XM1156 PGK, will be compatible with the 155mm M795 and M549A1 HE projectiles which are fired from the M109A6 Paladin and M777A2 Lightweight 155mm Howitzer. Future increments could have improved resistance to threat GPS jammers and expand this capability to projectiles containing an insensitive munitions explosive fill.

SYSTEM INTERDEPENDENCIES
None

PROGRAM STATUS
- **Current:** Increment 1 program is in Engineering and Manufacturing Development

PROJECTED ACTIVITIES
Increment 1
- **2QFY13:** Milestone C

Precision Guidance Kit (PGK)

FOREIGN MILITARY SALES
None

CONTRACTORS
Increment 1
Prime:
Alliant Techsystems (Plymouth, MN)
Subcontractor:
L-3 Interstate Electronics Corp.
    (Anaheim, CA)

Prophet

MISSION

Provides a near-real-time picture of the battlespace through the use of signals intelligence sensors with the capability to detect, identify, and locate selected emitters.

DESCRIPTION

Prophet is a 24-hour, all weather, near-real-time, ground-based, tactical signals intelligence/electronic warfare capability organic to the Brigade Combat Team (BCT), Stryker BCT (SBCT), and Battlefield Surveillance Brigade (BfSB). Prophet is fielded with two Prophet Enhanced (PEs) sensors and one Prophet Control (PC) or Prophet Analytical Cell (PAC) to BCTs, three PEs sensors and one PC or PAC to SBCTs, and eight PEs sensors and four PCs or PACs to BfSBs. Prophet provides actionable intelligence, situational understanding, and force protection. It is interoperable on the

Global Signals Intelligence Enterprise, delivering collected data to common databases for access by the intelligence community. Prophet's tactical mobility allows supported units to easily reposition its collection capability on the battlefield to support evolving situations.

The Prophet Enhanced is a non-platform dependent modular system that will allow easy integration onto a vehicle. The Prophet Enhanced supports stationary, on-the-move (Mobile) and Manpack operations simultaneously. The Prophet Enhanced Mobile Mission Equipment was integrated on an XM1229 Medium Mine Protected Vehicle and Mobile-All Terrain Vehicle to provide better crew protection and was fielded to units deployed in support of Operation Enduring Freedom.

The Prophet Enhanced has a Wideband Beyond-Line-of-Site (WBLOS) capability, which is based on the present Project Manager Warfighter Information Network–Tactical (WIN-T) architecture. This capability allows operation without the constraints of

line-of-sight communication, increasing the system's capability to operate at extended distance and perform distributed operations.

PC/PAC is the analytical node that tasks the Prophet Enhanced for data collection and reporting. Each PC/PAC contains Satellite Communications (SATCOM). The PC has TROJAN-Lightweight Integrated Telecommunications Equipment (T-Lite) and PAC has a SATCOM Capability Set. PC is integrated on an armored M1165 High Mobility Multipurpose Wheeled Vehicle and PAC is a platform independent modular system that is easily integrated onto a vehicle/trailer.

SYSTEM INTERDEPENDENCIES
In this Publication
None

Other Major Interdependencies
Global Positioning System, Trojan-Lightweight Integrated Telecommunications Equipment, Tactical Radio Communications Systems, Light Tactical Vehicles and Assured Mobility Systems.

PROGRAM STATUS
- **1QFY12-4QFY12:** Fielded Prophet Enhanced Systems
- **1QFY12-4QFY12:** Defielded, as necessary, Prophet Block I systems as Prophet Enhanced Systems are fielded
- **2QFY12:** Prophet Enhanced/Prophet Analytical Cell fielded to 201st

PROJECTED ACTIVITIES
- **1QFY13-4QFY13:** Continue Prophet Enhanced Systems fieldings
- **1QFY13-4QFY13:** Continue to defield, as necessary, Prophet Block I systems as Prophet Enhanced Systems are fielded

Prophet

FOREIGN MILITARY SALES
None

CONTRACTORS
Prophet Enhanced Sensor/
Analysis Cell Production:
General Dynamics (Scottsdale, AZ)

PAC Capability Set

Network Server Module (5U)

Backup Battery Module (5U)

NSAnet WMI

DAGR

Shore Power Module (5U)

SIPRNET TDN Module (5U)

JWICS TDN Module (5U)

SIPRnet WMI

VOIP Phone

JWICS WMI

VOIP Phone

SATCOM Capability Set*

SATCOM Module (5U)

SATCOM 1.4M Antenna

SATCOM WMI

* Currently Required by CPD but may be satisfied by Organic Brigade SATCOM

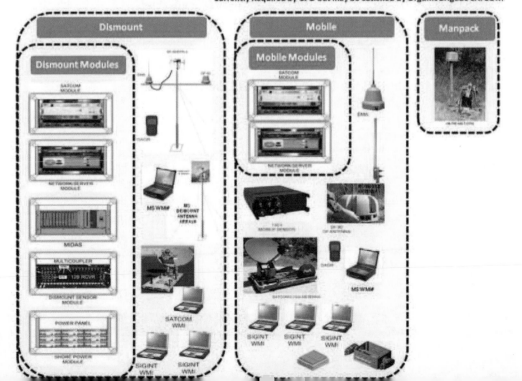

Dismount

Mobile

Manpack

Rocket, Artillery, Mortar (RAM) Warn

INVESTMENT COMPONENT

Modernization

Recapitalization

Maintenance

MISSION

Provides an organic, early, localized warning capability to support all Maneuver Brigade Combat Teams (BCTs).

DESCRIPTION

The Rocket, Artillery, Mortar (RAM) Warn program evolved from the Counter-Rocket, Artillery, Mortar (C-RAM) program and is a horizontal technology insertion, using current C-RAM warning equipment. RAM Warn employs the Air Defense Airspace Management (ADAM) Cell already resident in the BCT headquarters as the command and control (C2) element, uses the existing radars in the Target Acquisition Platoon of the Fires Battalion as the Sense element, and adds enhanced C2, warning equipment, controllers, and dedicated communications devices between the existing radars, ADAM Cell, and warning devices.

RAM Warn detects threat Indirect Fire (IDF) rounds, transmits the detection data to the C2 element for correlation and determination of predicted Point of Origin (POO) and Point of Impact

(POI), and if necessary activates audio and visual alarms for localized, or full area warning over the defended area. Timely warning will enable personnel in the hazard area of an inbound IDF threat to seek cover or a prone position prior to impact, thus reducing casualties.

SYSTEM INTERDEPENDENCIES

In this Publication

Forward Area Air Defense Command and Control (FAAD C2)

Other Major Interdependencies

FAAD C2/C-RAM C2 (at ADAM Cell), Firefinder Radar, Lightweight Counter Mortar Radar (LCMR)

PROGRAM STATUS

- **2QFY12:** Acquisition Decision Memorandum (ADM) signed establishing RAM Warn as an Aquisition Category III acquisition program under Program Executive Office, Missiles and Space
- **3QFY12:** RAM Warn Technical Manual and Training Support Plan Verification
- **4QFY12:** Logistics Demonstration

- **4QFY12:** New Equipment Training (NET) for Initial Operational Test (IOT)

PROJECTED ACTIVITIES

- **1QFY13:** IOT at Network Integration Evaluation (NIE) 13.1
- **2QFY13:** Milestone C Decision
- **3QFY13:** Production/Integration Contract Award
- **4QFY13:** First Unit Equipped

Technology Development | Engineering & Manufacturing Development | Production & Deployment | Operations & Support

Rocket, Artillery, Mortar (RAM) Warn

FOREIGN MILITARY SALES
None

CONTRACTORS
To be determined

Production contract award for RAM Warn hardware, integration, training, test, and installation/fielding is pending Milestone C decision.

Rough Terrain Container Handler (RTCH)

Modernization

Recapitalization

Maintenance

MISSION

Provides container handling and materiel handling capability in cargo transfer companies, transportation companies, quartermaster units, and ammunitions platoons.

DESCRIPTION

The Rough Terrain Container Handler (RTCH) is a commercial, non-developmental item acquired for cargo-handling missions worldwide. The vehicle lifts, moves, and stacks both 20-foot and 40-foot American National Standards Institute/International Organization for Standardization containers and shelters weighing up to 53,000 pounds.

RTCH improvements include the following: 1.) the capability to transport by rail, highway, or water in less than 2 1/2 hours. This reduces preparation time for air transport (C5A and C17) from 16 hours to less than one hour. 2.) Stacking nine-foot, six-inch containers three high achieves a forward speed of 23 miles per hour, and 3.) adding a full-range extendable boom and flexible top handler.

RTCHs operate worldwide, on hard-stand, over-sand terrain, and cross-country, executing ammunition handling and transportation operations. The system is capable of conducting operations in cold, basic, and hot climates. Additionally, RTCH can ford up to 60 inches of sea water.

SYSTEM INTERDEPENDENCIES

None

PROGRAM STATUS

- **Current:** Fielding complete

PROJECTED ACTIVITIES

- **Current:** contract deliveries are for foreign military sales. Contract expires in FY13. Contract may be utilized if cost to RESET is too high
- **FY13-FY15:** Research, Development, Test and Evaluation effort for sling load and armor

Rough Terrain Container Handler (RTCH)

FOREIGN MILITARY SALES
Australia, United Kingdom, Afghanistan, Iraq

CONTRACTORS
Kalmar Rough Terrain Center (KRTC) LLC (Cibolo, TX)

RQ-11B Raven Small Unmanned Aircraft System (SUAS)

INVESTMENT COMPONENT

Modernization

Recapitalization

Maintenance

MISSION

Provides reconnaissance, surveillance, target acquisition (RSTA), and force protection for the battalion commander and below during day/night operations.

DESCRIPTION

The RQ-11B Raven is a Small Unmanned Aircraft System (SUAS). It is a hand-launched, unmanned aircraft system capable of 90 minutes of flight time with an operational range of approximately ten kilometers. The Raven system comprises three air vehicles, a ground control station (GCS), a remote video terminal (identical to GCS), electro-optical (EO) / infrared (IR) payloads, aircraft and GCS batteries, a field repair kit, and a spares package. Normal operational altitude is 500 feet or lower. The system, aircraft, and ground control station are assembled by operators in approximately five minutes. The aircraft has a wingspan of 4.5 feet and weighs 4.2 pounds. Both color EO sensors and IR sensors are fielded for day and night capabilities with each system. A hand controller displays live video and aircraft status. Mission planning is performed on the hand controller or ruggedized laptop running Portable Flight Planning Software/ Falcon View flight planning software. Aircraft flight modes include fully autonomous navigation, altitude hold, loiter, and return home. In-flight re-tasking and auto-loiter at sensor payload point of interest are also available. Raven incorporates secure Global Positioning System navigation. The digital data link incorporates encryption, improves spectrum management allowing more air vehicles to be flown in an operational area, and provides range extension via data relay between two Raven aircraft.

The Raven is operated by two Soldiers and has a rucksack-portable design. No specific military occupational specialty is required. Operator training is ten days.

SYSTEM INTERDEPENDENCIES

None

PROGRAM STATUS

- **Current:** In production and deployment
- **Current:** Operational in both Operation New Dawn and Operation Enduring Freedom

PROJECTED ACTIVITIES

- **FY11-FY12:** Continue full-rate production and product improvements; integrate and field gimbaled payload (combined EO/IR/ Laser Illuminator)

RQ-11B Raven Small Unmanned Aircraft System (SUAS)

FOREIGN MILITARY SALES
Denmark, Estonia, Lebanon, Uganda

CONTRACTORS
Aerovironment, Inc. (Simi Valley, CA)
Indigo System Corp. (Goleta, CA)
All American Racers, Inc. (Santa Ana, CA)
L-3 Communications (San Diego, CA)
Bren-Tronics (Commack, NY)

RQ-7B Shadow Tactical Unmanned Aircraft System (TUAS)

INVESTMENT COMPONENT

Modernization

Recapitalization

Maintenance

MISSION

Provides reconnaissance, surveillance, target acquisition, and force protection for the Brigade Combat Team (BCT) in near-real-time during day/night and limited adverse weather conditions.

DESCRIPTION

The RQ-7B Shadow Tactical Unmanned Aircraft System (TUAS) has a wingspan of 20 feet and a payload capacity of approximately 60 pounds. Gross takeoff weight exceeds 440 pounds and endurance is more than eight hours on-station at a distance of 50 kilometers. The system is compatible with the All Source Analysis System, Advanced Field Artillery Tactical Data System, Joint Surveillance Target Attack Radar System Common Ground Station, Joint Technical Architecture–Army, the Defense Information Infrastructure Common Operating Environment, and the One System Ground Control Station (OSGCS). The RQ-7B Shadow can be transported by six U.S. Air Force C-130 aircraft. It is currently operational in both the U.S. Army and U.S. Marine Corps.

The RQ-7B Shadow configuration, fielded in platoon sets, consists of:
- Four air vehicles with day/night electro-optical (EO)/infrared (IR) with laser designator and IR illuminator payloads
- Two One System Ground Control Station on High Mobility Multipurpose Wheeled Vehicles (HMMWV)
- Four One System Remote Video Transceivers
- One hydraulic launcher
- Two ground data terminals
- Associated trucks, trailers, and support equipment

Shadow platoons are organic to the BTC. The Soldier platoon consists of a platoon leader, platoon sergeant, unmanned aerial vehicle (UAV) warrant officer, 12 Air Vehicle Operators/Mission Payload Operators, four electronic warfare repair personnel, and three engine mechanics supporting launch and recovery. The Maintenance Section Multifunctional is manned by Soldiers who also transport spares and provide maintenance support. The Mobile Maintenance Facility is manned by contractor personnel located with the Shadow platoon to provide logistics support to include "off system support" and "maintenance by repair."

The RQ-7B Shadow also has an early entry configuration of 15 Soldiers, one Ground Control Station (GCS), the air vehicle transport HMMWV, and the launcher trailer, which can be transported in three C-130s. All components can be slung under a CH-47 or CH-53 helicopter for transport.

SYSTEM INTERDEPENDENCIES
None

PROGRAM STATUS
- **Current:** In production and deployment; flown more than 480,000 hours in support of combat operations in Operation Iraqi Freedom and Operation Enduring Freedom since achieving initial operating capability. Total system flight hours is more than 650,000 hours

PROJECTED ACTIVITIES
- **FY11-FY12:** Field remaining production systems; procure and field laser designator, Tactical Common Data Link, and Universal Ground Control Station retrofits; develop and field reliability and product improvements

ACQUISITION PHASE

Technology Development	Engineering & Manufacturing Development	Production & Deployment	Operations & Support

RQ-7B Shadow Tactical Unmanned Aircraft System (TUAS)

FOREIGN MILITARY SALES
Australia

CONTRACTORS
Shadow System:
AAI Corp. (Textron Systems) (Hunt Valley, MD)
TCDL:
L-3 Communications (Salt Lake City, UT)
Shelter Integration:
CMI (Huntsville, AL)
GDT:
Tecom (Chatsworth, CA)
Shelters:
General Dynamics (Marion, VA)
ACE II/II+/III Flight:
Rockwell Collins (Warrenton, VA)
Mode IV IFF:
Raytheon (Baltimore, MD)
Amplifiers:
CTT (Santa Clara, CA)

Secure Mobile Anti-Jam Reliable Tactical Terminal (SMART-T)

INVESTMENT COMPONENT

Modernization

Recapitalization

Maintenance

MISSION
Provides range extension to the Army's current and future tactical communications networks.

DESCRIPTION
The Secure Mobile Anti-Jam Reliable Tactical–Terminal (SMART-T) is a mobile, military satellite communication terminal that provides worldwide, anti-jam, low probability of intercept and detection, secure voice and data capabilities for the Joint warfighter. The SMART-T provides range extension to the Army's current and future tactical communications networks through DoD Milstar and Advanced Extremely High Frequency (AEHF) communication satellites. SMART-Ts are being upgraded to interoperate with AEHF satellites and can now provide data rates up to 8.192 million bits per second (Mbps). The total Army procurement is 278 systems. The AEHF satellite system will dramatically increase the Army's end-to-end anti-jam communications throughput capability.

The Army's 278 SMART-Ts are being fielded at the brigade, division, and corps echelons in the Active Army, National Guard, and Reserve components.

SMART-T is part of Capability Set 13, the Army's first package of network components, associated equipment and software that provides integrated connectivity from the static tactical operations center to the commander on-the-move to the dismounted Soldier. CS 13 begins fielding to Brigade Combat Teams in October 2012.

SYSTEM INTERDEPENDENCIES
In this Publication
None

Other Major Interdependencies
The SMART-T communicates with Milstar military communication satellites and is being upgraded to communicate with AEHF communication satellites

PROGRAM STATUS
- **1QFY12-4QFY12:** Continued upgrade of SMART-Ts and conduct of New Equipment Training at the Fielding and Training Facility in Largo, FL

PROJECTED ACTIVITIES
- **2QFY13:** SMART-T will participate in Air Force AEHF multi-service operational test and evaluation
- **2QFY13-2QFY15:** Continued upgrade and fielding of AEHF SMART-Ts

ACQUISITION PHASE

| Technology Development | Engineering & Manufacturing Development | Production & Deployment | Operations & Support |

Secure Mobile Anti-Jam Reliable Tactical Terminal (SMART-T)

FOREIGN MILITARY SALES
SMART-Ts:
Canada (19), Netherlands (7)

CONTRACTORS
Production and Spares:
Raytheon (Largo, FL)
Engineering Support, Management:
Raytheon (Marlborough, MA)
Circuit Cards:
Teledyne (Lewisburg, TN)
Filters:
Transtector Systems (Hayden, ID)
Amplifier Assemblies:
Spectrum Microwave (Marlborough, MA)
COMSEC:
L-3 Communications (Camden, NJ)
New Equipment Training/Fielding:
EPS Corp. (Martinez, GA)
Satellite Simulator:
Lincoln Labs (Lexington, MA)
Technical/Fielding Support:
Linquest Corp. (Colorado Springs, CO)
Admin/Tech:
JANUS Research (Belcamp, MD)
Technical:
Booz Allen Hamilton (Belcamp, MD)

Sentinel

INVESTMENT COMPONENT

Modernization

Recapitalization

Maintenance

MISSION
Provides persistent surveillance and fire control quality data through external command and control platforms, enabling protection against cruise missiles, aircraft, unmanned aerial vehicles, and rocket, artillery, and mortar threats.

DESCRIPTION
Sentinel is used with the Army's Forward Area Air Defense Command and Control (FAAD C2) system and provides key target data to Stinger-based weapon systems and battlefield commanders through FAAD C2 or directly, using an Enhanced Position Location Reporting System or the Single Channel Ground and Airborne Radio System.

Sentinel consists of the M1097A1 High Mobility Multipurpose Wheeled Vehicle (HMMWV), the antenna transceiver group mounted on a high-mobility trailer, the identification friend-or-foe system (IFF), and the FAAD C2 interface. The sensor is an advanced three-dimensional battlefield X-band air defense phased-array radar with a 75-kilometer instrumented range.

Sentinel can operate day and night, in adverse weather conditions, and in battlefield environments of dust, smoke, aerosols, and enemy countermeasures. It provides 360-degree azimuth coverage for acquisition and tracking of targets (cruise missiles, unmanned aerial vehicles, rotary and fixed-wing aircraft) moving at supersonic, to hovering speeds, and at positions from the nap of the earth to the maximum engagement altitude of short-range air defense weapons. Sentinel detects targets before they can engage, thus improving air defense weapon reaction time and allowing engagement at optimum ranges. Sentinel's integrated IFF system reduces the potential for engagement of friendly aircraft.

Sentinel modernization efforts include increased detection and acquisition range of targets and enhanced situational awareness. The system provides integrated air tracks with classification and recognition of platforms that give an integrated air and cruise missile defense solution for the Air and Missile Defense System of Systems Increment 1 architecture and subsequent increments. Sentinel provides critical air surveillance of the National Capital Region and other areas as part of ongoing homeland defense efforts, and is a component of the counter rocket, artillery and mortar batteries in the area of responsibility.

SYSTEM INTERDEPENDENCIES
In this Publication
None

Other Major Interdependencies
Integration efforts with the Army Integrated Air and Missile Defense (AIAMD) architecture

PROGRAM STATUS
- **FY11-FY12:** Procurement of 31 Improved Sentinel A1 Kits

- **2QFY12:** Signed Indefinite Delivery/ Indefinite Quantity Contract with CECOM for AN/TPX-57 IFFs
- **4QFY12:** Prototype Family of Medium Tactical Vehicles (FMTV) Sentinel A3 variant
- **4QFY12:** Delivery of Finland and Netherlands F1 Software
- **4QFY12:** Procure A1 to A3 conversion kits for current Sentinel fleet

PROJECTED ACTIVITIES
- **FY13:** Procurement Production/ Delivery of 48 FMTV Sentinel A3 systems and 8 A3 Repair Cycle Floats
- **FY13-FY14:** Fielding/installation of Improved Sentinel A1 Kits

ACQUISITION PHASE

Technology Development	Engineering & Manufacturing Development	Production & Deployment	Operations & Support

Sentinel

FOREIGN MILITARY SALES
Egypt, Lithuania, Turkey

CONTRACTORS
Thales Raytheon Systems (Fullerton, CA;
    El Paso, TX; Forest, MS; Largo, FL)
Various SETA Contractors (Huntsville, AL)

Single Channel Ground and Airborne Radio System (SINCGARS)

INVESTMENT COMPONENT

Modernization

Recapitalization

Maintenance

MISSION

Provides Joint commanders with a highly reliable, low-cost, secure, and easily maintained Combat Net Radio (CNR) that has both voice and data handling capability in support of tactical command and control operations.

DESCRIPTION

The Single Channel Ground and Airborne Radio System (SINCGARS) Advanced SINCGARS System Improvement Program (ASIP) radio is the DoD/U.S. Army multi-service fielded solution for voice communication for platoon level and above, operating over the 30.000 to 87.975 megahertz frequency range. This radio provides the capability of establishing two-way communications (including jam-resistance) using the SINCGARS waveform and provides multimode voice and data communications supporting ground, air-to-ground, and ground-to-air line-of-sight communications links.

The ASIP radio is the newer version of the SINCGARS radio. It is smaller than the System Improvement Program (SIP) and weighs significantly less, while still maintaining all the functionalities of the SIP for backward compatibility. Enhancements include the Embedded Global Positioning System (GPS) Receiver (EGR) and the radio-based combat identification/radio-based situational awareness (RBCI/RBSA) capability, which provides the warfighter with enhanced situational awareness and identification of friendly forces in targeted areas.

RBCI serves as a gap filler for combat identification, providing an interrogation/responder capability to satisfy the air-to-ground positive identification of platforms prior to release of weapons to prevent fratricide. RBSA adds a radio beaconing capability for every ASIP-equipped platform to enhance the Blue Force situational awareness picture. The Internet controller enhancements add improved addressing capabilities in support of tactical Internet enhancements being provided by Joint Battle Command–Platform for Joint interoperability. Crypto modernization is a programmable communications security capability for SINCGARS that will allow the radios to continue to provide secure communications to the secret and top-secret level of security.

SINCGARS is part of Capability Set (CS) 13, the Army's first package of network components, associated equipment, and software that provides integrated connectivity from the static tactical operations center to the commander on-the-move to the dismounted Soldier. CS 13 begins fielding to Brigade Combat Teams in 2012.

SYSTEM INTERDEPENDENCIES

None

PROGRAM STATUS

- **1QFY12-4QFY12:** Continue to field in accordance with Headquarters Department of the Army guidance to support the Army Campaign Plan; National Guard, Army Reserve, and Active Army, Operation Enduring Freedom requirements and urgent Operational Needs Statement

PROJECTED ACTIVITIES

- **2QFY13-2QFY15:** Continued fielding of SINCGARS

ACQUISITION PHASE

| Technology Development | Engineering & Manufacturing Development | Production & Deployment | Operations & Support |

Single Channel Ground and Airborne Radio System (SINCGARS)

FOREIGN MILITARY SALES
Australia, Bahrain, Croatia, Egypt, Estonia, Finland, Georgia, Greece, Hungary, Ireland, Italy, Korea, Kuwait, Morocco, New Zealand, Portugal, Saudi Arabia, **Supreme Headquarters Allied Powers Europe (SHAPE) Tech Center:** Slovakia, Taiwan, Thailand, Ukraine, Uzbekistan, Zimbabwe

CONTRACTORS
Radio design/production:
ITT (Fort Wayne, IN)
Hardware Installation Kits:
UNICOR (Washington, DC)
Engineering Support and Testing:
ITT (Clifton, NJ)
Total Package Fielding:
USFalcon/EPS Corp. (Morrisville, NC, and Tinton Falls, NJ)

Small Arms–Crew Served Weapons

INVESTMENT COMPONENT

Modernization

Recapitalization

Maintenance

MISSION
Enables warfighters and small units to engage targets with lethal fire to defeat or deter adversaries.

DESCRIPTION
The M249 Squad Automatic Weapon (SAW) replaced the M16A1 Automatic Rifle at the squad level, as well as some M60 multipurpose machine guns in non-infantry units. The M249 delivers greater range and rates of fire than the M16 or M4. Improvements include a collapsible buttstock, a shorter barrel, and an improved bipod, providing increased combat efficiency.

The M240B 7.62mm Medium Machine Gun has been reconfigured for ground applications with buttstock, bipod, iron sights, and forward-rail assemblies.

The M240L 7.62mm Medium Machine Gun (Light) incorporates titanium construction and alternative manufacturing methods to achieve significant weight savings, reducing the Soldier's combat load while allowing easier handling and movement of the weapon. The M240L is approximately five pounds lighter than the M240B.

The M2 .50 Caliber Machine mounts on the M3 tripod and on most vehicles while also serving as an anti-personnel and anti-aircraft weapon. It is highly effective against light armored vehicles, low- and slow-flying aircraft, and small boats. It is capable of single-shot (ground M2 Machine Gun) and automatic fire.

The M2A1 with Quick Change Barrel (QCB) is an enhancement to the M2 .50 Caliber Machine Gun offering Soldiers increased performance as well as new features and design improvements. The M2A1 provides a fixed headspace and timing configuration, flash hider, and removable carrying handle. The M2A1 speeds target engagement and improves survivability and safety by reducing the time required to change the barrel and eliminating the timely procedure of setting headspace and timing.

The MK19 Grenade Machine Gun supports the Soldier by delivering heavy, accurate, and continuous firepower against enemy personnel and lightly armored vehicles. The MK19 can be mounted on a tripod or on multiple vehicle platforms and is the primary suppression weapon for combat support and combat service support units.

SYSTEM INTERDEPENDENCIES
In this Publication
Common Remotely Operated Weapon Station (CROWS)

PROGRAM STATUS
M249:
- Fielding and sustainment; Army production ended in July 2012
- Production for other Services, Agencies and Foreign Military Sales continue

M240B:
- Production and fielding

- Product Qualification/Verification activities on-going for the Colt M240B Weapons

M240L:
- Production and fielding

M2/M2A1:
- 4QFY11: M2A1 First Unit Equipped
- U.S. Ordnance successfully completed M2 first article testing

MK19
- Production and fielding
- Support Foreign Military Sales (FMS) requirements

PROJECTED ACTIVITIES
- In sustainment

Technology Development | Engineering & Manufacturing Development | Production & Deployment | Operations & Support

Small Arms–Crew Served Weapons

FOREIGN MILITARY SALES
M249 SAW:
Afghanistan, Colombia, Croatia, Philippines, Bangladesh, Tunisia
M240B:
Azerbaijan, Barbados, Bangladesh, Jordan
M2:
Afghanistan, Lebanon, Oman
MK19 Grenade Machine Gun:
Croatia, Kenya, Colombia, Bahrain, Chili, Czech Republic, Afghanistan, Saudi Arabia

CONTRACTORS
M249 SAW:
Fabrique National Manufacturing, LLC
    (Columbia, SC)
M240B Machine Gun:
Fabrique National Manufacturing, LLC
    (Columbia, SC)
Colt Defense LLC (West Hartford, CT)
M2:
General Dynamics Armament and Technical
    Products (GDATP) (Williston, VT)
GDATP manufacturing facility (Saco, ME)
U.S. Ordnance (McCarran, NV)
MK19 Grenade Machine Gun:
General Dynamics Armament and Technical
    Products (Saco, ME)
Alliant Techsystems (Mesa, AZ)

Small Arms–Individual Weapons

INVESTMENT COMPONENT

Modernization

Recapitalization

Maintenance

MISSION

Enables warfighters and small units to engage targets with lethal fire to defeat or deter adversaries.

DESCRIPTION

The M4 Carbine is designed for lightness, speed, mobility and firepower and is standard issue for Brigade Combat Teams in theater. The M4 series of carbines can also be mounted with the M203A2 Grenade Launcher, M320 Grenade Launcher, or M26 Modular Accessory Shotgun System (MASS). The weapon has incorporated 62 refinements since its inception. In post-combat surveys, 94 percent of Soldiers rate the M4 as an effective weapons system.

In September 2010, Headquarters Department of the Army G3/5/7 authorized the M4A1 as the standard carbine for the U.S. Army. Compared to the M4, the M4A1 has a heavier barrel and is fully automatic, improvements that deliver greater sustained rates of fire. The Army is upgrading both M4 configurations with an ambidextrous fire control assembly.

The M14 Enhanced Battle Rifle (EBR) has a new adjustable buttstock, cheek rest, and M4-style pistol grip. Five thousand M14 EBRs were assembled at TACOM Lifecycle Management Command at Rock Island Arsenal in response to Operational Need Statements requesting a longer range capability. The upgraded weapons are currently in service with select Army units.

The M320 Grenade Launcher is the replacement to all M203 series grenade launchers on M16 Rifles and M4 Carbines. A modular system, it attaches under the barrel of the rifle or carbine and can convert to a stand-alone weapon. The M320 features an integral day/night sighting system and improved safety features. It also has a side-loading unrestricted breech that allows the system to fire longer 40mm low-velocity projectiles (NATO standard and non-standard).

The M26 12-Gauge MASS attaches to the M4 Carbine and enables soldiers to transition between lethal and less-than-lethal fires and adds the capability of a separate shotgun without carrying a second weapon. It is also designed to operate as a stand-alone system, and comes with a recoil-absorbing, collapsible buttstock. The Picatinny rail mounted on top allows accessory equipment to be mounted on the shotgun. Additional features include a box magazine, flip-up sights, ambidextrous configurations, and an extendable stand-off device for door breaching.

SYSTEM INTERDEPENDENCIES

Common Remotely Operated Weapon Station (CROWS)

PROGRAM STATUS

M4 Carbine:
- Delivery and fielding complete

M4A1:
- Deliveries ongoing

M14 EBR:
- **3QFY09:** Army Requirements and Resourcing Board approved 5,000 systems
- **1QFY10:** 1,200 additional systems authorized

- **3QFY10:** First 5,000 complete

M320 Grenade Launcher Module:
- Production and fielding

M26 Modular Accessory Shotgun System:
- **4QFY09:** Limited user test and evaluation with Military Police units
- **2QFY10:** Low-rate initial production approved
- **4QFY10:** First article testing complete

PROJECTED ACTIVITIES

M4 Carbine:
- In sustainment

M4A1:
- New M4A1 competitive award pending for 24,000 carbines
- Continue M4A1 production, deliveries, and fielding

M14 EBR:
- In sustainment

M320 Grenade Launcher Module:
- Competitive solicitation planned to select improved day/night sighting system
- Competitive solicitation planned to provide additional M320 Grenade Launchers

M26 Modular Accessory Shotgun System:
- Continue fielding

ACQUISITION PHASE

| Technology Development | Engineering & Manufacturing Development | Production & Deployment | Operations & Support |

Small Arms–Individual Weapons

FOREIGN MILITARY SALES
M4 Carbine:
Afghanistan, Philippines, Colombia, Jamaica, Thailand
M4A1:
Bahrain, Thailand, Pakistan, Azerbaijan, Djibouti

CONTRACTORS
M4 Carbine:
Colt Defense, LLC. (West Hartford, CT)
M320 Grenade Launcher Module:
Heckler and Koch Defense Inc.
   (Ashburn, VA)
M26 Modular Accessory Shotgun System:
Vertu Corp. (Warrenton, VA)

Small Arms–Precision Weapons

INVESTMENT COMPONENT

Modernization

Recapitalization

Maintenance

MISSION

Enables sniper teams to employ destructive force against light materiel and personnel targets at longer ranges.

DESCRIPTION

The M107 Semi-Automatic Long Range Sniper Rifle (LRSR) fires .50 caliber ammunition and is capable of delivering precise, rapid fire on targets out to 2,000 meters, greatly exceeding the terminal effect capability of the XM2010, M110 or M24 sniper rifles. It is especially valuable during military operations in urban terrain where greater firepower and standoff ranges provide counter-sniper capability while enhancing sniper survivability.

The XM2010 Enhanced Sniper Rifle (ESR) is a bolt action, magazine-fed weapon system that utilizes .300 WinMag ammunition. The rifle is built around a rechambered M24

Sniper Weapon System receiver and is a capable of providing precision fire on targets out to 1,200 meters. The XM2010 is equipped with a fully adjustable right-folding chassis system featuring accessory cable routing channels and Military Standard 1913 Picatinny rails that mount a Leupold 6-20x50mm variable power Day Optic Scope with advanced scalable H-58 ranging and targeting reticle. The XM2010 is also equipped with a sound suppressor and adjustable bipod. The shooter interface can be tailored to accommodate a wide range of shooter preferences and its folding stock provides Soldiers flexibility in transporting the weapon during operations.

The M110 7.62mm Semi-Automatic Sniper System (SASS) is an anti-personnel and light materiel weapon that fires 7.62mm ammunition out to a maximum effective range of 800 meters. It is also the first U.S. Army weapon system that integrates a quick attach/detach suppressor to reduce the weapon's firing signature. The weapon system exceeds the rate of fire and lethality of the M24, the previous

medium-caliber sniper rifle. The M110's Leupold Mark IV 3.5–10x scope provides both a wide field of view at low magnification for close-in engagements and a narrow field of view for precision long shots at high magnification. The SASS leverages a rapid fire/rapid reload design, variable-power day optic sight, and 10 or 20-round detachable magazines.

SYSTEM INTERDEPENDENCIES

None

PROGRAM STATUS
M107:
- **FY12:** In sustainment

XM2010:
- **4QFY12:** Follow-on Urgent Materiel Release in support of Operation Enduring Freedom
- **Current:** Sustainment activities brought on-line to provide weapon repair and maintenance
- **Current:** Production underway to support additional fieldings

M110:
- **3QFY12:** Award for 700 additional M110s and accompanying advanced sniper accessory kits

PROJECTED ACTIVITIES
M107:
- Fielding to residual Explosive Ordnance Disposal units
- Residual maintenance work order/ upgrade remaining

XM2010:
- **1QFY13:** Type Classification Standard and Milestone C
- **1QFY13:** Full Materiel Release
- **2QFY13:** Full-rate production

M110:
- Fielding

ACQUISITION PHASE

| Technology Development | Engineering & Manufacturing Development | Production & Deployment | Operations & Support |

Small Arms–Precision Weapons

FOREIGN MILITARY SALES
None

CONTRACTORS
M107 LRSR:
Barrett Firearms Manufacturing
    (Murfreesboro, TN)
XM2010 ESR:
Remington Arms Co. Inc. (Illion, NY)
M110 SASS:
Knight's Armament Co. (Titusville, FL)

Small Caliber Ammunition

INVESTMENT COMPONENT

Modernization

Recapitalization

Maintenance

MISSION
Provides warfighters with the highest quality, most capable small caliber ammunition for training and combat.

DESCRIPTION
Small Caliber Ammunition consists of 5.56mm, 7.62mm, 9mm, 10- and 12-gauge, .22 Cal., .30 Cal., .300 WinMag, .50 Cal., and Grenade Rifle Entry Munition (GREM). The M16 Rifle, M249 Squad Automatic Weapon (SAW), and the M4 Carbine all use 5.56mm cartridges. The 7.62mm cartridge is used by the M240 Machine Gun, and M60 Machine Gun, as well as the M24, M110 and M14 Enhanced Battle Rifle Sniper Rifles. The 9mm cartridge is fired by the M9 Pistol. The XM2010 Sniper Rifle uses the .300 WinMag cartridge. The M2 Machine Gun and the M107 Sniper Rifle use .50 Cal. cartridges. The remaining Small Caliber Ammunition is used in a variety of pistols, rifles and shotguns.

Three categories of Small Caliber Ammunition are currently in use. War Reserve Ammo is modern ammunition that supports individual and crew served weapons during combat operations. Training Standard Ammunition is dual-purpose, and is used to support training or operational requirements. Training Unique Ammunition is designed specifically for use in training and is not for combat use, i.e., blank, dummy-inert, close combat mission capability kit man marking, and short range training ammunition.

SYSTEM INTERDEPENDENCIES
In this Publication
Small Arms–Crew Served Weapons, Small Arms–Individual Weapons

PROGRAM STATUS
- **2QFY12:** Began Production on Small Caliber Ammunition Second Source Production Contract
- **4QFY12:** Awarded Lake City Army Ammunition Plant Operating (LCAAP) and Production Contract

PROJECTED ACTIVITIES
- **3QFY13:** 7.62mm M80A1 Enhanced Performance Round (EPR) in production
- **1QFY14:** Transition to new LCAAP contract complete

Technology Development | Engineering & Manufacturing Development | Production & Deployment | Operations & Support

5.56MM, 7.62MM, 9MM, CCMCK, .50 CAL

5.56mm

M1037 · M862 SRTA · M200 · M193 · M855 Ball · M856A1 · M995 · M855A1

9mm

CCMCK

7.62mm

M993 · M62A1 · M80 Ball · M82 · M80 Ball Lnkd · M276 · M118 Ball · M80A1

.50 Cal.

M1A1 Blank · M33 Ball · M17 Trace · M20 · M8 · M211 · M962 · M903 · M860 Tracer · M858 Ball

Small Caliber Ammunition

FOREIGN MILITARY SALES
5.56mm, 7.62mm, .50 Caliber:
Afghanistan, Barbados, Colombia, France, Japan, Jordan, Lebanon, Philippines, Saudi Arabia

CONTRACTORS
Rifle/Machine Gun Ammunition:
Alliant Techsystems (Independence, MO)
Rifle/Machine Gun, pistol and Shotgun Ammunition:
Olin Corp. (East Alton, IL and Oxford, MS)
Small Caliber Ammunition Propellant (sub-contractor)
General Dynamics (Saint Marks, FL)
CCMCK Man Marking 5.56mm Ammunition:
Ultimate Training Munitions (Branchburg, NJ)
Ammunition Storage Containers:
Bway Corp. (Atlanta, GA)

Stryker Family of Vehicles

Modernization

Recapitalization

Maintenance

MISSION
Enables the Army to immediately respond to urgent operational requirements anywhere in the world using rapidly deployable, agile, and strategically responsive support vehicles.

DESCRIPTION
As the primary combat and combat support platform of the Stryker Brigade Combat Team (SBCT), the Stryker Family of Vehicles fulfills an immediate requirement for a strategically deployable (C-17/C-5) brigade capable of rapid movement anywhere on the globe in a combat-ready configuration. The Stryker Family of Vehicles is built on a common chassis, each with a different Mission Equipment Package. There are ten variants, including the Infantry Carrier Vehicle (ICV), the Mobile Gun System (MGS), the Reconnaissance Vehicle (RV), the Mortar Carrier (MC), the Commanders

Vehicle (CV), the Fire Support Vehicle (FSV), the Engineer Squad Vehicle (ESV), Medical Evacuation Vehicle (MEV), the Anti-tank Guided Missile (ATGM) Vehicle, and the Nuclear Biological Chemical Reconnaissance Vehicle (NBCRV). Additionally, there are Double V-Hull variants for the following: ICV, CV, MEV, MC, ATGM, FSV, ESV.

The Stryker Variants (excluding the MEV, ATGM, FSV, RV and MGS) are armed with a Remote Weapon Station supporting an M2 .50 caliber machine gun or MK19 automatic grenade launcher, the M6 grenade launcher, and a thermal weapons sight. Stryker supports communication suites that integrate the Single-Channel Ground-and-Air Radio System (SINCGARS) radio family; Enhanced Position Location Reporting System (EPLRS); Force XXI Battle Command Brigade-and-Below (FBCB2) or Blue Force Tracker (BFT); Global Positioning System (GPS); high-frequency (HF) and multi-band very-high and ultra-high frequency (VHF/UHF) radio systems. Stryker provides 360-degree protection against armor-piercing

threats. Stryker is powered by a 350-horsepower diesel engine, runs on eight wheels that possess a run-flat capability, and has a central tire-inflation system. It also incorporates a vehicle-height management system.

The Stryker program leverages non-developmental items with common subsystems and components to allow rapid acquisition and fielding. Stryker integrates government furnished materiel subsystems as required and stresses performance and commonality to reduce the logistics footprint and minimize costs.

SYSTEM INTERDEPENDENCIES
In this Publication
None

Other Major Interdependencies
DAGR, DVE, EPLRS, FH MUX, FS3, KNIGHT, LRAS3, MCS, MFCS, RWS, SHADOWFIRE, SPITFIRE, STORM, VIS VIC, Sensor Processing Group, Sensor Suite

PROGRAM STATUS
- **1QFY12:** NBCRV Full Materiel Release and Full-rate production approved
- **1QFY12:** Army decision to deploy second DVH SBCT to Afghanistan
- **3QFY12:** Army Systems Acquisition Review Council approves Phase 1 for Stryker Engineering Change Proposal effort

PROJECTED ACTIVITIES
- **2QFY13:** SBCT 8 reaches Initial Operational Capability
- **2QFY13:** ASARC to approve Phase 2 Stryker ECP effort (tent)
- **3QFY13:** SBCT 9 completes Stryker fielding
- **1QFY14:** SBCT 9 reaches Initial Operational Capability
- **FY12-FY14:** NBCRV fielding to select Active component Heavy BCT (HBCT) and Chemical Companies
- **FY14-FY17:** NBCRV fielding to select Reserve component HBCT and Chemical Companies

Technology Development	Engineering & Manufacturing Development	Production & Deployment	Operations & Support

Stryker Family of Vehicles

FOREIGN MILITARY SALES
None

CONTRACTORS
General Dynamics Land Systems
(Sterling Heights, MI)
Manufacturing/Assembly:
General Dynamics Land Systems-Canada
(Ontario, Canada)
Joint Systems Manufacturing Center
(JSMC) (Lima, OH)
General Dynamics Assembly Operations
(Anniston, AL)
Engineering:
General Dynamics (Sterling Heights, MI)
Kits
Manifold/Alternator:
North American Controls (Shelby Twp, MI)
Sensors/CCA:
Raytheon (El Segundo, CA)
Fire System Assembly:
Kidde Dual Spectrum (Goleta, CA)
Rock Island Arsenal (IL)

Sustainment System Mission Command (SSMC)

INVESTMENT COMPONENT

Modernization

Recapitalization

Maintenance

MISSION

Provides sustainment mission planning and execution in classified, unclassified, and coalition environments to enable commanders and staffs to view, tailor, and scale actionable sustainment information in a modular COP environment to meet critical information requirements across the full spectrum of operations.

DESCRIPTION

Sustainment System Mission Command (SSMC) is an integral part of Mission Command (MC), and is a component of the future Command Post Computing Environment (CP/CE) architecture. It provides commanders and staffs of Joint, inter-agency, inter-governmental, and multi-national organizations with logistical/sustainment situational awareness across the full spectrum of operations, from the company level to echelons above Corps, to support training exercises, mission planning, rehearsal, and operations worldwide.

Core competencies are asset visibility, in-transit visibility, and bottoms-up reporting. It displays comprehensive logistical/sustainment mission analysis data with map-centric operational views.

Battle Command Sustainment Support System (BCS3) provides a real-time logistical map-based operational view using customized filters to meet the full spectrum of logistics/sustainment mission command requirements.

It is aligned with the Mission Command Convergence Strategy and is migrating its existing functionality to the Mission Command Workstation and web services.

SYSTEM INTERDEPENDENCIES
In this Publication:
Battle Command Sustainment Support System (BCS3)

Other Major Interdependencies:
Publish and Subscribe Services (PASS), Data Distribution Services (DDS), Command Web services, MC Workstation (Logistics Dashboards), LIW/LOGSA, ILAP, SARSS, SASS-MOD, PBUSE, EMILPO, RFID, Joint-Automated Identification Technology (JAIT), Radio Frequency (RF), Satellite Transponders and Enhanced ITV data feeds (Orbit One, Global Track, and SUPREME (Hawkeye) Class I shipments), Container Intrusion and Detection Devices (CIDD) Radio Frequency Tag capability, Integrated Data Environment / Asset Visibility and Global Transportation Network Convergence (IGC), Global Air Transportation and Execution Management System / Worldwide Port System (GATES/WPS), In-Transit Visibility (ITV) for Surface Deployment and Distribution Cargo (ISDDC), U.S. Marine Corps Last Tactical Mile System (LTM), and Sustain Business System Modernization – Energy (BSM-E)

PROGRAM STATUS
- **3QFY12:** Network Integration Evaluation (NIE) 12.2
- **Continue:** Fielding of BCS3

PROJECTED ACTIVITIES
- **1QFY13:** Network Integration Evaluation (NIE) 13.1
- **1QFY13:** COE V1 Fielding SC210.0.3 (CS 11/12 Capability)
- **3QFY13:** NIE 13.2
- **1QFY14:** COE v2 Fielding 14/15
- **1QFY14:** NIE 14.1
- **3QFY14:** NIE 14.2
- **1QFY15:** NIE 15.1

ACQUISITION PHASE

| Technology Development | Engineering & Manufacturing Development | Production & Deployment | Operations & Support |

Sustainment System Mission Command (SSMC)

FOREIGN MILITARY SALES
None

CONTRACTORS
Software Development/Engineering Services:
IBM (Armonk, NY)
Tech Flow (San Diego, CA)
General Dynamics (Scottsdale, AZ)
JB Management (Eatontown, NJ)
Field Support/Engineering & New Equipment Training Services:
Raytheon (Waltham, MA)
Tapestry (Yorktown, VA)
INTECON (Colorado Springs, CO)
Software Professional Solutions (Tinton Falls, NJ)
Program Support:
CACI (Arlington, VA)
Engility (Chantilly, VA)
LMI (McLean, VA)
Hardware:
Dell Computer Corp. (Round Rock, TX)

T-9 Medium Dozer

INVESTMENT COMPONENT

Modernization

Recapitalization

Maintenance

MISSION
Performs earth moving operations (cutting, moving, and finishing grading) through a forward/reverse motion, to support various construction tasks; build and maintain roads, airfields, and shelters; and Reduce rubble obstacles and fortified buildings in urban terrain.

DESCRIPTION
The T-9/D7R dozer model is a medium drawbar, air transportable by C-5 and C-17, diesel-engine-driven crawler tractor with a dozer blade and optional winch or ripper.

The medium dozer is a commercial vehicle with military modifications to include NATO start, arctic kit, rifle rack and armor C-Kit capability.

SYSTEM INTERDEPENDENCIES
In this Publication
None

Other Major Interdependencies
M870 truck trailer for highway transportability; C-17/C-5 for fixed wing transport

PROGRAM STATUS
- **3QFY12:** System received Full Materiel Release and Type Classification Standard
- **3QFY12-4QFY12:** Systems being fielded.

PROJECTED ACTIVITIES
- **Continue:** Fielding of 480 systems

ACQUISITION PHASE

Technology Development

Engineering & Manufacturing Development

Production & Deployment

Operations & Support

T-9 Medium Dozer

FOREIGN MILITARY SALES
Uganda

CONTRACTORS
Caterpillar Inc. (Peoria, IL)
BAE Systems (Cincinnati, OH)

Tactical Electric Power (TEP)

Modernization

Recapitalization

Maintenance

MISSION
Provides a standardized family of tactical electric power sources to the DoD in accordance with DoDD 4120.11.

DESCRIPTION
The Tactical Electric Power (TEP) program consists of a variety of generator set sizes. Small generators include 2kW Military Tactical Generators (MTG), 3kW Tactical Quiet Generators (TQG), and Small Tactical Electric Power (STEP). Medium generators include 5, 10, 15, 30 and 60kW TQGs, Advanced Medium Mobile Power Sources* (AMMPS), and trailer-mounted Power Units and Power Plants (PU/PP). Large generators include 100-200kW TQGs, Large Advanced Mobile Power Sources (LAMPS); 840kW Deployable Power Generation and Distribution System (DPGDS); and Power Distribution: Power Distribution Illumination System Electrical (PDISE) – microgrid Intelligent Power Management and Distribution System (IPMDS).

* The AMMPS is the third generation of mobile electric power generation systems, which will replace the Tactical Quiet Generators (TQG) over time.

Tactical Electric Power Systems offer:
- Maximize fuel efficiency; diesel/JP8 based, eliminating gasoline on battlefield
- Increased reliability (750 hours mean time between failure), maintainability, and transportability via skid- or trailer-mount
- Improved sustainability, operating at rated loads in all military environments
- Minimized weight and size while meeting all user requirements with military ruggedized commercial components
- Reduced infrared signature and noise (less than 70 decibels at seven meters)
- Survivability in chemical, biological, and nuclear environments
- Advanced Technology, including digital controls and permanent magnet alternators
- A standard DoD military tactical generator fleet that meets power generation and conditioning standards in accordance with Military Standards-1332
- Man-portability with 2kW MTG and 3kW TQG

SYSTEM INTERDEPENDENCIES
None

PROGRAM STATUS
- Continued production and/or fielding the following systems: 2kW MTG, 3kW TQG, 5kW TQG, 10kW TQG, 15kW TQG, 30kW TQG, 60kW TQG, 100kW TQG; 5kW AMMPS, 10kW AMMPS, 15kW AMMPS, 30kW AMMPS, 60kW AMMPS, and the Power Distribution Illumination System Electrical (PDISE)
- **4QFY12:** 3kW TQG Re-procurement Contract Awarded
- **4QFY12:** Large Advanced Mobile Power Sources (LAMPS) Contract Award for Engineering and Manufacturing Development (EMD)

PROJECTED ACTIVITIES
- **1QFY13-3QFY13:** Power Distribution Illumination System Electrical (PDISE) Re-procurement
- **FY13-FY15:** Continue production and fielding of Advanced Medium Mobile Power Sources (AMMPs), 3kW TQG, 100kW TQG, and Power Distribution Illumination System Electrical (PDISE)

ACQUISITION PHASE

Technology Development	Engineering & Manufacturing Development	Production & Deployment	Operations & Support

Tactical Electric Power (TEP)

FOREIGN MILITARY SALES
Tactical Quiet Generators (TQGs) have
been purchased by 38 countries

CONTRACTORS
2kW MTG:
Dewey Electronics (Oakland, NJ)
**3kW, 5kW, 10kW, 15kW, 100kW, and
200kW TQG:**
DRS Fermont (Bridgeport, CT)
**Advanced Medium Mobile Power
Sources (AMMPS) 5 – 60kW:**
Cummins Power Generation
　(Minneapolis, MN)
30kW, 60kW TQG:
L-3 Westwood (Tulsa, OK)
**Deployable Power Generation and
Distribution System (DPGDS):**
DRS Fermont (Bridgeport, CT)

Tactical Mission Command (TMC)/ Maneuver Control System (MCS)

INVESTMENT COMPONENT

Modernization

Recapitalization

Maintenance

MISSION

Supports the core mission command computing environments and applications for the Army's operating forces.

DESCRIPTION

Tactical Mission Command (TMC)/ Maneuver Control System (MCS) suite of products and services provide commanders and staffs collaborative, executive decision-making capabilities using Common Operational Picture (COP) management, planning, information and knowledge management, and other maneuver functional tools. TMC products support Army modernization and migration to a DoD net-centric environment and Common Operating Environment (COE).

Command Post Web: Web-based strategic and tactical operational and intelligence data sharing, visualization,

and Command Post of the Future (CPOF) collaboration through Web applications. Implements Army-directed operations-intelligence convergence.

CPOF: The Army's primary Mission Command system that allows commanders and their staff to collaborate on the COP through integrated data feeds from other Army, Joint, and coalition systems. CPOF is the core of the Mission Command Workstation, which provides enhanced airspace, fires, sustainment, and air defense data in the collaborative environment.

SharePoint: Browser-based collaboration and documentation platform with offline access to document libraries for tactical users.

SYSTEM INTERDEPENDENCIES
In this Publication
Advanced Field Artillery Tactical Data System (AFATDS), Global Command and Control System–Army (GCCS-A), Force XXI Battle Command Brigade and Below (FBCB2), Distributed Common Ground System–Army (DCGS-A),

Battle Command Sustainment Support System (BCS3), Warfighter Information Network Tactical (WIN-T) Increment 2

Other Major Interdependencies
JADOCS, BFT, JCR, JBC-P, ENFIRE, AMDWS, AMPS, FIST, Joint Warning and Reporting Network (JWARN), JEM, CIRS, JTCW/C2PC, NCES, TAIS, TBMCS, SIRIS, BCT Modernization, JPMG, MIP, CIDNE, TIGR, Gorgon Stare, and JOCWATCH

PROGRAM STATUS
- **1QFY12:** Beta release of Command Web in Operation Enduring Freedom (OEF); CPOF and MC WS demonstration of CP CE Convergence and support for NIE 12.1
- **2QFY12:** Army-wide fielding initiated for CS11-12 baseline for CPOF 10.0; OEF upgrade completed
- **3QFY12:** CPOF and MC WS demonstration of CP CE Convergence and OTM capabilities and Command Post Web demonstration of operations-intelligence convergence, Maneuver Widget and CBRN capabilities and support for NIE 12.2
- **4QFY12:** Certification Test and Fielding decision for Command Post

Web Maneuver Widget and CIRS
- **FY12:** Development and integration of COE v1.0 baseline

PROJECTED ACTIVITIES
- **1QFY13:** CPOF 13.1 and Command Post Web 13.0.5 Support for NIE 13.1
- **1QFY13:** Quarterly release decision for MC13.0 baseline (Convergence)
- **2QFY13:** Certification Test for Command Post Web EMS and OH Widgets
- **3QFY13:** CPOF next-generation architecture demonstration and Command Post Web demonstration of all Maneuver Function capabilities at NIE 13.2
- **3QFY13:** Fielding decision for Command Post Web EMS and OH Widgets
- **2QFY13-3QFY13:** COE 1.0 I2E and AIC for CPOF and Command Post Web
- **3QFY14:** CPOF and Command Post Web demonstration and support for NIE 14.2
- **3QFY14-4QFY14:** COE 2.0 I2E and AIC for CPOF and Command Post Web
- **1QFY15:** COE 2.0 Fielding Decision for CPOF and Command Post Web
- **1QFY15:** Support for NIE 15.1

ACQUISITION PHASE

| Technology Development | Engineering & Manufacturing Development | Production & Deployment | Operations & Support |

Tactical Mission Command (TMC)/Maneuver Control System (MCS)

FOREIGN MILITARY SALES
Canada

CONTRACTORS
General Dynamics (Taunton, MA;
  Scottsdale, AZ)
ManTech (Chantilly, VA)
Lockheed Martin (Tinton Falls, NJ)
CACI (Chantilly, VA)

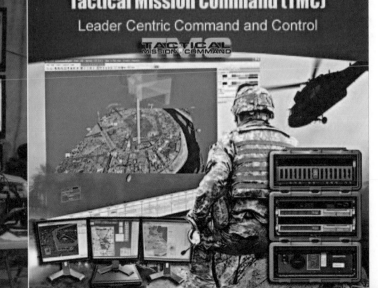

Tactical Mission Command (TMC)
Leader Centric Command and Control

Tank Ammunition

INVESTMENT COMPONENT

Modernization

Recapitalization

Maintenance

MISSION

Provides overwhelming lethality overmatch in tank ammunition.

DESCRIPTION

The current 120mm family of tank ammunition consists of fourth-generation kinetic energy (KE), multipurpose, and canister ammunition. KE ammunition lethality is optimized by firing a maximum-weight sub-caliber projectile at the greatest velocity possible, defeating advanced threat armor. The M829A3 kinetic energy cartridge provides armor-defeat capability. The M829E4, the next-generation KE cartridge, is currently in the Engineering, Manufacturing, Design (EMD) Phase of the acquisition cycle and Milestone C is scheduled for FY14. Multipurpose ammunition uses a high-explosive warhead to provide blast, armor penetration, and fragmentation effects. The shotgun shell-like M1028 canister cartridge provides the Abrams tank with effective, rapid, lethal fire against massed assaulting infantry, and is also used in training. The 120mm family has dedicated training cartridges in

production: The M865, with its reduced range, simulates tactical trajectory to 2,500 meters; and the M1002 simulates the M830A1 size, weight, and nose switch. To support the Stryker force, the 105mm Mobile Gun System uses M1040 canister cartridges. The M1040 canister cartridge provides rapid, lethal fire against massed assaulting infantry at close range, and is also used in training. The 105mm M724A1E1 is a reduced range training cartridge intended to provide the Soldier the training capability to maximize the effectiveness of the tactical 105mm M900 KE cartridge, which provides armor defeat capability. The M724A1E1 is a ballistic match for the M900 KE cartridge. The cartridge will be used in the Stryker Mobile Gun System (MGS). The MGS employs the M68A1/A2 105mm rifled gun tube with a Muzzle Reference System (MRS) and an autoloader for storage and handling of its 105mm ammunition. The 105mm M467A1 training cartridge is a ballistic match to the M393A3 tactical round, both of which completed production in FY10.

SYSTEM INTERDEPENDENCIES

In this Publication
None

Other Major Interdependencies
The Abrams Main Battle Tank fires 120mm ammunition, the Stryker Mobile Gun System fires 105mm ammunition.

PROGRAM STATUS
- **Cuurent:** M829A3, M830, M830A1, M1002 and M908, M1028, M1040, M393A3, M467A1 are fielded

PROJECTED ACTIVITIES
- **FY13:** M1002 in production; M865, M831A1 in recapitalization
- **FY13:** M724A1E1 Milestone C
- **FY14:** M829E4 Milestone C

| Technology Development | Engineering & Manufacturing Development | Production & Deployment | Operations & Support |

Large Caliber Ammunition
120mm, 105mm Training & Tactical Ammunition

| M724A1E1 | M829E4 | M1002 | M865 | M1040 | M1028 | M829A3 | M830A1 | M831A1 | M908 | M829A1 | M829A2 | M830 | M393A3 | M467A1 | M900 | M724A1 | M456A2 | M490A1 |

CONCEPT | DEVEL. | PRODUCTION | SUSTAINMENT

105mm

FOREIGN MILITARY SALES
M831A1 and M865:
Iraq
KE-WA1:
Kuwait

CONTRACTORS
M1002, M865 and M829E4:
Alliant Techsystems (Plymouth, MN)
M1002, M865, and KEW:
General Dynamics Ordnance and Tactical
    Systems (St. Petersburg, FL)
M1040:
L-3 Communications (Lancaster, PA)

Test Equipment Modernization (TEMOD)

INVESTMENT COMPONENT

Modernization

Recapitalization

Maintenance

MISSION

Improves readiness of Army weapon systems; minimizes test, measurement, and diagnostic equipment proliferation and obsolescence; and reduces operations/support costs.

DESCRIPTION

The Test Equipment Modernization (TEMOD) program modernizes the Army's existing inventory with new equipment, which reduces the proliferation of test equipment and is essential to the continued support of systems and weapon systems. Acquisitions are commercial items that have significant impact on readiness, power projection, safety, and training operations of the Army, Army Reserve, and National Guard. The TEMOD program has procured 38 products replacing over 334 models.

- **High Frequency Signal Generator (SG- 1366/U)** is a signal source to test electronic receivers and transmitters of all types throughout the Army and provide standards to compare signals, ensuring that battlefield commanders can communicate in adverse conditions.
- **Radar Test Set Identification Friend or Foe (IFF) Upgrade Kit and Radar Test Set with Mode S enhanced and Mode 5 cryptography (TS-4530A/UPM)** is used to perform pre-flight checks on aviation and missile transponders and interrogators to alleviate potential fratricide concerns. It is also required to ensure all Army platforms are in compliance with European and Federal Aviation Administration airspace mandates.
- **Multimeter (AN/GSM-437)** enables quick, reliable troubleshooting, which positively affects operational availability.
- **Radio Test Set (AN/PRM-36)** will be used to quickly and effectively diagnose various radios at the field maintenance level.
- **Ammeter (ME-572/U)** measures and displays alternating current and direct current without interrupting the measured circuit. It is used for testing power generators cables, installation wiring, and high current weapon system interfaces.
- **Telecommunication System Test Set (TS-4544/U)** measures and displays various bit-data information as related to digital transmission.

SYSTEM INTERDEPENDENCIES
None

PROGRAM STATUS
High Frequency Signal Generator:
- **2QFY12:** Full-rate production (FRP)
- **3QFY12:** Engineering Change Proposal

IFF Radar Test Set Mode S (Enhanced) Mode 5:
- **4QFY12:** Low-rate initial production (LRIP)

Multimeter:
- **1QFY12:** Contract Award

Radio Test Set:
- **1QFY12:** Contract Award
- **1QFY12:** Protest Filed
- **3QFY12:** Protest Resolved
- **4QFY12:** Contract Award

Ammeter:
- **4QFY11:** Issue Letter Request For Bid Samples (LRFBS)

Telecommunication System Test Set:
- **4QFY12:** Bid Sample Testing
- **4QFY12:** Contract Award

Oscilloscope (bench top):
- **3QFY12:** Market Research Oscilloscope (portable)
- **4QFY12:** Maket Research

PROJECTED ACTIVITIES
Radio Test Set:
- **2QFY13:** LRIP
- **3QFY13:** Product Verification Testing (PVT)
- **4QFY13:** FRP

Ammeter:
- **1QFY13:** Contract Award
- **2QFY13:** LRIP
- **3QFY13:** PVT
- **4QFY13:** FRP

Telecommunication System Test Set:
- **1QFY13:** LRIP
- **3QFY13:** PVT
- **1QFY14:** FRP

Oscilloscope (Bench Top):
- **2QFY13:** Contract Award
- **1QFY14:** PVT
- **4QFY14:** FRP

Oscilloscope (Portable):
- **2QFY15:** Contract Award

ACQUISITION PHASE

Technology Development | Engineering & Manufacturing Development | Production & Deployment | Operations & Support

ELECOMMUNICATION SYSTEM TEST SET
(PHOTO OF TS-4281/G SHOWN ABOVE
WHICH WILL BE REPLACED BY THE TS-4544/U
ACTUAL PHOTO NOT AVAILABLE)

SIGNAL GENERATOR SG-1366/U

RADIO TEST SET AN/PRM-36
(PHOTO OF AN/PRM-34 SHOWN ABOVE
WHICH WILL BE REPLACED BY THE AN/PRM-36
ACTUAL PHOTO NOT AVAILABLE)

RADAR TEST SET TS-4530A/UPM

AMMETER ME-572/U
(PHOTO OF ME-563/U SHOWN ABOVE
WHICH WILL BE REPLACED BY THE ME-572/U
ACTUAL PHOTO NOT AVAILABLE)

MULTIMETER AN/GSM-437

Test Equipment Modernization (TEMOD)

FOREIGN MILITARY SALES
IFF Radar Test Set Mode S (Enhanced) Mode 5:
Azerbaijan, Greece, Hungary, Kuwait, Netherlands, Norway, Portugal, Saudi Arabia, Singapore, United Kingdom

CONTRACTORS
High Frequency Signal Generator:
Agilent Technologies (Englewood, CO)
IFF Radar Test Set Mode S (Enhanced) Mode 5:
Tel-Instrument Electronics Corp. (Carlstadt, NJ)
Multimeter:
To be determined
Radio Test Set:
To be determined
Ammeter:
To be determined
Telecommunication System Test Set:
To be determined

Transportation Coordinators'–Automated Information for Movement System II (TC-AIMS II)

INVESTMENT COMPONENT

Modernization

Recapitalization

Maintenance

MISSION

Facilitates movement, management and control of personnel, equipment and supplies from home station to a theater of operations and back, and provides in-theater support for onward movement, sustainment planning requirements and source in-transit visibility data.

DESCRIPTION

The Transportation Coordinators'–Automated Information for Movements System II (TC-AIMS II) provides enterprise-level deployment planning and execution tools in support of worldwide military operations, both in peace and wartime. TC-AIMS II is configurable for use as a standalone or client/server system when connectivity to the enterprise is unavailable. The system provides:

- Unit Movement and Deployment Management (asset management, movement planning, deployment planning, load planning and manifesting)
- Theater Operations (movement requests, convoy planning and mode operations)
- Automatic Identification Technology for writing data to Radio Frequency Identification tags and sending content information to Army and Joint In-Transit Visibility (ITV) systems

SYSTEM INTERDEPENDENCIES

None

PROGRAM STATUS

- **1QFY12:** TC-AIMS II reached Full Operational Capacity
- **1QFY12:** TC-AIMS II displaced legacy National Guard Mobilization Movement Control (MOBCON) system
- **2QFY12:** Delivered first increment of new functionality directed by a Joint Urgent Operational Needs Statement (JUONS) to improve ITV within Afghanistan
- **2QFY12:** Released TC-AIMS II v6.3, providing Windows 7/Vista compatibility
- **4QFY12:** Deliver final increment of new JUONS functionality

PROJECTED ACTIVITIES

- **1QFY13:** Release TC-AIMS II v7.0
- **1QFY13:** Begin migration of Continental United States installations to Web-enabled TC-AIMS II enterprise
- **3QFY13:** Release software to support Joint Requirements Oversight Council (JROC)-directed Transportation Tracking Number (TTN) initiative

ACQUISITION PHASE

Technology Development | Engineering & Manufacturing Development | Production & Deployment | Operations & Support

Transportation Coordinators'–Automated Information for Movement System II (TC-AIMS II)

FOREIGN MILITARY SALES
None

CONTRACTORS
Systems Integration:
Engineer Research and Development Center (Vicksburg, MS)
Future Research Corp. (Huntsville, AL)
Apptricity Corp. (Dallas, TX)
Program and Fielding/Training support:
Engility Corp. (Alexandria, VA)
LMI (McLean, VA)

Tube-Launched, Optically-Tracked, Wire-Guided (TOW) Missiles

INVESTMENT COMPONENT

Modernization

Recapitalization

Maintenance

MISSION

Provides long-range, heavy anti-tank and precision assault fire capabilities to Army and Marine forces.

DESCRIPTION

The Close Combat Missile System–Heavy (CCMS-H) TOW (Tube-Launched, Optically Tracked, Wire-Guided) is a heavy anti-tank/precision assault weapon system, consisting of a launcher and a missile. The missile is six inches in diameter (encased, 8.6 inches) and 49 inches long. The gunner defines the aim point by maintaining the sight crosshairs on the target. The launcher automatically steers the missile along the line-of-sight toward the aim point via a pair of control wires or a one-way radio frequency (RF) link, which links the launcher and missile.

TOW missiles are employed on the High Mobility Multipurpose Wheeled Vehicle (HMMWV)-mounted Improved Target Acquisition System (ITAS), HMMWV-mounted M220A4 launcher (TOW 2), Stryker Anti-Tank Guided Missile (ATGM) Vehicles, and Bradley Fighting Vehicles (A2/A2ODS/A2OIF/A3) within the Infantry, Stryker, and Heavy Brigade Combat Teams, respectively. TOW missiles are also employed on the Marine HMMWV mounted ITAS, HMMWV-mounted M220A4 launcher (TOW 2), LAV-ATGM Vehicle, and AH-1W Cobra attack helicopter. TOW is also employed by allied nations on a variety of ground and airborne platforms.

The TOW 2B Aero is the most modern and capable missile in the TOW family, with an extended maximum range to 4,500 meters. The TOW 2B Aero has an advanced counteractive protection system capability and defeats all current and projected threat armor systems. The TOW 2B Aero flies over the target (offset above the gunner's aim point) and uses a laser profilometer and magnetic sensor to detect and fire two downward-directed, explosively formed penetrator warheads into the target. The TOW 2B Aero's missile weighs 49.8 pounds (encased, 65 pounds).

The TOW Bunker Buster (TOW BB) is optimized for performance against urban structures, earthen bunkers, field fortifications, and light-skinned armor threats. The missile impact is at the aim point. It has a 6.25 pound, 6-inch diameter high-explosive, bulk-charge warhead, and its missile weighs 45.2 pounds. The TOW BB has an impact sensor (crush switch) located in the main-charge and gives a pyrotechnic detonation delay to enhance warhead effectiveness. The PBXN-109 explosive is housed in a thick casing for maximum performance. The TOW BB can produce a 21- to 24-inch diameter hole in an 8-inch thick, double reinforced concrete wall at a range of 65 to 3,750 meters.

SYSTEM INTERDEPENDENCIES

In this Publication
None

Other Major Interdependencies
M1121/1167 High Mobility Multipurpose Wheeled Vehicle (HMMWV), Stryker Anti-Tank Guided Missile (ATGM)

PROGRAM STATUS

- **Current:** TOW 2B Aero RF and BB RF in production

PROJECTED ACTIVITIES

- **FY12-FY16:** TOW multi-year contract

ACQUISITION PHASE

Technology Development | Engineering & Manufacturing Development | Production & Deployment | Operations & Support

Tube-Launched, Optically-Tracked, Wire-Guided (TOW) Missiles

FOREIGN MILITARY SALES
The TOW weapon system has been sold to more than 43 allied nations over the life of the system

CONTRACTORS
TOW 2B Aero and TOW BB
Prime:
Raytheon Missile Systems (Tucson, AZ)
Control Actuator, Shutter Actuator:
Moog Inc. (Salt Lake City, UT)
Warheads:
Aerojet General (Socorro, NM)
Gyroscope:
BAE Systems (Cheshire, CT)
Sensor (TOW 2B only):
Thales (Basingstoke, United Kingdom)
Flight Motor:
ATK (Rocket Center, WV)
Machined/Fabricated Parts:
Klune (Spanish Fork, UT)

Unified Command Suite (UCS)

INVESTMENT COMPONENT

Modernization

Recapitalization

Maintenance

MISSION
Provides a command and control vehicle to civil support teams (CST) that perform on-site analysis of contaminants in support of first responders.

DESCRIPTION
The Unified Command Suite (UCS) vehicle is a self-contained, stand-alone, C-130 air mobile communications platform that provides both voice and data communications capabilities to CST commanders.

The UCS consists of a combination of commercial and existing government off-the-shelf communications equipment (both secure and non-secure data) to provide the full range of communications necessary to support the CST mission. It is the primary means of reach-back communications for the Analytical Laboratory System (ALS) and acts as a command and control hub to deliver a common operational picture for planning and fulfilling an incident response.

It provides:
- Digital voice and data over satellite network
- Secure Internet Protocol Router Network (SIPRNET) and Non-Secure (NIPRNET)
- Radio remote and intercom with cross-banding
- Over-the-horizon communication interoperable interface with state emergency management and other military units

SYSTEM INTERDEPENDENCIES
None

PROGRAM STATUS
- **Current:** Modernization of communication on the move system, radio cross-banding system, and secondary reach-back system
- **Current:** Platform integration/ modernization

PROJECTED ACTIVITIES
- Production and deployment

Technology Development	Engineering & Manufacturing Development	Production & Deployment	Operations & Support

Unified Command Suite (UCS)

FOREIGN MILITARY SALES
None

CONTRACTORS
UCS Vehicle:
Wolf Coach, Inc., an L-3 Communications
    Co. (Auburn, MA)
UCS Communications System
Integrator:
Naval Air Warfare Center Aircraft Division
    (Patuxent River, MD)

Unit Water Pod System (Camel II)

INVESTMENT COMPONENT

Modernization

Recapitalization

Maintenance

MISSION
Receives, stores, and dispenses potable water to units at all echelons throughout the battlefield.

DESCRIPTION
The Unit Water Pod System (Camel II) is the Army's primary water distribution system. Camel II replaces the M107, M149, and M1112 series water trailers. It consists of an 800–900 gallon capacity baffled water tank with integrated freeze protection and all hoses and fittings necessary to dispense water by gravity flow. The acquisition strategy consists of two increments: Increment 1 is the basic system with freeze protection; Increment 2 will provide modular component(s) to give the Camel II water chilling, pumping, circulation, and on-board power generation as add-on capabilities.

The Camel II is mounted on a M1095 Trailer, allowing for better transportability on and off the road by utilizing the Family of Medium Tactical Vehicle Trucks. It holds a minimum of 800 gallons of water and provides a one-day supply of potable water for drinking and other purposes. If the unit has another source of drinking water, such as bottled water, then the Camel II can provide two days of potable water for other purposes. It is operational from -25 to +120 degrees Fahrenheit. The system also contains six filling positions for filling canteens and five-gallon water cans.

SYSTEM INTERDEPENDENCIES
In this Publication
None

Other Major Interdependencies
M1095 Medium Tactical Vehicle Trailer

PROGRAM STATUS
- **FY12:** Deliver test units
- **FY12:** Conduct government testing

PROJECTED ACTIVITIES
- **FY13:** Full materiel release; type classification standard
- **FY14:** First Unit Equipped

Unit Water Pod System (Camel II)

FOREIGN MILITARY SALES
None

CONTRACTORS
Choctaw Defense Manufacturing
Contractors (CDMC) (Durant, OK)

Warfighter Information Network–Tactical (WIN-T) Increment 1

INVESTMENT COMPONENT

Modernization

Recapitalization

Maintenance

MISSION
Provides the warfighter seamless, assured, mobile communications, along with advanced network management tools.

DESCRIPTION
Warfighter Information Network–Tactical (WIN-T) Increment 1 is the Army's current and future tactical network, representing a generational leap forward in enabling widely dispersed, highly maneuverable units to communicate. Increment 1 is a converged tactical communications network providing voice, data, and video capability to connect the battalion-level warfighter. It allows greater flexibility of troop movement and is scalable to meet the mission commander's requirements. It is divided into two sub-increments defined as Increment 1a, "extended networking at-the-halt," and Increment 1b, "enhanced networking-at-the-halt." Increment 1 is a rapidly deployable, early entry system housed in a Lightweight Multipurpose Shelter (LMS) and mounted on an Expanded Capacity High Mobility Multipurpose Wheeled Vehicle for roll-on/rolloff mobility.

WIN-T Increment 1a upgrades the former Joint Network Node satellite capability to access the Ka-band defense Wideband Global SATCOM (WGS) system, reducing the reliance on commercial Ku-band satellites.

WIN-T Increment 1b introduces the Network Centric Waveform, a dynamic wave form that optimizes bandwidth and satellite utilization. It also introduces a colorless core security architecture, which meets Global Information Grid Information Assurance security compliance requirements and incorporates industry standards for network operations and intrusion detection.

WIN-T Increment 1 is a Joint compatible communications package that allows the warfighter to use advanced networking capabilities, interface to legacy systems, retain interoperability with Current Force systems, and keep in step with future increments of WIN-T.

SYSTEM INTERDEPENDENCIES
In this Publication
Warfighter Information Network Tactical (WIN-T) Increment 2, Warfighter Information Network Tactical (WIN-T) Increment 3

Other Major Interdependencies
Area Common User System Modernization (ACUS Mod), Army Battle Command System, Command and Control On The Move (C2OTM), Global Positioning System (GPS), Teleport, Secure Mobile Anti-Jam Reliable Tactical-Terminal (SMART-T), Tactical Operations Center System (TOCS)

PROGRAM STATUS
- **3QFY12:** Increment 1b Modification Work Order (MWO)verification test
- **4QFY12:** Increment 1a completed fielding

PROJECTED ACTIVITIES
- **2QFY13-2QFY15:** Increment 1b MWO Fielding
- **2QFY13-2QFY15:** Increment 1a End of Life (EOL) Tech Refreshments

ACQUISITION PHASE

Technology Development	Engineering & Manufacturing Development	Production & Deployment	Operations & Support

UNITED STATES ARMY

Joint Network Node (JNN)

Tactical Hub

Satellite Terminal Trailer (STT)

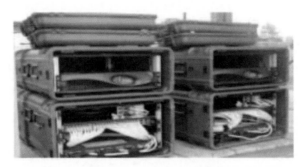

Battalion Command Post Node

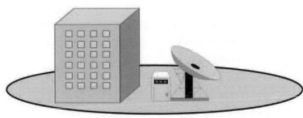

Fixed Regional Hub Node (FRHN)

Warfighter Information Network–Tactical (WIN-T) Increment 1

FOREIGN MILITARY SALES
None

CONTRACTORS
Increment 1b colorless core kits production and integration:
General Dynamics C4 Systems, Inc.
  (Taunton, MA)
Transportable Terminals:
General Dynamics SATCOM Technologies
(Duluth, GA)
PM Support:
Engineering Solutions and Products
  (Aberdeen, MD)
JANUS (Aberdeen, MD)

Warfighter Information Network–Tactical (WIN-T) Increment 2

INVESTMENT COMPONENT

Modernization

Recapitalization

Maintenance

MISSION

Provides initial networking on-the-move, enabling Joint land forces to engage enemy forces deeper and more effectively.

DESCRIPTION

Warfighter Information Network–Tactical (WIN-T) Increment 2 accelerates the delivery of a self-forming, self-healing mobile network infrastructure via commercial off-the-shelf and government off-the-shelf technologies. As a converged tactical communications and transport layer network, Increment 2 leverages an early release of the objective Highband Networking Waveform running on the Highband Networking Radio to provide high-throughput, line-of-sight communications. It also leverages an early release of the objective Net Centric Waveform on a ruggedized R-MPM-1000 modem

for on-the-move (OTM) satellite communications enabling greater situational awareness, and command and control. Multiple configuration items tailor capability from the division level down to the company level. It provides an accelerated delivery of network operations capability that allows management, prioritization, and protection of information while reducing organizational and operational support.

Increment 2 network operations include automated planning, on-the-move node planning, automated link planning for currently fielded systems, initial automated spectrum management, initial quality of service planning and monitoring, and over-the-air network management and configuration of WIN-T radios. Additionally, Increment 2 network operations automate the initial Internet Protocol planning and routing configurations.

WIN-T Increment 2 is part of Capability Set 13, the Army's first package of network components, associated equipment, and software that provides integrated connectivity from the

static tactical operations center to the commander on-the-move to the dismounted Soldier.

SYSTEM INTERDEPENDENCIES

In this Publication

Bradley Fighting Vehicle Systems Upgrade, Distributed Common Ground System–Army (DCGS-A), Integrated Air and Missile Defense (IAMD), Joint Tactical Radio System Handheld, Manpack, Small Form Fit (JTRS HMS), Mine Resistant Ambush Protected Vehicles (MRAP), Single Channel Ground and Airborne Radio System (SINCGARS), Stryker Family of Vehicles, Warfighter Information Network–Tactical (WIN-T) Increment 1, Warfighter Information Network Tactical (WIN-T) Increment 3, Ground Combat Vehicle (GCV)

Other Major Interdependencies

Battle Command Servers (BCS), Enhanced Position Location and Reporting System (EPLRS) (Stryker Brigade Combat Teams only), Joint Light Tactical Vehicle (JLTV), Tactical NW Operations Security Center (TNOSC), Wideband Global SATCOM (WGS)

PROGRAM STATUS

- **2QFY12:** Cold Region Test; New Equipment Training (NET)
- **3QFY12:** Initial Operational Test (IOT); Force Development Test/Evaluation (FDT/E); Production Readiness Review
- **4QFY12:** Full-rate production decision review; First Unit Equipped
- **1QFY13:** Begin fielding CS 13 to brigade combat teams

PROJECTED ACTIVITIES

- **3QFY13:** Initial operational capability

ACQUISITION PHASE

Technology Development | Engineering & Manufacturing Development | Production & Deployment | Operations & Support

WIN-T Inc 2 Tactical
Communications
Node (While M...)

RMPM-1000

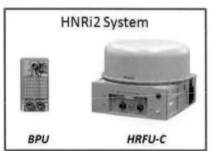

-The-Move
SATCOM
Antenna

HNRi2 System

BPU   HRFU-C

WIN-T Inc 2 Tactical
Communications
Node (While
Stationary)

Initial Network
Operations

WIN-T Inc 2 Point of
Presence

Warfighter Information Network–Tactical (WIN-T) Increment 2

FOREIGN MILITARY SALES
None

CONTRACTORS
WIN-T System
Prime:
General Dynamics (Taunton, MA)
Subcomponent:
Lockheed Martin (Gaithersburg, MD)
Subcontractors:
Harris Corp. (Melbourne, FL)
L-3 Communications (San Diego, CA)
General Dynamics (Richardson, TX)

Warfighter Information Network–Tactical (WIN-T) Increment 3

INVESTMENT COMPONENT

Modernization

Recapitalization

Maintenance

MISSION

Provides full networking on-the-move as a mobile, multi-tiered, tactical communications/transport layer network, enabling Joint land forces to engage enemy forces effectively.

DESCRIPTION

The Warfighter Information Network–Tactical (WIN-T) Increment 3 enables the full-objective mobile, tactical network distribution of command, control, communications, computers, intelligence, surveillance, and reconnaissance information via voice, data, and real-time video. Building on previous increments, Increment 3 provides more robust connectivity and greater network access via military specification radios, higher bandwidth satellite communications (SATCOM) and line-of-sight (LOS) waveforms, an aerial tier (LOS airborne relay), and integrated network operations. It manages, prioritizes, and protects information through network operations (network management, quality of service, and information assurance) while reducing organizational and operational

support. WIN-T Increment 3 ensures communications interoperability with Joint, allied, coalition, Current Force, and commercial voice and data networks. Using communications payloads mounted on Unmanned Aerial Systems, Increment 3 increases network reliability and robustness. It extends connectivity and provides increased Soldier mobility, providing constant mobile communications.

SYSTEM INTERDEPENDENCIES

In this Publication

Bradley Fighting Vehicle Systems Upgrade, Integrated Air and Missile Defense (IAMD), Joint Tactical Radio System Ground Mobile Radios (JTRS GMR), Joint Tactical Radio System Handheld, Manpack, Small Form Fit (JTRS HMS), Distributed Common Ground System–Army (DCGS-A), Stryker Family of Vehicles, Warfighter Information Network–Tactical (WIN-T) Increment 1, Warfighter Information Network Tactical (WIN-T) Increment 2, Mine Resistant Ambush Protected Vehicles (MRAP), Single Channel Ground and Airborne Radio System (SINCGARS), Ground Combat Vehicle (GCV)

Other Major Interdependencies

BCT Modernization, Enhanced Position Location and Reporting System (EPLRS) (for Stryker Brigade Combat Teams only), Light Tactical Vehicle (LTV), Unmanned Aircraft Systems (UAS), Tactical NW Operations and Security Center (TNOSC), Wideband Global SATCOM (WGS)

PROGRAM STATUS

- **3QFY12:** Transmission Subsystem (TSS) In Process Review

PROJECTED ACTIVITIES

- **3QFY13:** Transmission Subsystem Critical Design Review (CDR)
- **4QFY14:** Full Critical Design Review

ACQUISITION PHASE

Technology Development | **Engineering & Manufacturing Development** | Production & Deployment | Operations & Support

Antenna
(High Band Radio Frequency Unit)

Ground Only
HRFU-MT

Air Only
HRFU -Ek

2 Channel JC4ISR Air & Ground

Soldier Network Extension

Addition of Airborne Tier

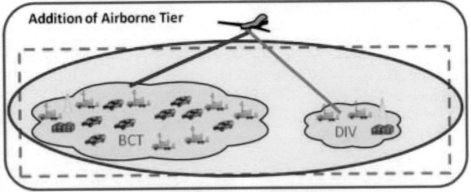

BCT

DIV

Full Network Operations

Tactical Communications Node (White ...)

FOREIGN MILITARY SALES
None

CONTRACTORS
Prime:
General Dynamics (Taunton, MA, and Sunrise, FL)
Subcomponent:
Lockheed Martin (Gaithersburg, MD)
Subcontractors:
Harris Corp. (Melbourne, FL)
BAE Systems (Wayne, NJ)
L-3 Communications (San Diego, CA)

XM1216 & XM1216 E1 Small Unmanned Ground System (SUGV)

INVESTMENT COMPONENT

Modernization

Recapitalization

Maintenance

MISSION

The XM1216 & XM1216 E1 Small Unmanned Ground Vehicle (SUGV) is a lightweight, Soldier-portable unmanned ground vehicle capable of conducting Military Operations in Urban Terrain (MOUT), tunnels, sewers and caves. The SUGV provides Situational Awareness, Situational Understanding (SA/SU) and Intelligence, Surveillance and Reconnaissance (ISR) to dismounted Soldiers. Enables the performance of manpower intensive or high-risk functions without exposing Soldiers directly to the hazards while providing intelligence, information and increase situational awareness at the squad level.

DESCRIPTION

The 34.4 pound SUGV consists of the following: Day/Night/Thermal cameras, laser range finder, Infra-Red (IR) illuminator to facilitate night combat operations, Global Positioning System (GPS), two-way speaker and microphone, and a ruggedized handheld controller. The XM1216 E1 SUGV includes a militarized EO/IR

sensor head, adding target location and improves day/night recognition to 300 meters. The improved XM1216 E1 adds the Joint Tactical Radio System (JTRS) Handheld, Manpack and Small Form Fit (HMS) radio, and ability to mount payloads including: a tether/spooler; tactical engagement simulator and a manipulator arm. Mission duration up to six hours on two BB-2590 batteries, operating in sewers, tunnels and caves, travelling at speeds up to 10 kph, climbing stairs and 10 inches vertical obstacles and traversing slopes up to 17 degrees lateral and 31 degrees vertical. It is capable of carrying up to four pounds of payload weight and fording 6 inches of water.

SYSTEM INTERDEPENDENCIES

None

PROGRAM STATUS

- **4QFY12:** 3rd Brigade, 1st Armored Division successfully employed XM1216 SUGVs in Operation Enduring Freedom deployment and completed Operational Assessment Theater
- **4QFY12:** Fielding 38 XM1216 SUGV systems to 4th Brigade, 101st Airborne Division (Air Assault)

PROJECTED ACTIVITIES

- **1QFY14:** XM1216 E1 Milestone C Decision

ACQUISITION PHASE

Technology Development | Engineering & Manufacturing Development | Production & Deployment | Operations & Support

XM1216 & XM1216 E1
Small Unmanned
Ground System (SUGV)

FOREIGN MILITARY SALES
None

CONTRACTORS
Prime:
iRobot (Burlington, MA)
Electronics:
Benchmark Electronics (Nashua, NH)
Communications/Navigation Units:
BAE (GEC) (Wayne, NJ)
Radios:
General Dynamics (Scottsdale, AZ)
Operator Control Unit Software:
Boeing (Anaheim, CA)

XM7 Spider

INVESTMENT COMPONENT

Modernization

Recapitalization

Maintenance

MISSION
Provides the commander with a new capability to shape the battlefield, protect the force, and respond to changing battlefield environments in a graduated manner while minimizing risk to friendly troops and non-combatants.

DESCRIPTION
The XM7 Spider is a hand-emplaced, remotely-controlled, Man-in-the-Loop (MITL), anti-personnel munition system. Spider provides munition field effectiveness, but does so without residual life-threatening risks after hostilities end. The fielding of this system with its sensors, communications, and munitions changes the way Soldiers operate in an otherwise unpredictable battlefield. Each munition is controlled by a remotely stationed Soldier who monitors its sensors, allowing for more precise (non-lethal to lethal)

responses—a significant advancement and advantage. The Spider Networked Munitions System enables the MITL to detect, track, classify, count, and destroy the enemy.

The Spider system contains three main components: the remote control unit, residing within a computer interface; the repeater, extending the remote control range; and a munition control unit for sending and receiving commands as well as activating the munitions. Spider can be used as a force-protection-reinforcing obstacle to delay, disrupt, and channel enemy forces as well as restrict their use of critical routes of terrain, thereby reducing civil casualties and the exposure of personnel to hostile fire. It can also be integrated into a base defense system, providing protection to Soldiers in forward operating bases and combat outposts.

The system's design allows for safe, flexible, and rapid deployment, reinforcement, and recovery as well as safe passage of friendly forces. Spider eliminates the possibility of an unintended detonation through early

warning and selective engagement of enemy forces, and has a self-destruct capability. Spider is designed for storage, transport, rough handling, and use in worldwide military environments.

SYSTEM INTERDEPENDENCIES
In this Publication
None

Other Major Interdependencies
Interface with Tactical Internet through Force XXI Battle Command Brigade-and-Below and obstacle positioning through Global Positioning System (GPS)

PROGRAM STATUS
- **2QFY12:** Conditional Materiel Release of 43 Systems to U.S. Army Europe
- **2QFY12:** Software V4 Contract Award
- **3QFY12:** LRIP4 Undefinitized Contract Action Awarded for 120 Sytstems
- **3QFY12:** Level III Tech Data Package Contract Award

PROJECTED ACTIVITIES
- **1QFY13:** Follow On Operational Test #3
- **1QFY13:** Network Integration Evaluation Participation
- **2QFY13:** LRIP4 Modification Contract Award for up to 135 Systems
- **2QFY13:** Materiel Release/Type Classification Standard
- **3QFY13:** Full-rate production decision
- **2QFY14:** Full-rate production Contract Award

XM7 Spider

FOREIGN MILITARY SALES
None

CONTRACTORS
Prime:
A joint venture between
Textron Defense Systems
 (Wilmington, MA)
Alliant Techsystems (Plymouth, MN)
Subcontractors:
Alliant Techsystems (Rocket Center, WV)
BAE Systems/Holston (Kingsport, TN)
American Ordnance (Milan, TN)

Science and Technology

The Army's Science and Technology (S&T) mission is to "foster invention, innovation and demonstration of technologies to enable Future Force capabilities while exploiting opportunities to transition technology-enabled capabilities to the Current Force." The U.S. Army depends on its S&T Program to research, develop, and demonstrate high-payoff technological solutions to hard problems faced by Soldiers in the ever-changing, complex environments across the full spectrum of conflict. In order to prevent, shape, and win future conflicts in an uncertain, complex world, Army S&T delivers timely technology solutions that address the Army's top priority capability gaps, while investing in developing technology solutions that Soldiers will need in the future.

The Army's S&T Enterprise includes Research, Development, and Engineering Centers (RDECs) and laboratories, depicted in Figure 1, with more than 12,000 scientists and engineers who are committed to ensuring Soldiers have the technological edge in any environment, against any possible adversary. S&T programs and projects also include participation from academia, industry, and other organizations such as University Affiliated Research Centers (UARCs) and Federally Funded Research and Development Centers (FFRDCs). Our vision is to provide technology-enabling capabilities that empower, unburden, and protect our Soldiers and warfighters in an environment of persistent conflict

ARMY S&T MISSION

FOSTER INVENTION, INNOVATION AND DEMONSTRATION OF TECHNOLOGIES TO ENABLE FUTURE FORCE CAPABILITIES WHILE EXPLOITING OPPORTUNITIES TO TRANSITION TECHNOLOGY-ENABLED CAPABILITIES TO THE CURRENT FORCE

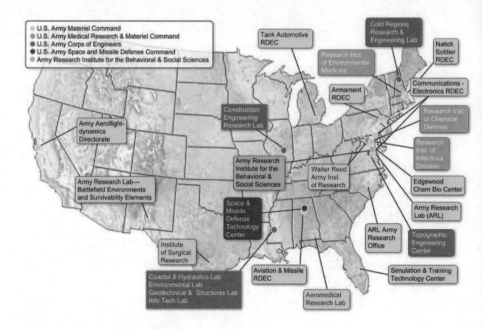

Figure 1: The Army S&T Enterprise

SCIENCE AND TECHNOLOGY TENETS

It is the objective of Army S&T to maintain Army-critical enabling technologies (Army Campaign Plan, 5-6). The Army S&T program balances investments in innovative, game-changing "revolutionary" research with other investments in more "evolutionary" research to advance performance of existing and developing warfighting systems. Underpinning all S&T research is a commitment to strong business practices, which ensures efficient and appropriate investment across the S&T enterprise. The Army's S&T program is guided and directed by Army leadership and is based on the following tenets:

Innovative, Revolutionary S&T
- Create future Army mission capabilities via technical innovation, especially in areas with the greatest potential utility to Soldiers and small units
- Prepare for an uncertain future by researching and developing leap-ahead and disruptive technologies that can be matured and demonstrated to provide advanced capabilities in accordance with identified Army warfighter technology gaps

Advancing Performance of Warfighting Systems
- Provide research and engineering that can rapidly provide solutions for urgent needs in current operations and/or to assist Acquisition Programs to achieve threshold and objective requirements
- Demonstrate advanced technologies and manufacturing methods that will reduce lifecycle costs of future systems and enable more effective, efficient, affordable systems through upgrades

Effective Business Practices
- Invest in areas where Army must invest because no one else will (i.e., take the lead); leverage other service laboratories, Department of Defense (DoD) laboratories, academia, industry, and international partners for everything else
- Maintain critical in-house Army research facilities (in 22 organizations in 5 commands), workforce, and capabilities in areas where the Army must lead in invention and innovation

- Synchronize our S&T programs and major efforts with fiscal processes and Army Force Generation (ARFORGEN) timelines determined by needs of the warfighter
- Develop strong partnerships throughout the Army, especially with the Training & Doctrine Command (TRADOC), Acquisition, and Threat Communities, as well as with the Soldiers in the Active and Reserve Components, so that there is a greater understanding of the value of Army S&T and its endeavors

RESOURCING S&T

The Army S&T budget is apportioned within the Research Development Test & Evaluation (RDT&E) appropriation, as follows:

≈20 percent to Basic Research, which seeks to:
- Obtain knowledge for an uncertain future through invention and discovery
- Understand theories and phenomena that may impact Army needs

≈35 percent to Applied Research, which seeks to:
- Conduct research and apply knowledge and understanding to specific Army problems and challenges
- Conceptualize, design, and experiment with components, subsystems, models (discovery and innovation)

≈40 percent to Advanced Technology Development, which seeks to:
- Mature, develop, and integrate technologies at sub-system and system level
- Demonstrate feasibility of technology-enabled capabilities
- Define transition paths to accelerate introduction of technology-enabled capabilities to the warfighter

≈4 percent in Technology Maturation Initiatives which are designed to take selected technologies above Technology Readiness Level 6 (TRL6) in order to facilitate transition, or to conduct competitive prototyping in accordance with the principles of the Weapons Systems Acquisition Reform Act for technologies that have high probability of transitioning to Army Programs of Record (PORs) soon

≈1 percent in Manufacturing Technology, which is focused on advancing the ability of the U.S. industrial base to manufacture affordable key advanced technologies

Army S&T will invest to provide effective, affordable, and supportable solutions for Army needs. For FY 13, the Army has dedicated more than $2 billion to its S&T programs: $444 million in Basic Research, $875 million in Applied Research, $891 million in Technology Development, $25 million in Technology Maturation, and $60 million in Manufacturing Technology.

ARMY S&T IN ACTION

The Army S&T community is organizing its investments into programs which address major investment priorities. Near-term integrated capabilities are developed with Advanced Technology Development and some late stage Applied Research funds to address the aforementioned capability gaps. Today, the Army S&T community is addressing these capability gaps in Technology-Enabled Capability Demonstration (TECD) programs, which either measurably enhance performance and effectiveness of an existing capability, or enable a new and necessary capability for the Soldier. The culminating event for the TECD is an integrated demonstration of the enhanced capability in a relative environment. In early FY12, Army senior leadership approved nine TECD programs.

In addition, several major S&T programs have been developed based on their visibility across the Army and the DoD. Each of these programs is managed with well-defined deliverables and transition commitments.

The major S&T programs and nine TECDs are detailed in the portfolio descriptions included in this section.

S&T PORTFOLIOS – DEFINING THE ARMY'S CAPABILITIES OF TOMORROW

Army S&T investment portfolios support Army modernization goals to develop and field affordable equipment in a rapidly changing technological environment by fostering invention, innovation, maturation, and the demonstration of technologies for the current and future fight. The Army S&T program is organized into investment portfolios that address challenges in six capability areas: four Army-wide areas (Air; Soldier; Ground; Command, Control, Communications, and Intelligence (C3I)) and two areas unique to S&T (Basic Research and Innovation Enablers).

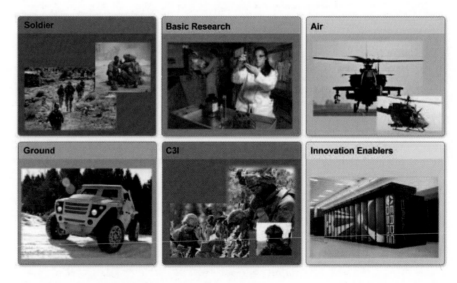

Figure 2: The six Army S&T investment portfolios

SOLDIER S&T PORTFOLIO

The Soldier S&T portfolio researches the science of human performance and matures and demonstrates technologies for the Soldier and squad across a host of supporting sub-portfolios detailed in Figure 3. The efforts of this portfolio are designed to maximize the effectiveness of squad performance as a collective formation.

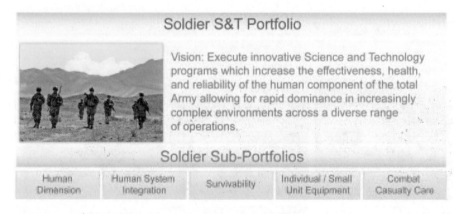

Figure 3: The Soldier S&T Portfolio and its six sub-portfolios

Technology-Enabled Capability Programs within the Solider portfolio include:

- Force Protection – Soldier and Small Unit which will develop and demonstrate technologies that increase protective gear performance while reducing weight and volume—protection from weapon threats, blast, fire, insect-borne diseases, weather conditions, and chemical/biological threats by FY16.
- Medical Assessment and Treatment which will develop and demonstrate capability to assess, diagnose, treat, and rehabilitate Soldiers who have been exposed to ballistic and blast events or other insults by FY16.
- Individual Training to Tactical Tasks which will develop and demonstrate self-training mechanisms that can monitor and track Soldier learning needs, assess and diagnose problems, and guide Soldiers through training events, provide effective performance feedback, select appropriate instructional strategies, anticipate and seek out information and learning content tailored to the learner's needs, and provide other assistance as needed by FY17.
- Overburdened – Physical Burden which will demonstrate technologies that reduce the weight and volume of items that individual Soldiers in a small unit must physically carry to accomplish their missions, while maintaining or increasing the ability of the unit to perform tasks, whether dismounted or in vehicles, by FY16.
- Sustainability/Logistics – Basing which will develop technologies to reduce supply demands, and reduce waste at small bases to sustain the small unit for the duration of the mission at lower cost and lower risk without adversely impacting primary mission accomplishment by FY17.

Major Efforts

The Soldier Portfolio also includes the following major efforts:

A major program within the Soldier Survivability sub-portfolio is the Warrior Injury Assessment Manikin (WIAMan) Medical Research Program (Figure 4). WIAMan provides medical program management for the effort to develop an improved blast manikin test device for the Live Fire Test and Evaluation (LFTE) Program. This program seeks to provide an enhanced capability to measure and predict combat vehicle occupant injuries during underbody blast events. The project will evaluate skeletal injuries to occupants during vehicle underbody blast events, which will be used with the medically validated set of skeletal injury criteria and the improved blast test manikin to improve LFTE evaluation capability, vehicle design, and soldier survivability.

Figure 4: The WIAMan program

The Military Infectious Diseases Research Program (MIDRP) manages research for the DoD on infectious diseases with a focus on protecting the warfighter by developing vaccines, drugs, diagnostics, and vector control. The U.S. military has been notably successful in this undertaking; since World War I, deaths from naturally occurring infections have not exceeded deaths due to combat injury in wartime. MIDRP's role is of continuing importance because diseases such as malaria, dengue fever, diarrhea, and leishmaniasis continue to have an adverse impact on military operations and the health of service members. MIDRP has supported HIV vaccine research and development since 1985 as HIV remains a significant threat to service members deployed overseas and is a major source of regional instability in areas of U.S. force protection. The MIDRP HIV research program is heavily engaged with efforts of other U.S. government agencies. The National Institute of Allergy and Infectious Disease (NIAID)is a major partner of MIDRP-supported HIV vaccine development activities. MIDRP also develops preventive medicine products to reduce insect and vector-borne disease transmission, such as improved repellents suitable for the military operational environment, bed nets and other products that enhance medical officers' ability to minimize disease threats in the field. DEET is a successful example. This repellent was developed in collaboration with the U.S. Department of Agriculture (USDA), introduced to the public in the 1950s, and has become by far the most common insect repellent used throughout the world today.

The are several Combat Casualty Care efforts related to Traumatic Brain Injury (TBI). The Combat Casualty Care Research Program Neurotrauma research portfolio is organized as a continuum of care model that includes eight phases: Basic Science/Epidemiology, Prevention/Education/Training, Head Impact/Blast, Screening, Assessment, Treatment, Recovery, and Return to Duty. The portfolio includes 472 projects, with 347 of these projects including a mild TBI component and 354 of these projects including a moderate to severe TBI component. Although we have improved diagnostic and treatment guidelines and therapies, recent research indicates that TBI and especially repeated mild exposures may lead to chronic neurodegeneration, mild cognitive impairments, mood disorders, and chronic traumatic encephalopathy. Combat Casualty Care also conducts research in rehabilitative medicine to enhance the ability to diagnose, stabilize, and accelerate wound healing and repair for casualties.

GROUND S&T PORTFOLIO

As depicted in Figure 5, the Ground S&T Portfolio includes technologies across four sub-portfolios that address survivability, weapon systems, active and passive protection systems for ground vehicles, manned and unmanned ground platforms and mobility systems, countermine/counter-IED efforts, and deployable small base protection.

Ground S&T Portfolio

Vision: U.S. overmatch of offensive and defensive capabilities in weapons and military vehicles. The strategy is to invest in technologies that increase performance and affordability of Army Ground Systems.

Ground Sub-Portfolios

| Survivability | Weapons | Ground Platforms | Mobility / Countermobility |

Figure 5: The Ground Portfolio vision and sub-portfolios

Major Efforts

The Ground Portfolio is developing protection from underbody blast threats. In the past we designed vehicles with little consideration for accommodating Soldiers—focusing more on mission capabilities. Today we are beginning to explore ways to design vehicles around Soldiers.

Technology-Enabled Capability Programs within the Ground S&T portfolio include:

• The Occupant Centric TECD program, which provides the mechanism to develop, design, demonstrate, and document an occupant-centered Army ground vehicle design that improves vehicle and Soldier survivability by mitigating Soldier injury due to underbody IED and mine blasts, vehicle rollover, and vehicle crash events. This design philosophy considers the Soldier first, integrates occupant protection technologies, and builds the vehicle to surround and support Soldiers and their mission. To this end, Army S&T is developing an occupant-centric survivability concept design demonstrator, as well as platform-specific demonstrators with unique occupant protection technologies tailored to the platform design constraints. We are also publishing standards for occupant-centric design guidelines, test procedures, and safety specifications as well as improving Modeling and simulation capabilities and toolsets.

We have developed tools and methods which have led to system-level evaluations through modeling and simulation resulting in improved Live Fire Test and Evaluation, faster delivery of technologies to theater/customers, and necessary characterizations of threats, systems and environment. We continue to look at a full range of technologies to address underbody blast events, from modeling and simulation and physiological studies to seats, restraints and energy-absorbing materials. This TECD will complete in FY15.

- Force Protection – Basing which will demonstrate an ability to construct and protect a 300-person combat outpost (COP) or patrol base (PB) in 30 days with integrated sensing and defense capabilities. Additional payoffs expected from this TECD include increased Soldier availability for mission tasks vice set-up and security tasks; force protection levels comparable to larger forward operating bases; decreased combat outpost tear-down time to less than four days and up to a fifty percent increase in reusable materials. This TECD will complete in FY17.

Figure 6: Occupant-Centric TECD

Within the Survivability Sub-Portfolio:

Another major effort is Deployable Force Protection (DFP), which is exploring technologies to improve survivability of small patrol bases and forward operating bases. In 2012, this S&T Program developed designs and methods to quickly establish mortar pits and provide overhead cover using modular systems that incorporate blast and ballistic protection. Members of the DFP team worked with troops on design and employment options. The 82nd Airborne Division will deploy in 2012 with a number of modular protective mortar pit and overhead cover systems; these will be used in an operational assessment in theater. Use of these systems will save of dozens of hours typically spent establishing mortar pits and protection and will increase the associated level of protection for Soldiers.

The Extended Area Protection & Survivability (EAPS) program addresses range, coverage, and performance limitations in protection against rocket, artillery, and mortar (RAM) attacks. The EAPS program encompasses high-risk, robust performance missile and gun technology developments required to provide a capability for 360-degree protection against asymmetric, multiple-threat, simultaneous RAM attacks. The Aviation & Missile Research, Development, and Engineering Center manages the development of two EAPS missile-technology programs; one, a miniature hit-to-kill interceptor with a semi-active radio frequency seeker; the other, a command-guided interceptor with a forward-firing warhead. The EAPS gun-technology program, managed by the Armaments Research, Development, and Engineering Center, manages the development of a command-guided, course-correcting 50mm artillery-round and a rapid-fire auto cannon. In 2011, the U.S. Army received approval to acquire a robust next-generation counter-RAM system to replace what has been fielded, the Indirect Fire Protection Capability Increment 2-Intercept program. The program strategy is to leverage the EAPS technology if the output from the recommendations of the Analysis of Alternatives is to pursue development of either a gun or missile solution based on the EAPS designs. A program Milestone A decision is anticipated in 2013.

AIR PORTFOLIO

The Air Portfolio includes technologies for manned and unmanned systems; air-delivered lethality; and air-platform safety, survivability and protection. These and other efforts are depicted in Figure 7.

Figure 7: The Air Portfolio and its 7 sub-portfolios

Major Efforts

The major effort within the Air Portfolio, and shown in Figure 8, is the Joint Multi-Role (JMR) Technology Demonstration program, a key facilitator of Future Vertical Lift (FVL). In FY11, the Army, Navy, and NASA agreed to use a common toolset and database and are collaboratively sharing in the definition of a model performance specification for FVL medium, an aircraft intended to replace our Blackhawk/Seahawk and Apache fleets. Three different configurations of representative FVL aircraft have been designed as concepts by the government: a conventional helicopter, a large-wing slowed rotor compound, and a tilt rotor. Seven design excursions have been investigated, fully exploring the size and environmental characteristics of interest, including shipboard operations. In FY13, the design and fabrication phase of two flight demonstrators will begin with flight testing scheduled to begin in FY17.

Figure 8: Future Vertical Lift/Joint Multi-Role Program

While many of our rotorcraft research efforts are focused on the development of technologies for transition to new platforms in 2025 and beyond, Army S&T is also investing to keep the current fleet effective. One recent transition success has been the Advanced Affordable Turbine Engine (AATE), a 3000-shaft horsepower engine with 25% greater fuel efficiency and 35% reduced lifecycle costs. In FY12, AATE transitioned to Program Manager (PM) - Utility for Engineering and Manufacturing Development under the Improved Turbine Engine Program, which will re-engine our Blackhawk and Apache fleet.

COMMAND, CONTROL, COMMUNICATIONS & INTELLIGENCE (C3I) PORTFOLIO

The C3I Portfolio, detailed in Figure 9, includes technologies for ground, air, and Soldier communications devices and networks, air and space sensors, intelligence, electronic warfare (EW), cyber and network payloads, and mission command. The key to successful operations in an increasingly complex battle space is secure, seamless, and timely communications across all echelons, from headquarters to the Soldier. As such, the C3I portfolio aims to provide Soldiers at the tactical edge with trusted and responsive sensors, communications, and information adaptable in dynamic, austere environments to support battlefield operations and non-kinetic warfare.

Figure 9: C3I Portfolio and sub-portfolios

Major Efforts

A major effort in the C3I portfolio combines enhanced mission command capabilities for the Soldier and small unit with improved actionable intelligence that is pertinent to the squad's mission using mobile networks to connect the squad to higher echelons. The Mission Command and Actionable Intelligence TECDs will reduce tactical surprise and achieve overmatch at the squad level by providing the intelligence and mission command (MC) tools to the tactical edge that allow leaders to synchronize action, seize the initiative, and maintain situational awareness. These TECDs will complete in FY15.

Within the Sensors sub-portfolio a major effort is the Infrared (IR) Revitalization program, an innovative effort to help maintain Department of Defense proficiency in IR sensors. The program is focused on developing new semiconductor materials for higher temperature focal plane arrays and advanced digital read out integrated circuits for the next generation of sensors.

Within the Mission Command sub-portfolio, Army S&T is providing solutions to improve command and control, situational awareness, and dynamic communications, while maintaining appropriate military security not found in commercial devices. In order to exploit the full range of capabilities that smart devices offer the Soldier, we need an adaptive network in an on-the-move (OTM) environment; handheld devices that provide Soldiers with the necessary decision and communications capabilities with an intuitive interface; and appropriate security protection against potential cyber threats on the battlefield.

Army mobile network research efforts are increasing network efficiency and reliability, increasing OTM connectivity and bandwidth use, and allowing for reliable message delivery in difficult communications environments. Our mission command efforts are aimed at providing Soldiers and small units with the kinds of data-driven decision tools once available only to higher echelons. As our defense strategy moves to a smaller, more agile force, it is critical that small units and individual Soldiers have access to accurate and relevant situation awareness information including geospatial and meteorological data, combat identification and battle space awareness, as well as full-spectrum decision support tools.

Within the intelligence/electronic warfare sub-portfolio, Army research is focusing on identifying, locating, neutralizing, and protecting against electronic threats on the battlefield while coordinating EW and cyber capabilities while maintaining interoperability between EW transmissions and Blue Force communications.

The most useful tools are worthless if they are not properly secured. Security must address capabilities such as approved encryption for Secret and below, identity management, security policy management, exploitable applications, bi-directional cross-domain data transfer (Secret to UNCLAS) and security for computing, network and wireless infrastructure. Army S&T efforts in this area include authentication of approved applications and preventing the installation of rogue applications, providing Secret voice and data connections across disparate technologies including handheld devices, and developing a mutual authentication mechanism between users, handheld devices, and the network core.

Beyond the specific security efforts for mobile battlefield communications, the C3I portfolio also directs our broader cyber security S&T efforts. Our work in a resilient cyber security framework will provide a more secure foundation for participants, including cyber devices and software, to work together in near real time to anticipate and prevent cyber attacks, limit the spread of attacks across participating devices, minimize the consequences of attacks, and recover systems and networks to trusted states.

INNOVATION ENABLERS PORTFOLIO

The Innovation Enablers Portfolio includes technology development associated with environmental quality and installations, in the areas of sustainable ranges and lands, pollution prevention, military materials in the environment, and adaptive and resilient installations. This portfolio also includes the Department of Defense High Performing Computer (HPC) Modernization efforts and is detailed in Figure 10.

The High Performance Computing Modernization Program (HPCMP) devolved from the Office of the Secretary of Defense to the Army in FY12. This program supports research and development in Army, Navy, Air Force, Marine Corps, and DoD agencies, by enabling advanced computational capabilities as a solution of first resort to explore and evaluate new theories. The HPCMP, shown in Figure 11, provides the hardware, software, and expertise to solve the complex problems faced by the RDT&E community and also reduces the time and cost of acquiring weapons systems and platforms through advanced computing, simulations, and calculations in support of military operations.

Innovation Enablers Portfolio

Environmental Quality & Installations

Vision: Army fully understands our built and natural environment to ensure sustainability and success of military operations in support of the Army mission.

High Performance ComputingQuality & Installations

Vision: High Performance Computing – Be the global leader in developing specialized and unique HPC resources to solver DoD's most demanding computational and critical problems with ready access to HPC resources across the research, development, acquisition, test and operational communities.

Figure 10: The Innovation Enablers Portfolio and its 2 sub-portfolios

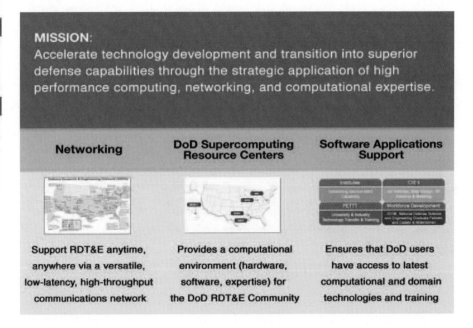

MISSION:
Accelerate technology development and transition into superior defense capabilities through the strategic application of high performance computing, networking, and computational expertise.

Networking	DoD Supercomputing Resource Centers	Software Applications Support
Support RDT&E anytime, anywhere via a versatile, low-latency, high-throughput communications network	Provides a computational environment (hardware, software, expertise) for the DoD RDT&E Community	Ensures that DoD users have access to latest computational and domain technologies and training

Figure 11: The High Performance Computing Modernization Program

BASIC RESEARCH PORTFOLIO

Underpinning all Army S&T efforts is a strong basic research program. Figure 12 details the vision and sub-portfolios of the Basic Research portfolio. Investments in Basic Research are critical to acquiring new knowledge in areas that hold great promise in advancing new and technically challenging Army capabilities and concepts—concepts that will enable revolutionary advances and paradigm-shifting future operational capabilities. Long-term exploration efforts look to discover or invent new technologies and capabilities relevant to the Army mission—we explore with a purpose. Our long-term disruptive technology investments are researching technologies that will change the rules of the playing field for our Soldiers. Long-term enabling research looks for innovative ways to move the inventions and discoveries into components and subcomponents and technologies that our labs and research partners can exploit. By doing this we enable future S&T applied research, advanced tech development, and capabilities. Taken together, this basic research provides the solid foundation for Army S&T.

Basic Research Portfolio

Vision: Advance the frontiers of fundamental science and technology and drive long-term, game-changing capabilities for the Army through a multi-disciplinary portfolio that teams our technically skilled and agile in-house researchers with the global academic community.

Basic Research Sub-Portfolios

| Human Centric | Information Centric | Material Centric | Platform Centric | Enrichment Initiatives |

Figure 12: The Basic Research Portfolio and its 5 sub-portfolios

Major Efforts

The following is a brief summary of select areas of investment noted above, the synergy among them, and some of the capabilities they may provide by sub-portfolio.

Within the Human Centric sub-portfolio:

Immersive Technology—the path to virtual reality training

The evolving threat environment continues to put increasing demands on the diversity and effectiveness of Soldier skills. To meet these demands, superior training tools and methods are needed. Virtual worlds can provide this capability; however, we are currently at primitive stages in their realization. With advances in computational processing and steady progress in understanding the brain's "software" comes the possibility of creating highly realistic virtual training environments inhabited by humanlike avatars. Such environments will create a paradigm shift in the way we provide training, while achieving low-cost, safe, low-environmental impact, highly variable simulation environments for the future training of our Soldiers.

Neuroscience—Understanding how the human brain works

Fundamental to the conduct of military operations is superior Soldier performance. Understanding how the human brain works—determining the brain's "software"—is the key to developing these capabilities. When embedded in a wide range of military platforms, this "software" will provide superior training methods and human system interfaces that will be tuned to an individual's characteristics, thereby resulting in superior Soldier performance. Research in this area will also dramatically advance our ability to prevent and treat those suffering from various types of battlefield brain injury.

Within the Information Centric sub-portfolio:

Network Science—managing complex military operations with greater speed and precision

Networks tie together the following: highly distributed sensor systems for reconnaissance and surveillance, information for decision-making, Soldiers, and the execution of fast distributed precision fires. Better-functioning networks advance our ability to conduct complex military operations with greater speed and precision. However, we know relatively little about network science and are not yet realizing the potential that networks can provide on current and future battlefields. We are implementing a new multidisciplinary approach that combines communications, information, and the social/human component of networks. This approach changes the way we think about optimizing the use of networks. Advances in network science will allow us to predict and optimize network performance before we build them through the creation of wholly new design tools.

Within the Material Centric sub-portfolio:

Materials Modeling research develops fundamental scientific principles to develop underpinning, cross-cutting, and transferable physics-based modeling capabilities. Research focuses on two-way multi-scale modeling for predicting performance and designing materials. We also investigate analytical and theoretical analyses to effectively define the interface physics across length scales. In addition we aim to advance experimental capabilities for verification and validation of multi-scale physics, and modeling and strategies for the synthesis of high loading rate tolerant materials. The intent is to provide the Army with next generation multi-functional materials for ballistic and electronic applications, for light-weight vehicle and facility protection, and for energy storage and electronic devices, and to provide new materials to address the extreme challenges associated with understanding and modeling materials subject to Army operational environments. This research supports the development of computational tools, software, and new methods for material characterization to speed the process of discovery and development of advanced materials and make it less expensive and more predictable.

Biotechnology—leveraging four billion years of evolution

The increasing importance of and demands for wide-area persistent surveillance create significant challenges for sensor systems, real-time processing of vast amounts of data, the real-time interpretation of information for decision-making and power and energy requirements to support such demanding systems. Through four billion years of evolution, biological systems have engineered solutions to some of these challenges. We seek to leverage research in these areas to improve the performance of our Soldiers. Major investments in this area through reverse engineering will lead to totally new sensing systems, new ways to rapidly process data into information, the development of novel sense and response systems, and biologically inspired power and energy solutions.

Nanotechnology—dramatically changing our ability to manufacture new material by design

The last century was dominated by advances in the physical sciences through the discovery of the atom, its structure, and the laws that govern its behavior. This century will be dominated by the complex world of nanoscience, whose mysteries will be unraveled by our understanding of systems of atoms and molecules. Nanotechnology is the manipulation of matter on a near-atomic scale to produce new structures, materials, and devices. Nanotechnology research makes it possible to explore the emerging biotech field and dramatically change our ability to create new materials by design. This technology has the ability to transform many industries in discovering and creating new materials with properties that will revolutionize military technology and make Soldiers less vulnerable to the enemy and to environmental threats. Research in nanoscale technologies is growing rapidly worldwide. By 2015, the National Science Foundation estimates that nanotechnology will have a one trillion dollar impact on the global economy and employ two million workers, one million of whom may be in the United States.

Within the Platform centric sub-portfolio:

Autonomous Systems—extending the operational effectiveness of Soldiers through robotic systems

A major military objective is to totally frustrate and defeat our adversaries across a wide spectrum of conflicts while dramatically increasing the survivability of our Soldiers by keeping them out of harm's way. Autonomous systems of extraordinary capability can fulfill this objective; however, they must be completely safe and secure while operating in highly complex operational environments. Achieving such levels of capability will require significant investments in highly sophisticated sense, response, and processing systems to approach the capabilities of biological systems; major advances in artificial intelligence; the development of intelligent agents approaching human performance levels; and advances in machine learning, swarming, and actuation and control.

Quantum Effects—overcoming the limitations of Moore's Law

Increasing demand for information to support decision making on the battlefield requires advanced sensor systems to collect relevant data, as well as the means to process it into actionable forms. Major advancements in processing power are required to cope with the demand to process ever-larger amounts of data.

Investments in this area will exploit the massive parallelism of the quantum world to create computers that will dwarf the capabilities of the most powerful computers today, making them look like pocket calculators. The development of such computational systems will enable the embedding of high-performance computing in all military platforms, including the Soldier's uniform.

In 2012, Army S&T developed a process similar to TECDs to define a set of priorities for Basic Research and identify challenge statements against which programs can be proposed and approved. Moving forward, Army S&T will continue to refine these statements and develop long-term basic research programs and investments to meet these challenges for the Army of the future.

TECHNOLOGY TRANSITION – A KEY METRIC OF PERFORMANCE

Neither a technology nor a collection of technologies are useful unless accepted by someone and provided to the warfighter. A concept or technology can be matured and demonstrated to a Technology Readiness Level (TRL) 6 with S&T funds (Budget Activity 2 or 3). Products or deliverables of Army S&T programs can be knowledge products; experimental data on performance; hardware (devices, components, sub-systems, or prototype systems) or software; specifications, or concepts.

There are many ways that transition from S&T can happen.

- Transition can be to an Army Program of Record (POR) managed by a Program Executive Office (PEO). It is best practice to have a Technology Transition Agreement (TTA) between the S&T program and the PEO that specifies the S&T products or deliverables. A TTA is negotiated at the beginning of the S&T program and revisited every year to ensure that the effort remains on track and in alignment with the modernization strategy of the Army.
- Transition can be to a field element through a PEO or rapid fielding initiative— usually in cases of retrofits or solutions to problems with equipment identified in the field.
- Transitions can be through Joint Capability Technology Demonstrations (JCTDs), which provide capability demonstrations to combatant commanders along with leave-behind assets.
- Transitions can be to field units through the response to Operational Needs Statements (ONS) or Joint Urgent Operational Needs (JUONS).

- Transitions can be to an Army organization such as TRADOC, ATEC, a member of the Army S&T enterprise, the G-1, another DoD entity, or industry. These transitions are typically in the form of "knowledge products" that inform requirements such as analysis of alternatives; inform acquisition, such as a proof of concept or information to enable a milestone decision or a technology data package. These knowledge products can take the form of TTP or even science and engineering support for follow-on development, demonstration, experimentation, and assessments. Finally, these knowledge products can include standards, certification, accreditation, or test methods that will enable subsequent fielding of capabilities to the Army.

Many times transition of S&T is indirect because the research and technology maturation funded by S&T and executed by industry or university partners often becomes the technology solution provided years later to the government in response to competitive solicitations for systems.

The Army technology transition strategy is to address the principal causes of the chasm between S&T and acquisition. Army S&T will seek to improve:

- Systems integration: People performing on both sides of Budget Activity 3/Budget Activity 4 or Budget Activity 3/Manufacturing Technology hand-off must have knowledge of the target system and understand system interfaces of target system/platform; we will strive to hire, retain, and utilize more systems engineers in S&T.
- Collaboration: We will seek to improve partnerships and maintain a more collaborative environment among all entities (S&T laboratories and centers, industry, PEOs/PMs, TRADOC).
- Accountability: S&T will strive to improve its program planning and execution—stressing accountability for the cost, schedule, and performance of technology products and solutions in all our efforts.
- Matching technology maturation and insertion cycle times: S&T will, where appropriate, use prototyping to facilitate timely insertion of rapidly developing technologies into PORs within constraints imposed by deliberate requirements, budgeting and acquisition processes.

- Time horizon balance: We will seek to revitalize "advanced concepting" and wargaming activities to provide game-changing concepts to inform requirements development and guide the maturation and transition of innovative technologies in the mid- to long-term.

SUMMARY

The Army Science and Technology (S&T) investments support Army modernization goals to develop and field affordable equipment in a rapidly changing technological environment by fostering invention, innovation, maturation, and demonstration of technologies for the current and future fight. As we continue to diligently identify and harvest technologies suitable for transition to our force, we aim to remain ever vigilant of potential and emerging threats. We will sharpen our research efforts to focus upon those core capabilities we will need to sustain for the future and also identify promising "leap-ahead" technologies able to change the existing paradigms of understanding. Ultimately, our focus remains upon Soldiers; we consistently seek new avenues to increase their capability and ensure their technological superiority—today, tomorrow and decades from now.

Appendices

Army Combat Organizations

Glossary of Terms

Systems by Contractors

Contractors by State

Points of Contact

Army Combat Organizations

Army organizations are inherently built around people and the tasks they must perform. Major combat organizations are composed of smaller forces, as shown here.

Squad
- Leader is a sergeant
- Smallest unit in Army organization
- Size varies depending on type: Infantry (9 Soldiers), Armor (4 Soldiers), Engineer (10 Soldiers)
- Three or four squads make up a platoon

Platoon
- Leader is a lieutenant
- Size varies: Infantry (40 Soldiers), Armor (4 tanks, 16 Soldiers)
- Three or four platoons make up a company

Company
- Leader is a captain
- Usually up to 220 Soldiers
- Artillery unit of this size is called a battery
- Armored Cavalry or Air Cavalry unit is called a troop
- Basic tactical element of the maneuver battalion or cavalry squadron
- Normally five companies make up a battalion

Battalion
- Leader is a lieutenant colonel
- Tactically and administratively self-sufficient
- Armored Cavalry and Air Cavalry equivalents are called squadrons
- Two or more combat battalions make up a brigade

Brigade
- Leader is a colonel
- May be employed on independent or semi-independent operations
- Combat, combat support, or service support elements may be attached to perform specific missions
- Normally three combat brigades are in a division

Division
- Leader is a major general
- Fully structured division has own brigade-size artillery, aviation, engineer, combat support, and service elements
- Two or more divisions make up a corps commanded by a lieutenant general

To better confront current and future threats, the Army is transforming its force structure into Brigade Combat Teams (BCTs). The goal is to provide more flexible and self contained forces with the capability of rapid deployment and the ability to engage in the full spectrum of warfare without sacrificing lethality and staying power on the battlefield. These BCTs will be organized as Infantry (IBCTs), Heavy (HBCTs) and Stryker (SBCTs).

Glossary of Terms

Acquisition Categories (ACAT)

ACAT I programs are Milestone Decision Authority Programs (MDAPs) or programs designated ACAT I by the Milestone Decision Authority (MDA).

Dollar value: estimated by the Under Secretary of Defense (Acquisition and Technology) (USD [A&T]) to require an eventual total expenditure for research, development, test and evaluation (RDT&E) of more than $365 million in fiscal year (FY) 2000 constant dollars or, for procurement, of more than $2.190 billion in FY 2000 constant dollars. ACAT I programs have two subcategories:

1. **ACAT ID**, for which the MDA is USD (A&T). The "D" refers to the Defense Acquisition Board (DAB), which advises the USD (A&T) at major decision points.
2. **ACAT IC**, for which the MDA is the DoD Component Head or, if delegated, the DoD Component Acquisition Executive (CAE). The "C" refers to Component. The USD (A&T) designates programs as ACAT ID or ACAT IC.

ACAT IA programs are MAISs (Major Automated Information System Acquisition Programs), or programs designated by the Assistant Secretary of Defense for Command, Control, Communications, and Intelligence (ASD [C3I]) to be ACAT IA.

Estimated to exceed: $32 million in FY 2000 constant dollars for all expenditures, for all increments, regardless of the appropriation or fund source, directly related to the AIS definition, design, development, and deployment, and incurred in any single fiscal year; or $126 million in FY 2000 constant dollars for all expenditures, for all increments, regardless of the appropriation or fund source, directly related to the AIS definition, design, development, and deployment, and incurred from the beginning of the Materiel Solution Analysis Phase through deployment at all sites; or $378 million in FY 2000 constant dollars for all expenditures, for all increments, regardless of the appropriation or fund source, directly related to the AIS definition, design, development, deployment, operations and maintenance, and incurred from the beginning of the Materiel Solution Analysis Phase through sustainment for the estimated useful life of the system.

ACAT IA programs have two subcategories:

1. **ACAT IAM**, for which the MDA is the Chief Information Officer (CIO) of the DoD, the ASD (C3I). The "M" refers to Major Automated Information System Review Council (MAISRC). (Change 4, 5000.2-R)
2. **ACAT IAC**, for which the DoD CIO has delegated milestone decision authority to the CAE or Component CIO. The "C" refers to Component.

ACAT II programs are defined as those acquisition programs that do not meet the criteria for an ACAT I program, but do meet the criteria for a major system, or are programs designated ACAT II by the MDA. The dollar value is estimated to require total expenditure for RDT&E of more than $140 million in FY 2000 constant dollars, or for procurement of more than $660 million in FY 2000 constant dollars.

ACAT III programs are defined as those acquisition programs that do not meet the criteria for an ACAT I, an ACAT IA, or an ACAT II. The MDA is designated by the CAE and shall be at the lowest appropriate level. This category includes less-than-major AISs. The dollar values are under the threshold for ACAT II.

Acquisition Phase

All the tasks and activities needed to bring a program to the next major milestone occur during an acquisition phase. Phases provide a logical means of progressively translating broadly stated mission needs into well-defined system-specific requirements and ultimately into operationally effective, suitable, and survivable systems. The acquisition phases for the systems described in this handbook are defined below:

Technology Development Phase

The purpose of this phase is to reduce technology risk, determine and mature the appropriate set of technologies to be integrated into a full system, and to demonstrate critical technology elements on prototypes. Technology Development is a continuous technology discovery and development process reflecting close collaboration between the Science and Technology (S&T) community, the user, and the system developer. It is an iterative process designed to assess the viability of technologies while simultaneously refining user requirements. Entrance into this phase depends on the completion of the Analysis of Alternatives (AOA), a proposed materiel solution, and full funding for planned Technology Development Phase activity.

Engineering and Manufacturing Development (EMD) Phase

The purpose of the EMD phase is to develop a system or an increment of capability; complete full system integration (technology risk reduction occurs during Technology Development); develop an affordable and executable manufacturing process; ensure operational supportability with particular attention to minimizing the logistics footprint; implement human systems integration (HSI); design for producibility; ensure affordability; protect critical program information by implementing appropriate techniques such as anti-tamper; and demonstrate system integration, interoperability, safety, and utility. The Capability Development Document, Acquisition Strategy, Systems Engineering Plan, and Test and Evaluation Master Plan (TEMP) shall guide this effort. Entrance into this phase depends on technology maturity (including software), approved requirements, and full funding. Unless some other factor is overriding in its impact, the maturity of the technology shall determine the path to be followed.

Production and Deployment Phase

The purpose of the Production and Deployment phase is to achieve an operational capability that satisfies mission needs. Operational test and evaluation shall determine the effectiveness and suitability of the system. The MDA shall make the decision to commit DoD to production at Milestone C and shall document the decision in an Acquisition Decision Memorandum. Milestone C authorizes entry into low rate initial production (for MDAPs and major systems), into production or procurement (for non-major systems that do not require LRIP) or into limited deployment in support of operational testing for MAIS programs or software-intensive systems with no production components. Entrance into this phase depends on the following criteria: acceptable performance in developmental test and evaluation and operational assessment (OSD OT&E oversight programs); mature software capability; no significant manufacturing risks; manufacturing processes under control (if Milestone C is full-rate production); an approved Initial Capabilities Document (ICD) (if Milestone C is program initiation); an approved Capability Production Document (CPD); a refined integrated architecture; acceptable interoperability; acceptable operational supportability; and demonstration that the system is affordable throughout the life cycle, fully funded, and properly phased for rapid acquisition. The CPD reflects the operational

requirements, informed by EMD results, and details the performance expected of the production system. If Milestone C approves LRIP, a subsequent review and decision shall authorize full-rate production.

Operations and Support Phase

The purpose of the Operations and Support phase is to execute a support program that meets materiel readiness and operational support performance requirements, and sustains the system in the most cost-effective manner over its total life cycle. Planning for this phase shall begin prior to program initiation and shall be documented in the Life-Cycle Sustainment Plan (LLSP). Operations and Support phase has two major efforts: life-cycle sustainment and disposal. Entrance into the Operations and Support Phase depends on meeting the following criteria: an approved CPD; an approved LCSP; and a successful Full-Rate Production (FRP) Decision.

Acquisition Program

A directed, funded effort designed to provide a new, improved or continuing weapons system or AIS capability in response to a validated operational need. Acquisition programs are divided into different categories that are established to facilitate decentralized decision-making, and execution and compliance with statutory requirements.

Advanced Concept Technology Demonstrations (ACTDs)

ACTDs are a means of demonstrating the use of emerging or mature technology to address critical military needs. ACTDs themselves are not acquisition programs, although they are designed to provide a residual, usable capability upon completion. If the user determines that additional units are needed beyond the residual capability and that these units can be funded, the additional buys shall constitute an acquisition program with an acquisition category generally commensurate with the dollar value and risk of the additional buy.

Automated Information System (AIS)

A combination of computer hardware and software, data, or telecommunications, that performs functions such as collecting, processing, transmitting, and displaying information. Excluded are computer resources, both hardware and software, that are physically part of, dedicated to, or essential in real-time to the mission performance of weapon systems.

Commercial and Non-Developmental Items

Market research and analysis shall be conducted to determine the availability and suitability of existing commercial and non-developmental items prior to the commencement of a development effort, during the development effort, and prior to the preparation of any product description. For ACAT I and IA programs, while few commercial items meet requirements at a system level, numerous commercial components, processes, and practices have application to DoD systems.

Demilitarization and Disposal

At the end of its useful life, a system must be demilitarized and disposed of. During demilitarization and disposal, the program manager shall ensure materiel determined to require demilitarization is controlled and shall ensure disposal is carried out in a way that minimizes DoD's liability due to environmental, safety, security, and health issues.

Developmental Test and Evaluation (DT&E)

DT&E shall identify potential operational and technological capabilities and limitations of the alternative concepts and design options being pursued; support the identification and description of design technical risks; and provide data and analysis in support of the decision to certify the system ready for operational test and evaluation.

Full Materiel Release

The process that ensures all Army materiel is safe, operationally suitable, and is supportable before release of issue to users. The PM determines necessary activities to certify materiel release readiness. This decision should be accomplished prior to full-rate production.

Joint Program Management

Any acquisition system, subsystem, component or technology program that involves a strategy that includes funding by more than one DoD component during any phase of a system's life cycle shall be defined as a Joint program. Joint programs shall be consolidated and collocated at the location of the lead component's program office, to the maximum extent practicable.

Live Fire Test and Evaluation (LFT&E)

LFT&E must be conducted on a covered system, major munition program, missile program, or product improvement to a covered system, major munition program, or missile program before it can proceed beyond low-rate initial production. A covered system is any vehicle, weapon platform, or conventional weapon system that includes features designed to provide some degree of protection to users in combat and that is an ACAT I or II program. Depending upon its intended use, a commercial or non-developmental item may be a covered system, or a part of a covered system. Systems requiring LFT&E may not proceed beyond low-rate initial production until realistic survivability or lethality testing is completed and the report required by statute is submitted to the prescribed congressional committees.

Low-Rate Initial Production (LRIP)

The objective of this activity is to produce the minimum quantity necessary to provide production-configured or representative articles for operational tests; establish an initial production base for the system; and permit an orderly increase in the production rate for the system, sufficient to lead to full-rate production upon successful completion of operational testing. LRIP quantity may not exceed 10 percent of the total production quantity without an approved waiver by the Acquisition Executive and documented in the Acquisition Decision Memorandum.

Major Automated Information System (MAIS) Acquisition Program

An AIS acquisition program that is (1) designated by ASD (C3I) as a MAIS, or (2) estimated to require program costs in any single year in excess of $32 million in FY 2000 constant dollars, total program costs in excess of $126 million in FY 2000 constant dollars, or total life cycle costs in excess of $378 million in FY 2000 constant dollars. MAISs do not include highly sensitive classified programs.

Major Defense Acquisition Program (MDAP)

An acquisition program that is not a highly sensitive classified program (as determined by the Secretary of Defense) and that is: (1) designated by the USD (A&T) as an MDAP, or (2) estimated by the USD (A&T) to require an eventual total expenditure for research, development, test and evaluation of more than $365 million in FY 2000 constant dollars or, for procurement, of more than $2.190 billion in FY 2000 constant dollars.

Major Milestone

A major milestone is the decision point that separates the phases of an acquisition program. MDAP milestones include, for example, the decisions to authorize entry into the engineering and manufacturing development phase or full rate production. MAIS milestones may include, for example, the decision to begin program definition and risk reduction.

- **Milestone A:** Entry into the Technology Development Phase
- **Milestone B:** Entry into the Engineering and Manufacturing Development Phase
- **Milestone C:** Entry into the Production and Deployment Phase

Major Systems

Dollar value: estimated by the DoD Component Head to require an eventual total expenditure for RDT&E of more than $140 million in FY 2000 constant dollars, or for procurement of more than $660 million in FY 2000 constant dollars. (Lowest category for major system designation is ACAT II.)

Materiel Solution Analysis Phase

The purpose of this phase is to assess potential materiel solutions and to satisfy the phase-specific entrance criteria for the next program milestone designated by the MDA. Entrance into this phase depends upon an approved ICD resulting from the analysis of current mission performance and an analysis of potential concepts across the DoD components, international systems from allies, and cooperative opportunities.

Milestone (MS)

The point at which a recommendation is made and approval sought regarding starting or continuing an acquisition program, i.e., proceeding to the next phase.

Milestone Decision Authority (MDA)

The individual designated in accordance with criteria established by the USD (AT&L), or by the ASD (C3I) for AIS acquisition programs, to approve entry of an acquisition program into the next phase.

Modifications

Any modification that is of sufficient cost and complexity that it could itself qualify as an ACAT I or ACAT IA program shall be considered for management purposes as a separate acquisition effort. Modifications that do not cross the ACAT I or IA threshold shall be considered part of the program being modified, unless the program is no longer in production. In that case, the modification shall be considered a separate acquisition effort.

Operational Support

The objectives of this activity are the execution of a support program that meets the threshold values of all support performance requirements and sustainment of them in the most life cycle cost effective manner. A follow-on operational testing program that assesses performance and quality, compatibility, and interoperability, and identifies deficiencies shall be conducted, as appropriate. This activity shall also include the execution of operational support plans, to include the transition from contractor to organic support, if appropriate.

Operational Test and Evaluation (OT&E)

OT&E shall be structured to determine the operational effectiveness and suitability of a system under realistic conditions (e.g., combat) and to determine if the operational performance requirements have been satisfied. The following procedures are mandatory: threat or threat representative forces, targets, and threat countermeasures, validated in coordination with Defense Intelligence Agency (DIA), shall be used; typical users shall operate and maintain the system or item under conditions simulating combat stress and peacetime conditions; the independent operational test activities shall use production or production representative articles for the dedicated phase of OT&E that supports the full-rate production decision, or for ACAT IA or other acquisition programs, the deployment decision; and the use of modeling and simulation shall be considered during test planning.

For additional information on acquisition terms, or terms not defined, please refer to AR 70-1, Army Acquisition Policy, available on the Internet at http://www.army.mil/usapa/epubs/pdf/r70_1.pdf; or DA PAM 70-3, Army Acquisition Procedures, available on the Internet at http://www.dtic.mil/whs/directives/corres/pdf/500002p.pdf.

Systems by Contractors

AAI Corp.
MQ-1C Gray Eagle Unmanned Aircraft System (UAS)
RQ-7B Shadow Tactical Unmanned Aircraft System (TUAS)

Aardvark Technical
Improvised Explosive Device Defeat/Protect Force (IEDD/PF)

Aberdeen Test Center/Proving Ground
M109 Family of Vehicles (FOV) (Paladin/FAASV, PIM SPH/CAT)

Acambis PLC
Chemical Biological Medical Systems (CBMS)–Prophylaxis

Accenture Federal Services
General Fund Enterprise Business Systems (GFEBS)
Global Command and Control System–Army (GCCS-A)
One Semi-Automated Force (OneSAF)

Acme Electric
Javelin

Action Manufacturing
2.75-Inch Rocket Systems (Hydra-70)

ADSI
High Mobility Engineer Excavator (HMEE) I and III

Aegis
One Semi-Automated Force (OneSAF)

AEPCO
Longbow Apache (AH-64D) (LBA)

Aerial Machine and Tool Corp.
Air Warrior (AW)

Aerojet General
Guided Multiple Launch Rocket System (GMLRS) DPICM/Unitary/Alternative Warhead (Tactical Rockets)
Javelin
PATRIOT Advanced Capability–3 (PAC-3)
Tube-Launched, Optically Tracked, Wire-Guided (TOW) Missiles

Aerovironment, Inc.
RQ-11B Raven Small Unmanned Aircraft System (SUAS)

Agilent Technologies, Inc.
Calibration Sets Equipment (CALSETS)
Test Equipment Modernization (TEMOD)

Airborne Systems North America
Joint Precision Airdrop System (JPADS)

All American Racers, Inc.
RQ-11B Raven Small Unmanned Aircraft System (SUAS)

Alliant Techsystems
Counter Defilade Target Engagement (CDTE)–XM25

Guided Multiple Launch Rocket System (GMLRS) DPICM/Unitary/Alternative Warhead (Tactical Rockets)
HELLFIRE Family of Missiles
Medium Caliber Ammunition (MCA)
Precision Guidance Kit (PGK)
Small Arms–Crew Served Weapons
Small Caliber Ammunition
Tank Ammunition
XM7 Spider

Allison Transmission
Abrams Tank Upgrade
Family of Medium Tactical Vehicles (FMTV)
Heavy Expanded Mobility Tactical Truck (HEMTT)/HEMTT Extended Service Program (ESP)
Palletized Load System (PLS) and PLS Extended Service Program (ESP)

ALTA IT Services
Integrated Personnel and Pay System–Army (IPPS-A)

AM General
High Mobility Multipurpose Wheeled Vehicle (HMMWV) Recapitalization (RECAP) Program
Improved Ribbon Bridge (IRB)
Joint Light Tactical Vehicle (JLTV)

American Eurocopter
Lakota/UH-72A

American Ordnance
Artillery Ammunition
XM7 Spider

American Science & Engineering, Inc.
Non-Intrusive Inspection Systems (NIIS)
AMT
Mortar Systems

AMTEC Corp.
Medium Caliber Ammunition (MCA)

Anniston Army Depot (ANAD)
Abrams Tank Upgrade
Assault Breacher Vehicle (ABV)

Apptricity Corp.
Transportation Coordinators' – Automated Information for Movements System II (TC-AIMS II)

ArgonST Radix - Part of the Boeing Co.
Guardrail Common Sensor (GR/CS)

Armaments R&D Center
M109 Family of Vehicles (FOV) (Paladin/FAASV, PIM SPH/CAT)

ARMTEC
Artillery Ammunition

Artel Inc.
Combat Service Support Communications (CSS Comms)

AT&T Government Solutions
Installation Information Infrastructure Modernization Program (I3MP)

ATK
Tube-Launched, Optically Tracked, Wire-Guided (TOW) Missiles

Attain LLC
Army Enterprise Systems Integration
    Program (AESIP)

Avaya Government Solutions
Installation Information Infrastructure
    Modernization Program (I3MP)

Avenge
Enhanced Medium Altitude
    Reconnaissance and Surveillance
    System (EMARSS)

Aviation and Missile Solutions, LLC
Longbow Apache (AH-64D) (LBA)

Avon Protection Systems
Joint Service Aircrew Mask–Rotary Wing
    (JSAM RW) (MPU-5)

Avox Systems
Joint Service Aircrew Mask–Rotary Wing
    (JSAM RW) (MPU-5)

AVT
Close Combat Tactical Trainer (CCTT)

BAE Systems
120M Motor Grader
2.75-Inch Rocket Systems (Hydra-70)
Advanced Threat Infrared
    Countermeasures (ATIRCM) and
    Common Missile Warning System
    (CMWS) programs and Pre-MDAP
    Common Infrared Countermeasure
    (CIRCM)
Air Warrior (AW)
Airborne and Maritime/Fixed Station Joint
    Tactical Radio System (AMF JTRS)
Airborne Reconnaissance Low (ARL)

Bradley Fighting Vehicle Systems Upgrade
Enhanced Medium Altitude
    Reconnaissance and Surveillance
    System (EMARSS)
Heavy Loader
High Mobility Artillery Rocket System
    (HIMARS) M142
High Mobility Engineer Excavator (HMEE)
    I and III
Interceptor Body Armor (IBA)
Joint Effects Targeting System (JETS)
    Target Location Designation System
    (TLDS)
Joint Tactical Ground Station (JTAGS)
Joint Tactical Radio System Ground Mobile
    Radios (JTRS GMR)
Joint Tactical Radio System Handheld,
    Manpack, Small Form Fit (JTRS HMS)
Lightweight 155mm Howitzer System
    (LW155)
M109 Family of Vehicles (FOV) (Paladin/
    FAASV, PIM SPH/CAT)
Mine Protection Vehicle Family (MPVF),
    Area Mine Clearing System (AMCS),
    Interrogation Arm
Multiple Launch Rocket System (MLRS)
    M270A1
Night Vision Thermal Systems–Thermal
    Weapon Sight (TWS)
T-9 Medium Dozer
Tube-Launched, Optically Tracked, Wire-
    Guided (TOW) Missiles
Warfighter Information Network–Tactical
    (WIN-T) Increment 3
XM1216 & XM1216 E1 Small Unmanned
    Ground System (SUGV)
XM7 Spider

Barrett Firearms Manufacturing
Small Arms - Precision Weapons

Battelle Biomedical Research Center
Chemical Biological Medical Systems
    (CBMS)–Therapeutics

Battelle Memorial Institute
Chemical Biological Medical Systems
    (CBMS)–Therapeutics
Assembled Chemical Weapons
    Alternatives (ACWA)

Bechtel National, Inc.
Assembled Chemical Weapons
    Alternatives (ACWA)

Bell Helicopter Textron
Advanced Threat Infrared
    Countermeasures (ATIRCM) and
    Common Missile Warning System
    (CMWS) programs and Pre-MDAP
    Common Infrared Countermeasure
    (CIRCM)
Kiowa Warrior

Benchmark Electronics
XM1216 & XM1216 E1 Small Unmanned
    Ground System (SUGV)

Berg Companies, Inc.
Force Provider (FP)

Binary Group
General Fund Enterprise Business Systems
    (GFEBS)

Boeing
CH-47F Chinook
Enhanced Medium Altitude
    Reconnaissance and Surveillance
    System (EMARSS)

Joint Tactical Radio System Ground Mobile
    Radios (JTRS GMR)
Joint Tactical Radio System Network
    Enterprise Domain (JTRS NED)
Longbow Apache (AH-64D) (LBA)
PATRIOT Advanced Capability–3 (PAC-3)
XM1216 & XM1216 E1 Small Unmanned
    Ground System (SUGV)

Booz Allen Hamilton
Distributed Common Ground System–
    Army (DCGS-A)
Enhanced Medium Altitude
    Reconnaissance and Surveillance
    System (EMARSS)
Global Command and Control System–
    Army (GCCS-A)
Integrated Personnel and Pay System–
    Army (IPPS-A)
Joint Land Component Constructive
    Training Capability (JLCCTC)
Medical Communications for Combat
    Casualty Care (MC4)
Secure Mobile Anti-Jam Reliable Tactical
    Terminal (SMART-T)

Bound Tree Medical LLC
Medical Simulation Training Center
    (MSTC)

Bren-Tronics
RQ-11B Raven Small Unmanned Aircraft
    System (SUAS)

Bryant & Associates
Distributed Learning System (DLS)
Bway Corp.
Small Caliber Ammunition

CACI Technologies
Advanced Field Artillery Tactical Data System (AFATDS)
Airborne Reconnaissance Low (ARL)
Army Key Management System (AKMS)
Biometric Enabling Capability (BEC)
Common Hardware Systems (CHS)
Distributed Learning System (DLS)
Enhanced Medium Altitude Reconnaissance and Surveillance System (EMARSS)
Force XXI Battle Command Brigade and Below (FBCB2)
Global Command and Control System–Army (GCCS-A)
Joint Battle Command–Platform (JBC-P)
Joint Personnel Identification version 2 (JPIv2)
Nuclear Biological Chemical Reconnaissance Vehicle (NBCRV)–Stryker Sensor Suites
Sustainment System Mission Command (SSMC)
Tactical Mission Command (TMC)/Maneuver Control System (MCS)

CAE Inc.
One Semi-Automated Force (OneSAF)

Camber Corp.
One Semi-Automated Force (OneSAF)

Cangene, Corp.
Chemical Biological Medical Systems (CBMS)–Prophylaxis

Cap Gemini
Army Enterprise Systems Integration Program (AESIP)

Carleton Technologies Inc.
Air Warrior (AW)

Carter Enterprises, LLC
Interceptor Body Armor (IBA)

CAS, Inc.
Enhanced Medium Altitude Reconnaissance and Surveillance System (EMARSS)

Case New Holland
High Mobility Engineer Excavator (HMEE) I and III

Casteel Manufacturing
Line Haul Tractor

Caterpillar C-15
Palletized Load System (PLS) and PLS Extended Service Program (ESP)

Caterpillar Defense and Federal Products
Heavy Loader

Caterpillar Inc.
120M Motor Grader
Family of Medium Tactical Vehicles (FMTV)
Heavy Expanded Mobility Tactical Truck (HEMTT)/HEMTT Extended Service Program (ESP)
T-9 Medium Dozer

CEP Inc.
Air Warrior (AW)

Ceradyne, Inc.
Interceptor Body Armor (IBA)

CGI Federal
Meteorological Measuring Set–Profiler (MMS-P)/Computer Meteorological Data–Profiler (CMD-P)

Charleston Marine Containers
Force Provider (FP)

Choctaw Defense Manufacturing Contractors (CDMC)
Unit Water Pod System (Camel II)

Cloudera
Distributed Common Ground System–Army (DCGS-A)

CMI
RQ-7B Shadow Tactical Unmanned Aircraft System (TUAS)

Cogility
Distributed Common Ground System–Army (DCGS-A)

Cole Engineering Services Inc. (CESI)
One Semi-Automated Force (OneSAF)

Colt Defense LLC
Small Arms–Crew Served Weapons
Small Arms–Individual Weapons

Communications Security Logistics Activity
Cryptographic Systems

Computer Sciences Corp. (CSC)
Advanced Field Artillery Tactical Data System (AFATDS)
Advanced Threat Infrared Countermeasures (ATIRCM) and Common Missile Warning System (CMWS) programs and Pre-MDAP Common Infrared Countermeasure (CIRCM)
Defense Enterprise Wideband SATCOM System (DEWSS)
Distributed Common Ground System–Army (DCGS-A)
Medical Simulation Training Center (MSTC)

Comtech Mobile Datacom Corp. (CMDC)
Movement Tracking System (MTS)

CONCO
2.75-Inch Rocket Systems (Hydra-70)

CRGT
Distributed Learning System (DLS)

CSS
Army Key Management System (AKMS)

CTT
RQ-7B Shadow Tactical Unmanned Aircraft System (TUAS)

CUBIC Defense System
Instrumentable–Multiple Integrated Laser Engagement System (I-MILES)

Cummins Power Generation
Tactical Electric Power (TEP)

Curtiss-Wright
Bradley Fighting Vehicle Systems Upgrade

Daedalus Technologies
Aviation Combined Arms Tactical Trainer (AVCATT)

Daimler Trucks North America LLC/ Freightliner
Line Haul Tractor

DATA Inc.
Advanced Threat Infrared Countermeasures (ATIRCM) and Common Missile Warning System (CMWS) programs and Pre-MDAP Common Infrared Countermeasure (CIRCM)

David H. Pollock Consultants
Advanced Threat Infrared Countermeasures (ATIRCM) and Common Missile Warning System (CMWS) programs and Pre-MDAP Common Infrared Countermeasure (CIRCM)

Defense Information System Agency (DISA) Satellite Transmission Services-Global
Combat Service Support Communications (CSS Comms)

Dell Computer Corp.
Computer Hardware, Enterprise Software and Solutions (CHESS)
Distributed Common Ground System– Army (DCGS-A)

Sustainment System Mission Command (SSMC)

Detroit Diesel
Line Haul Tractor
Palletized Load System (PLS) and PLS Extended Service Program (ESP)

Dewey Electronics
Tactical Electric Power (TEP)

DHS Systems
Harbormaster Command and Control Center (HCCC)

Digital Reasoning
Distributed Common Ground System– Army (DCGS-A)

DMD
Army Integrated Air and Missile Defense (AIAMD)

DOK-ING d.o.o.
Anti-Personnel Mine Clearing System, Remote Control M160

Draper Laboratories
Joint Precision Airdrop System (JPADS)

DRS
Assault Breacher Vehicle (ABV)
Force XXI Battle Command Brigade and Below (FBCB2)
Helmet Mounted Night Vision Devices (HMNVD)

DRS Fermont
Tactical Electric Power (TEP)

DRS Optronics Inc.
Kiowa Warrior
Night Vision Thermal Systems–Thermal Weapon Sight (TWS)

DRS Sustainment Systems, Inc.
M1200 Armored Knight
Modular Fuel System (MFS)

DRS Tactical Systems
M1200 Armored Knight

DRS Technologies
Bradley Fighting Vehicle Systems Upgrade
Javelin
Joint Effects Targeting System (JETS) Target Location Designation System (TLDS)
Joint Service Transportable Small Scale Decontaminating Apparatus (JSTSS DA) M26
Movement Tracking System (MTS)

DRS-ES
Improved Environmental Control Unit (IECU)

DRS-TEM
Integrated Family of Test Equipment (IFTE)

DSE Corp.
Medium Caliber Ammunition (MCA)

Dynamics Research Corp.
One Semi-Automated Force (OneSAF)

Dynetics, Inc.
Calibration Sets Equipment (CALSETS)

DynPort Vaccine
Chemical Biological Medical Systems (CBMS)–Prophylaxis
Longbow Apache (AH-64D) (LBA)

E.D. Etnyre and Co.
Modular Fuel System (MFS)

EADS North America
Lakota/UH-72A

ECC International
Javelin

EDC Consulting, LLC
Integrated Personnel and Pay System– Army (IPPS-A)

EIBOT
Distributed Learning System (DLS)

Elbit Systems of America
Bradley Fighting Vehicle Systems Upgrade
Kiowa Warrior
Mortar Systems

EMC2
Distributed Common Ground System– Army (DCGS-A)

Emergent BioSolutions (Bioport)
Chemical Biological Medical Systems (CBMS)–Prophylaxis

EMS Safety Services, Inc.
Medical Simulation Training Center (MSTC)

Engility (formerly L3 MPRI)
Army Enterprise Systems Integration
  Program (AESIP)
Distributed Learning System (DLS)
Sustainment System Mission Command
  (SSMC)
Transportation Coordinators' – Automated
  Information for Movements System II
  (TC-AIMS II)

**Engineer Research and Development
Center**
Transportation Coordinators' – Automated
  Information for Movements System II
  (TC-AIMS II)

**Engineering Solutions and Products
(ESP)**
Common Hardware Systems (CHS)
Force XXI Battle Command Brigade and
  Below (FBCB2)
Global Command and Control System–
  Army (GCCS-A)
Movement Tracking System (MTS)
Warfighter Information Network–Tactical
  (WIN-T) Increment 1
eScience & Technology Solutions Inc.
Distributed Learning System (DLS)
Joint Land Attack Cruise Missile Defense
  Elevated Netted Sensor System (JLENS)

Esri
Distributed Common Ground System–
  Army (DCGS-A)

Fabrique National Manufacturing, LLC
Small Arms–Crew Served Weapons

FBM Babcock Marine
Improved Ribbon Bridge (IRB)

Firehouse Medical Inc.
Medical Simulation Training Center
  (MSTC)

Flight Suits
Air Soldier System (Air SS)

FLIR Systems
Chemical, Biological, Radiological, Nuclear
  Dismounted Reconnaissance Sets, Kits,
  and Outfits (CBRN DR SKO)

Fluke Corp.
Calibration Sets Equipment (CALSETS)

Future Research Corp.
Transportation Coordinators' – Automated
  Information for Movements System II
  (TC-AIMS II)

**General Atomics, Aeronautical Systems
Inc.**
MQ-1C Gray Eagle Unmanned Aircraft
  System (UAS)

General Dynamics
2.75-Inch Rocket Systems (Hydra-70)
Advanced Field Artillery Tactical Data
  System (AFATDS)
Common Hardware Systems (CHS)
Counter-Rocket, Artillery, Mortar (C-RAM)/
  Indirect Fire Protection Capability (IFPC)
Forward Area Air Defense Command and
  Control (FAAD C2)
Global Command and Control System–
  Army (GCCS-A)

Joint Tactical Radio System Handheld,
  Manpack, Small Form Fit (JTRS HMS)
Prophet
RQ-7B Shadow Tactical Unmanned
  Aircraft System (TUAS)
Small Caliber Ammunition
Stryker Family of Vehicles
Sustainment System Mission Command
  (SSMC)
Tactical Mission Command (TMC)/
  Maneuver Control System (MCS)
Warfighter Information Network–Tactical
  (WIN-T) Increment 2
Warfighter Information Network–Tactical
  (WIN-T) Increment 3
XM1216 & XM1216 E1 Small Unmanned
  Ground System (SUGV)

**General Dynamics Armament and
Technical Products (GDATP)**
Joint Biological Point Detection System
  (JBPDS)
Small Arms–Crew Served Weapons

General Dynamics C4 Systems, Inc.
Air Warrior (AW)
Airborne and Maritime/Fixed Station Joint
  Tactical Radio System (AMF JTRS)
Command Post Systems and Integration
  (CPS&I) Standardized Integrated
  Command Post Systems (SICPS)
Harbormaster Command and Control
  Center (HCCC)
Joint Battle Command–Platform (JBC-P)
Mortar Systems
Warfighter Information Network–Tactical
  (WIN-T) Increment 1

General Dynamics Canada
Lightweight 155mm Howitzer System
  (LW155)

**General Dynamics Communication
Systems**
Cryptographic Systems

**General Dynamics European Land
Systems–Germany**
Improved Ribbon Bridge (IRB)

**General Dynamics Information
Technology (GDIT)**
Medical Communications for Combat
  Casualty Care (MC4)

General Dynamics Land Systems
Abrams Tank Upgrade
Nuclear Biological Chemical
  Reconnaissance Vehicle (NBCRV)–
  Stryker Sensor Suites
Stryker Family of Vehicles

**General Dynamics Land Systems-
Canada**
Mine Protection Vehicle Family (MPVF),
  Area Mine Clearing System (AMCS),
  Interrogation Arm
Stryker Family of Vehicles

**General Dynamics Network
Systems, Inc.**
Installation Information Infrastructure
  Modernization Program (I3MP)

**General Dynamics Ordnance and
Tactical Systems**
2.75-Inch Rocket Systems (Hydra-70)

Artillery Ammunition
Excalibur (M982)
Javelin
Medium Caliber Ammunition (MCA)
Tank Ammunition

**General Dynamics Ordnance and
Tactical Systems-Scranton Operations**
Artillery Ammunition

**General Dynamics SATCOM
Technologies**
Warfighter Information Network–Tactical
(WIN-T) Increment 1

General Electric
Black Hawk/UH/HH-60

General Transmissions Products
High Mobility Multipurpose Wheeled
Vehicle (HMMWV) Recapitalization
(RECAP) Program

Genesis Concepts and Consultants LLC
Medical Simulation Training Center
(MSTC)

Gentex Corp.
Air Warrior (AW)

Georgia Tech Applied Research Corp.
Advanced Threat Infrared
Countermeasures (ATIRCM) and
Common Missile Warning System
(CMWS) programs and Pre-MDAP
Common Infrared Countermeasure
(CIRCM)

GEP
High Mobility Multipurpose Wheeled
Vehicle (HMMWV) Recapitalization
(RECAP) Program

Gibson and Barnes
Air Warrior (AW)

Gill Technology
Countermine

Goodrich
CH-47F Chinook

GP Strategies Corp.
Assembled Chemical Weapons
Alternatives (ACWA)

Group Home Foundation, Inc.
Joint Chem/Bio Coverall for Combat
Vehicle Crewman (JC3)

Gulfstream
Fixed Wing

H&K Gmbh
Counter Defilade Target Engagement
(CDTE)–XM25

Hamilton Sundstrand
Black Hawk/UH/HH-60

Harris Corp.
Cryptographic Systems
Defense Enterprise Wideband SATCOM
System (DEWSS)
High Mobility Artillery Rocket System
(HIMARS) M142

Multiple Launch Rocket System (MLRS)
M270A1
Warfighter Information Network–Tactical
(WIN-T) Increment 2
Warfighter Information Network–Tactical
(WIN-T) Increment 3

Hawker Beech Corp.
Enhanced Medium Altitude
Reconnaissance and Surveillance
System (EMARSS)
Fixed Wing

Heckler and Koch Defense Inc.
Small Arms–Individual Weapons

Helicopter Support, Inc.
Lakota/UH-72A

Honeywell
Abrams Tank Upgrade
M1200 Armored Knight
CH-47F Chinook
Guided Multiple Launch Rocket System
(GMLRS) DPICM/Unitary/Alternative
Warhead (Tactical Rockets)
Kiowa Warrior
Longbow Apache (AH-64D) (LBA)

Howmet Castings
Lightweight 155mm Howitzer System
(LW155)

HP
Distributed Common Ground System–
Army (DCGS-A)

Hunter Mfg.
Force Provider (FP)

IBM Global Business Services
Distributed Common Ground System–
Army (DCGS-A)
Distributed Learning System (DLS)
Integrated Personnel and Pay System–
Army (IPPS-A)
Sustainment System Mission Command
(SSMC)

iLuMinA Solutions Inc.
Army Enterprise Systems Integration
Program (AESIP)
General Fund Enterprise Business Systems
(GFEBS)

Independent Pipe Products
Javelin

Indigo System Corp.
RQ-11B Raven Small Unmanned Aircraft
System (SUAS)

Informatica
Distributed Common Ground System–
Army (DCGS-A)

Inmarsat Government
Combat Service Support Communications
(CSS Comms)

INTECON
Sustainment System Mission Command
(SSMC)

Integrated Global Solutions
Integrated Personnel and Pay System–
Army (IPPS-A)

Intelligent Decisions
Close Combat Tactical Trainer (CCTT)

Intercoastal Electronics
Improved Target Acquisition System (ITAS)

Intuitive Research and Technology
Medium Extended Air Defense System (MEADS)
PATRIOT Advanced Capability–3 (PAC-3)
iRobot Corp.
XM1216 & XM1216 E1 Small Unmanned Ground System (SUGV)

Ironbow
Computer Hardware, Enterprise Software and Solutions (CHESS)

ITT Corp.
Defense Enterprise Wideband SATCOM System (DEWSS)
Joint Tactical Radio System Network Enterprise Domain (JTRS NED)
Single Channel Ground and Airborne Radio System (SINCGARS)

ITT-Exelis
Helmet Mounted Night Vision Devices (HMNVD)

JANUS Research
Secure Mobile Anti-Jam Reliable Tactical Terminal (SMART-T)
Warfighter Information Network–Tactical (WIN-T) Increment 1

JB Management
AN/TPQ-53
Sustainment System Mission Command (SSMC)

JCB Inc.
High Mobility Engineer Excavator (HMEE) I and III

JLMI
Distributed Learning System (DLS)

Johns Hopkins University Applied Physics Laboratory
Defense Enterprise Wideband SATCOM System (DEWSS)

Joint Systems Manufacturing Center (JSMC)
Stryker Family of Vehicles.

Joint Technology Solutions, Inc.
Integrated Personnel and Pay System–Army (IPPS-A)

Juniper Networks
Combat Service Support Communications (CSS Comms)

Kalmar Rough Terrain Center (KRTC) LLC
Light Capability Rough Terrain Forklift (LCRTF)
Rough Terrain Container Handler (RTCH)

KDH Defense Systems
Interceptor Body Armor (IBA)

KForce Government Solutions
Medical Simulation Training Center (MSTC)

Kidde Dual Spectrum
Stryker Family of Vehicles

King Aerospace
Fixed Wing

Klune
Tube-Launched, Optically Tracked, Wire-Guided (TOW) Missiles

Knight's Armament Co.
Clip-on Sniper Night Sight (CoSNS), AN/PVS-30
Small Arms - Precision Weapons

Kongsberg Defense & Aerospace
Common Remotely Operated Weapon Station (CROWS)

L-3 Communications
Aviation Combined Arms Tactical Trainer (AVCATT)
Bradley Fighting Vehicle Systems Upgrade
Cryptographic Systems
Distributed Common Ground System–Army (DCGS-A)
Guardrail Common Sensor (GR/CS)
HELLFIRE Family of Missiles
Medical Communications for Combat Casualty Care (MC4)
MQ-1C Gray Eagle Unmanned Aircraft System (UAS)
RQ-11B Raven Small Unmanned Aircraft System (SUAS)
RQ-7B Shadow Tactical Unmanned Aircraft System (TUAS)
Secure Mobile Anti-Jam Reliable Tactical Terminal (SMART-T)
Tank Ammunition
Warfighter Information Network–Tactical (WIN-T) Increment 2
Warfighter Information Network–Tactical (WIN-T) Increment 3

L-3 Communications East
Force Protection Systems

L-3 Communications Electro-Optic Systems
Helmet Mounted Night Vision Devices (HMNVD)

L-3 Communications Space & Navigation
Multiple Launch Rocket System (MLRS) M270A1

L-3 Communications West
Enhanced Medium Altitude Reconnaissance and Surveillance System (EMARSS)

L-3 Communications/Brashear
Counter Defilade Target Engagement (CDTE)–XM25

L-3 CyTerra Corp.
Countermine

L-3 Global Communications Solutions Inc.
Combat Service Support Communications (CSS Comms)

L-3 Insight
Helmet Mounted Night Vision Devices (HMNVD)

L-3 Interstate Electronics Corp.
Precision Guidance Kit (PGK)

L-3 Vertex
Fixed Wing

L-3 Warrior Systems
Joint Effects Targeting System (JETS)
  Target Location Designation System
  (TLDS)

L-3 Westwood
Tactical Electric Power (TEP)

Letterkenny Army Depot
Force Provider (FP)
High Mobility Multipurpose Wheeled
  Vehicle (HMMWV) Recapitalization
  (RECAP) Program
Improvised Explosive Device Defeat/
  Protect Force (IEDD/PF)

Lex Products Corp.
Force Provider (FP)

Lincoln Labs
Secure Mobile Anti-Jam Reliable Tactical
  Terminal (SMART-T)

Linquest Corp.
Secure Mobile Anti-Jam Reliable Tactical
  Terminal (SMART-T)

Litton Advanced Systems
Airborne Reconnaissance Low (ARL)

LMI
Sustainment System Mission Command
  (SSMC)
Transportation Coordinators' – Automated
  Information for Movements System II
  (TC-AIMS II)

Lockheed Martin
Airborne and Maritime/Fixed Station Joint
  Tactical Radio System (AMF JTRS)
Airborne Reconnaissance Low (ARL)
Distributed Learning System (DLS)
Enhanced Medium Altitude
  Reconnaissance and Surveillance
  System (EMARSS)
AN/TPQ-53
Global Command and Control System–
  Army (GCCS-A)
Guardrail Common Sensor (GR/CS)
Guided Multiple Launch Rocket System
  (GMLRS) DPICM/Unitary/Alternative
  Warhead (Tactical Rockets)
HELLFIRE Family of Missiles
High Mobility Artillery Rocket System
  (HIMARS) M142
Instrumentable–Multiple Integrated Laser
  Engagement System (I-MILES)
Javelin
Joint Air-to-Ground Missile (JAGM)
Joint Light Tactical Vehicle (JLTV)
Joint Tactical Radio System Network
  Enterprise Domain (JTRS NED)
Longbow Apache (AH-64D) (LBA)
Medium Extended Air Defense System
  (MEADS)
Multiple Launch Rocket System (MLRS)
  M270A1
One Semi-Automated Force (OneSAF)
PATRIOT Advanced Capability–3 (PAC-3)
Tactical Mission Command (TMC)/
  Maneuver Control System (MCS)
Warfighter Information Network–Tactical
  (WIN-T) Increment 2
Warfighter Information Network–Tactical
  (WIN-T) Increment 3

**Lockheed Martin Global Training and
Logistics**
Close Combat Tactical Trainer (CCTT)
Joint Land Component Constructive
  Training Capability (JLCCTC)
Lockheed Martin Missiles and Fire Control
Guided Multiple Launch Rocket System
  (GMLRS) DPICM/Unitary/Alternative
  Warhead (Tactical Rockets)
PATRIOT Advanced Capability–3 (PAC-3)

Logistics Management Institute (LMI)
Army Enterprise Systems Integration
  Program (AESIP)

Longbow LLC
Longbow Apache (AH-64D) (LBA)

LTI DataComm Inc.
Combat Service Support Communications
  (CSS Comms)

M-7 Aerospace
Fixed Wing

MacAulay-Brown Inc.
Advanced Threat Infrared
  Countermeasures (ATIRCM) and
  Common Missile Warning System
  (CMWS) programs and Pre-MDAP
  Common Infrared Countermeasure
  (CIRCM)

Mainstream Engineering
Improved Environmental Control Unit
  (IECU)

MANTECH
Army Key Management System (AKMS)

Force XXI Battle Command Brigade and
  Below (FBCB2)
Tactical Mission Command (TMC)/
  Maneuver Control System (MCS)

Mantech Sensors Technology, Inc.
Meteorological Measuring Set–Profiler
  (MMS-P)/Computer Meteorological
  Data–Profiler (CMD-P)

Marsh Industrial
Force Provider (FP)

Maryn Consulting
Integrated Personnel and Pay System–
  Army (IPPS-A)

MaTech
Mortar Systems

McAlester Army Ammunition Plant
Artillery Ammunition

MEADS, Intl.
Medium Extended Air Defense System
  (MEADS)

Med-Eng Systems Inc.
Air Warrior (AW)

Medical Training Consultants, Inc.
Medical Simulation Training Center
  (MSTC)
Meridian Medical Technologies
Chemical Biological Medical Systems
  (CBMS)–Therapeutics

Meritor
Family of Medium Tactical Vehicles
(FMTV)
Line Haul Tractor

Michelin
Heavy Expanded Mobility Tactical Truck
(HEMTT)/HEMTT Extended Service
Program (ESP)
Palletized Load System (PLS) and PLS
Extended Service Program (ESP)

Microsoft
Computer Hardware, Enterprise Software
and Solutions (CHESS)

Mil-Mar Century, Inc.
Load Handling System Compatible Water
Tank Rack (Hippo)

MILVETS Systems Technology Inc.
Distributed Learning System (DLS)

MITRE
Army Enterprise Systems Integration
Program (AESIP)
Biometric Enabling Capability (BEC)
Distributed Common Ground System–
Army (DCGS-A)
Enhanced Medium Altitude
Reconnaissance and Surveillance
System (EMARSS)
Joint Battle Command–Platform (JBC-P)

Moog Inc.
HELLFIRE Family of Missiles
Tube-Launched, Optically Tracked, Wire-
Guided (TOW) Missiles

Moulage Sciences and Training, LLC
Medical Simulation Training Center
(MSTC)
Mountain High Equipment and Supply Co.
Air Warrior (AW)

**Naval Air Warfare Center Aircraft
Division**
Unified Command Suite (UCS)

Navistar Defense
Mine Protection Vehicle Family (MPVF),
Area Mine Clearing System (AMCS),
Interrogation Arm

NetApp
Distributed Common Ground System–
Army (DCGS-A)

NetCentric
Distributed Common Ground System–
Army (DCGS-A)

NextiraOne Federal LLC
Installation Information Infrastructure
Modernization Program (I3MP)

NIITEK
Countermine

N-Link
Distributed Learning System (DLS)

North American Controls
Stryker Family of Vehicles

Northrop Grumman
Advanced Field Artillery Tactical Data
System (AFATDS)

Advanced Threat Infrared
Countermeasures (ATIRCM) and
Common Missile Warning System
(CMWS) programs and Pre-MDAP
Common Infrared Countermeasure
(CIRCM)
Air/Missile Defense Planning and Control
System (AMDPCS)
Airborne and Maritime/Fixed Station Joint
Tactical Radio System (AMF JTRS)
Army Integrated Air and Missile Defense
(AIAMD)
Command Post Systems and Integration
(CPS&I) Standardized Integrated
Command Post Systems (SICPS)
Common Hardware Systems (CHS)
Counter-Rocket, Artillery, Mortar (C-RAM)/
Indirect Fire Protection Capability (IFPC)
Defense Enterprise Wideband SATCOM
System (DEWSS)
Force XXI Battle Command Brigade and
Below (FBCB2)
Guardrail Common Sensor (GR/CS)
Harbormaster Command and Control
Center (HCCC)
Integrated Family of Test Equipment (IFTE)
Joint Land Attack Cruise Missile Defense
Elevated Netted Sensor System (JLENS)
Joint Tactical Radio System Ground Mobile
Radios (JTRS GMR)
Longbow Apache (AH-64D) (LBA)
Movement Tracking System (MTS)
One Semi-Automated Force (OneSAF)

Northrop Grumman Electronic Systems
Joint Tactical Ground Station (JTAGS)

**Northrop Grumman Guidance and
Electronics, Laser Systems**
Joint Effects Targeting System (JETS)
Target Location Designation System
(TLDS)
Lightweight Laser Designator Rangefinder
(LLDR) AN/PED-1 & AN/PED-1A

**Northrop Grumman Information
Systems**
Biometric Enabling Capability (BEC)

**Northrop Grumman Information
Technology**
Joint Warning and Reporting Network
(JWARN)

Northrop Grumman Mission Systems
Joint Effects Model (JEM)

**Northrop Grumman Space and Mission
Systems Corp.**
Forward Area Air Defense Command and
Control (FAAD C2)

Northrop Grumman Technical Services
Fixed Wing

Oakland Consulting
Army Enterprise Systems Integration
Program (AESIP)

Olin Corp.
Small Caliber Ammunition

Oracle
Computer Hardware, Enterprise Software
and Solutions (CHESS)

Integrated Personnel and Pay System–
Army (IPPS-A)

Oshkosh Corp.
Dry Support Bridge (DSB)
Family of Medium Tactical Vehicles
(FMTV)
Heavy Expanded Mobility Tactical Truck
(HEMTT)/HEMTT Extended Service
Program (ESP)
Improved Ribbon Bridge (IRB)
Joint Light Tactical Vehicle (JLTV)
Mine Protection Vehicle Family (MPVF),
Area Mine Clearing System (AMCS),
Interrogation Arm
Palletized Load System (PLS) and PLS
Extended Service Program (ESP)

Osiris Therapeutics
Chemical Biological Medical Systems
(CBMS)–Therapeutics

OverWatch Systems
Distributed Common Ground System–
Army (DCGS-A)

**Oxygen Generating Systems
International**
Air Warrior (AW)

Parsons Government Services, Inc.
Assembled Chemical Weapons
Alternatives (ACWA)

Pennsylvania State University
Meteorological Measuring Set–Profiler
(MMS-P)/Computer Meteorological
Data–Profiler (CMD-P)

Physical Optics Corp.
Air Soldier System (Air SS)

Pixia
Distributed Common Ground System–
Army (DCGS-A)

PKMM
Forward Area Air Defense Command and
Control (FAAD C2)

Potomac Fusion
Distributed Common Ground System–
Army (DCGS-A)

Precision CastParts Corp.
Lightweight 155mm Howitzer System
(LW155)

Prototype Integration Facility
Force XXI Battle Command Brigade and
Below (FBCB2)

PULAU Corp.
Medical Simulation Training Center
(MSTC)

Quantum 3D
Close Combat Tactical Trainer (CCTT)

Rapiscan Systems
Non-Intrusive Inspection Systems (NIIS)

Raytheon
Advanced Field Artillery Tactical Data
System (AFATDS)
Air Soldier System (Air SS)
Air Warrior (AW)

Airborne and Maritime/Fixed Station Joint
Tactical Radio System (AMF JTRS)
M1200 Armored Knight
Army Integrated Air and Missile Defense
(AIAMD)
Bradley Fighting Vehicle Systems Upgrade
Distributed Common Ground System–
Army (DCGS-A)
Excalibur (M982)
Helmet Mounted Night Vision Devices
(HMNVD)
Improved Target Acquisition System (ITAS)
Javelin
Joint Air-to-Ground Missile (JAGM)
Joint Land Attack Cruise Missile Defense
Elevated Netted Sensor System (JLENS)
Kiowa Warrior
Night Vision Thermal Systems–Thermal
Weapon Sight (TWS)
PATRIOT Advanced Capability–3 (PAC-3)
RQ-7B Shadow Tactical Unmanned
Aircraft System (TUAS)
Secure Mobile Anti-Jam Reliable Tactical
Terminal (SMART-T)
Stryker Family of Vehicles
Sustainment System Mission Command
(SSMC)

Raytheon Missile Systems
Counter-Rocket, Artillery, Mortar (C-RAM)/
Indirect Fire Protection Capability (IFPC)
Javelin
Tube-Launched, Optically Tracked, Wire-
Guided (TOW) Missiles

Raytheon Solipsys
Joint Land Attack Cruise Missile Defense
Elevated Netted Sensor System (JLENS)

Raytheon Technical Services
Improvised Explosive Device Defeat/
Protect Force (IEDD/PF)

ReadyOne Industries
Joint Chem/Bio Coverall for Combat
Vehicle Crewman (JC3)

Red River Army Depot
High Mobility Multipurpose Wheeled
Vehicle (HMMWV) Recapitalization
(RECAP) Program

Redhat
Distributed Common Ground System–
Army (DCGS-A)

Remington Arms Co. Inc.
Small Arms - Precision Weapons

Ringtail Design
Distributed Common Ground System–
Army (DCGS-A)

Rini Corp.
Air Soldier System (Air SS)

Rock Island Arsenal
Stryker Family of Vehicles

Rockwell Collins
Advanced Threat Infrared
Countermeasures (ATIRCM) and
Common Missile Warning System
(CMWS) programs and Pre-MDAP
Common Infrared Countermeasure
(CIRCM)
Black Hawk/UH/HH-60
CH-47F Chinook

Close Combat Tactical Trainer (CCTT)

Enhanced Medium Altitude Reconnaissance and Surveillance System (EMARSS)

Joint Tactical Radio System Ground Mobile Radios (JTRS GMR)

Joint Tactical Radio System Handheld, Manpack, Small Form Fit (JTRS HMS)

Joint Tactical Radio System Network Enterprise Domain (JTRS NED)

NAVSTAR Global Positioning System (GPS)

RQ-7B Shadow Tactical Unmanned Aircraft System (TUAS)

Rolls Royce Corp.
Kiowa Warrior

SAAB Training USA LOC
Instrumentable–Multiple Integrated Laser Engagement System (I-MILES)

Saba
Distributed Learning System (DLS)

SAP
Army Enterprise Systems Integration Program (AESIP)

SCI Technology, Inc.
Command Post Systems and Integration (CPS&I) Standardized Integrated Command Post Systems (SICPS)
Harbormaster Command and Control Center (HCCC)

Science and Engineering Services Inc. (SESI)
Air Warrior (AW)

Science Applications International Corp. (SAIC)
Advanced Threat Infrared Countermeasures (ATIRCM) and Common Missile Warning System (CMWS) programs and Pre-MDAP Common Infrared Countermeasure (CIRCM)
Anti-Personnel Mine Clearing System, Remote Control M160
Army Key Management System (AKMS)
Aviation Combined Arms Tactical Trainer (AVCATT)
Enhanced Medium Altitude Reconnaissance and Surveillance System (EMARSS)
Installation Information Infrastructure Modernization Program (I3MP)
Joint Tactical Radio System Handheld, Manpack, Small Form Fit (JTRS HMS)
Non-Intrusive Inspection Systems (NIIS)
One Semi-Automated Force (OneSAF)

Secure Communication Systems Inc.
Air Warrior (AW)

Seiler Instruments
Lightweight 155mm Howitzer System (LW155)

Sensor Technologies
Common Hardware Systems (CHS)
Enhanced Medium Altitude Reconnaissance and Surveillance System (EMARSS)

Sierra Nevada Corp.
Airborne Reconnaissance Low (ARL)
Army Key Management System (AKMS)

Sigmatech, Inc.
Command Post Systems and Integration (CPS&I) Standardized Integrated Command Post Systems (SICPS)
Harbormaster Command and Control Center (HCCC)

Sikorsky
Black Hawk/UH/HH-60

SKEDCO, Inc.
Medical Simulation Training Center (MSTC)

Skillsoft Corp.
Distributed Learning System (DLS)

Smiths
Longbow Apache (AH-64D) (LBA)

Smiths Detection, Inc.
Chemical Biological Protective Shelter (CBPS) M8E1
Joint Chemical Agent Detector (JCAD) M4A1
Meteorological Measuring Set–Profiler (MMS-P)/Computer Meteorological Data–Profiler (CMD-P)

Software Engineering Center-Belvoir (SEC-B)
Combat Service Support Communications (CSS Comms)

Software Engineering Directorate, AMRDEC
Joint Battle Command–Platform (JBC-P)

Software Professional Solutions
Sustainment System Mission Command (SSMC)

Sotera Defense Systems
Force Provider (FP)

Southwest Research Institute
Chemical Biological Medical Systems (CBMS)–Therapeutics

SPAWAR Pacific
Harbormaster Command and Control Center (HCCC)

Spectrum Microwave
Secure Mobile Anti-Jam Reliable Tactical Terminal (SMART-T)

SRA International
Integrated Personnel and Pay System–Army (IPPS-A)
Joint Tactical Radio System Network Enterprise Domain (JTRS NED)

SRCTec Inc.
Counter-Rocket, Artillery, Mortar (C-RAM)/Indirect Fire Protection Capability (IFPC)
Lightweight Counter Mortar Radar (LCMR)

StackFrame
One Semi-Automated Force (OneSAF)

Susan L Berger, LLC
Integrated Personnel and Pay System–Army (IPPS-A)

Switlik Parachute Co.
Air Soldier System (Air SS)

Syracuse Research Corp. (SRC)
Lightweight Counter Mortar Radar (LCMR)

Systems Technologies (Systek) Inc.
Combat Service Support Communications
(CSS Comms)

Systems, Studies, and Simulation
Guided Multiple Launch Rocket System
(GMLRS) DPICM/Unitary/Alternative
Warhead (Tactical Rockets)

**Tank-Automotive and Armaments
Command (TACOM)**
M109 Family of Vehicles (FOV) (Paladin/
FAASV, PIM SPH/CAT)

Tapestry
Sustainment System Mission Command
(SSMC)

Tapestry Solutions Inc.
Joint Land Component Constructive
Training Capability (JLCCTC)

Taylor-Wharton
Air Warrior (AW)

TCOM
Joint Land Attack Cruise Missile Defense
Elevated Netted Sensor System (JLENS)

Tech Flow
Sustainment System Mission Command
(SSMC)

Technology Management Group (TMG)
Joint Personnel Identification version 2
(JPIv2)

Tecom
RQ-7B Shadow Tactical Unmanned
Aircraft System (TUAS)

Teledyne
Secure Mobile Anti-Jam Reliable Tactical
Terminal (SMART-T)

Telephonics Corp.
Air Warrior (AW)

Tel-Instrument Electronics Corp.
Test Equipment Modernization (TEMOD)

Telos Corp.
Combat Service Support Communications
(CSS Comms)

Textron Defense Systems
XM7 Spider
Textron Marine & Land Systems
M1200 Armored Knight

Thales Communications
Joint Tactical Radio System Handheld,
Manpack, Small Form Fit (JTRS HMS)
Tube-Launched, Optically Tracked, Wire-
Guided (TOW) Missiles

Thales Raytheon Systems
Sentinel

The Research Associates (TRA)
Biometric Enabling Capability (BEC)
Joint Personnel Identification version 2
(JPIv2)

Tobyhanna Army Depot
Combat Service Support Communications
(CSS Comms)
Command Post Systems and Integration
(CPS&I) Standardized Integrated
Command Post Systems (SICPS)
Forward Area Air Defense Command and
Control (FAAD C2)
Harbormaster Command and Control
Center (HCCC)
Lightweight Counter Mortar Radar (LCMR)

Transtector Systems
Secure Mobile Anti-Jam Reliable Tactical
Terminal (SMART-T)

Trioh Consulting
Integrated Personnel and Pay System–
Army (IPPS-A)

Tri-Tech USA Inc.
Force Provider (FP)

Triumph Structures
Lightweight 155mm Howitzer System
(LW155)

Tucson Embedded Systems
Distributed Common Ground System–
Army (DCGS-A)

**U.S. Army Information Systems
Engineering Command (USAISEC)**
Combat Service Support Communications
(CSS Comms)

Ultimate Training Munitions
Small Caliber Ammunition

Ultra, Inc.
Air/Missile Defense Planning and Control
System (AMDPCS)

URS Corp.
Assault Breacher Vehicle (ABV)
Assembled Chemical Weapons
Alternatives (ACWA)
Force Protection Systems

U.S. Divers
Air Warrior (AW)

USFalcon/EPS Corp.
Secure Mobile Anti-Jam Reliable Tactical
Terminal (SMART-T)
Single Channel Ground and Airborne Radio
System (SINCGARS)

Vallon GmbH
Countermine

Vertigo Inc.
Force Provider (FP)

Vertu Corp.
Small Arms–Individual Weapons

VIA SAT
Cryptographic Systems

ViaSat Inc.
Force XXI Battle Command Brigade and
Below (FBCB2)

Viking Air Limited
Fixed Wing

Vision Technology Miltope Corp.
Integrated Family of Test Equipment (IFTE)

Vmware
Distributed Common Ground System–
   Army (DCGS-A)

Watervliet Arsenal
Lightweight 155mm Howitzer System
   (LW155)
Mortar Systems

Websec Corp.
Army Key Management System (AKMS)

WESCAM
Airborne Reconnaissance Low (ARL)

Westwind Technologies Inc.
Air Warrior (AW)

Williams Fairey Engineering, Ltd.
Dry Support Bridge (DSB)

**Wolf Coach, Inc., an L-3
Communications Co.**
Unified Command Suite (UCS)

World Wide Technology
Computer Hardware, Enterprise Software
   and Solutions (CHESS)

XMCO
Assault Breacher Vehicle (ABV)
Dry Support Bridge (DSB)
Heavy Loader
High Mobility Engineer Excavator (HMEE)
   I and III

Yuma Test Center/Proving Ground
Lightweight Counter Mortar Radar (LCMR)
M109 Family of Vehicles (FOV) (Paladin/
   FAASV, PIM SPH/CAT)

ZETA
Guardrail Common Sensor (GR/CS)

Points of Contact

120M Motor Grader
Program Executive Office (PEO) Combat
Support and Combat Service Support
(CS&CSS)
Product Manager, Combat Engineer/
Material Handling Equipment
SFAE-CSS-FP-C
6501 E. 11 Mile Rd.
Mail Stop 401
Warren, MI 48397-5000

2.75 Inch Rocket Systems (Hydra-70)
PEO Missiles and Space
JAMS Project Office
SFAE-MSLS-JAMS
5250 Martin Road
Redstone Arsenal, AL 35898-8000

Abrams Tank Upgrade
PEO Ground Combat Systems (GCS)
HBCT
SFAE-GCS-CS-A
6501 E. 11 Mile Rd.
Warren, MI 48397-5000

Advanced Field Artillery Tactical Data System (AFATDS)
PEO Command, Control and
Communications-Tactical (C3T)
Product Director, Fire Support Command
and Control
SFAE-C3T-MC-FSC2
6007 Combat Drive
5th Floor
Aberdeen Proving Ground, MD 21005

Advanced Threat Infrared Countermeasures (ATIRCM) and Common Missile Warning System (CMWS) programs and Pre-MDAP Common Infrared Countermeasure (CIRCM)
PEO Intelligence Electronic Warfare &
Sensors (IEW&S)
Aircraft Survivability Equipment (ASE)
SFAE-IEW-ASE
6726 Odyssey Drive
Huntsville, AL 35806

Air Soldier System (Air SS)
PEO Soldier
Product Manager Air Warrior
SFAE-SDR-AW
6726 Odyssey Drive NW
Huntsville, AL 35806

Air Warrior (AW)
PEO Soldier
Product Manager Air Warrior
SFAE-SDR-AW
6726 Odyssey Drive NW
Huntsville, AL 35806

Air/Missile Defense Planning and Control System (AMDPCS)
PEO C3T
Counter-Rocket, Artillery, Mortar
(C-RAM) Program Directorate
SFAE-C3T-CR
Building 5250
Redstone Arsenal, AL 35898-5000

Airborne and Maritime/Fixed Station Joint Tactical Radio System (AMF JTRS)
PEO-C3T
Bldg 6007
Aberdeen Proving Ground, MD 21005

Airborne Reconnaissance Low (ARL)
PEO IEW&S
PM Airborne Reconnaissance and
Exploitation Systems (ARES)
SFAE-IEW-ARS
Building 6006, Room B1-125
Combat Drive
Aberdeen Proving Ground , MD 21005

Anti-Personnel Mine Clearing System, Remote Control M160
PEO GCS
SFAE-GCS
6501 E 11 Mile Rd.
Warren, MI

AN/TPQ-53
PEO IEW&S
LTC Robert Thomas, PM Radars
SFAE-IEW&S
6001 Combat Drive
Aberdeen Proving Ground, MD 21010

Armored Multi-Purpose Vehicle (AMPV)
Program Manager, Heavy Brigade
Combat Team
PEO GCS
6501 E. 11 Mile Rd.
Mail Stop 463
Warren, MI 48397-5000

Army Integrated Air and Missile Defense (AIAMD)
PEO Missiles and Space
Integrated Air and Missile Defense
Project Office
SFAE-MSLS-IAMD
Building 5250
Redstone Arsenal, AL 35898-8000

Army Key Management System (AKMS)
PEO C3T
Project Director, Communications
Security (PD COMSEC)
SFAE-C3T-COMSEC
6007 Combat Drive, 5th Floor
Aberdeen Proving Ground, MD 21005

Artillery Ammunition
PEO Ammunition
PM Combat Ammunition Systems
SFAE-AMO-CAS
Picatinny Arsenal, NJ 07806

Assault Breacher Vehicle (ABV)
PEO CS&CSS
PM Bridging Systems
SFAE-CSS-FP-H
6501 E. 11 Mile Rd.
Warren, MI 43897-5000

Assembled Chemical Weapons Alternatives
ATTN: SAAL-SFAE
5183 Blackhawk Road
APG-EA, MD 21010-5424

Aviation Combined Arms Tactical Trainer (AVCATT)
PEO for Simulation, Training and Instrumentation (STRI)
Project Manager Combined Arms Tactical Trainers
SFAE-STRI-PMCATT
12350 Research Parkway
Orlando, FL 32826-3276

Biometric Enabling Capability (BEC)
PEO Enterprise Information Systems (EIS)
PM DoD Biometrics
SFAE-PS-BI
9350 Hall Road
Building 1445
Ft. Belvoir, VA 22060

Black Hawk/UH/HH-60
PEO Aviation
Utility Helicopters Project Office (UHPO)
SFAE-AV-UH
Bldg 5308
Redstone Arsenal, AL 35898

Bradley Fighting Vehicle Systems Upgrade
PEO GCS
Program Manager, Heavy Brigade Combat Team
SFAE-GCS-CS-A
6501 E. 11 Mile Rd.
Warren, MI 48397-5000

Calibration Sets Equipment (CALSETS)
PEO CS&CSS
Test, Measurement, and Diagnostic Equipment Product Director
SFAE-CSS-JC-TM
Building 3651
Redstone Arsenal, AL 35898

Capability Set 13 (CS 13)
System of Systems Integration (SoSI) Directorate
6007 Combat Drive, Bldg. 6007
Aberdeen Proving Ground, MD 005-5001

CH-47F Chinook
PEO Aviation
SFAE-AV-CH-ICH
Building 5678
Redstone Arsenal, AL 35898

Chemical Biological Medical Systems (CBMS)–Prophylaxis
Joint Project Manager Chemical Biological Medical Systems
JPM CBMS
1564 Freedman Dr
Ft Detrick, MD 21702

Chemical Biological Medical Systems–Therapeutics
Joint Project Manager Chemical Biological Medical Systems
JPM CBMS
1564 Freedman Dr
Ft Detrick, MD 21702

Chemical Biological Medical Systems–Diagnostics
Joint Project Manager Chemical Biological Medical Systems
JPM CBMS
1564 Freedman Dr
Ft. Detrick, MD 21702

Chemical Biological Protective Shelter (CBPS) M8E1
Joint Project Manager Protection
JPM P
Suite 301
50 Tech Parkway
Stafford, VA 22556

Chemical, Biological, Radiological, Nuclear Dismounted Reconnaissance Sets, Kits, and Outfits (CBRN DR SKO)
JPEO CBD
Joint Project Manager NBC Contamination Avoidance
SFAE-CBD-NBC-R
Building 2800
5183 Blackhawk Road
Aberdeen Proving Ground, MD 21010-5425

Clip-on Sniper Night Sight (CoSNS), AN/PVS-30
PEO Soldier
PM Soldier Sensors and Lasers
SFAE-SDR-SW
10170 Beach Road
Building 325
Ft. Belvoir, VA 22060

Close Combat Tactical Trainer (CCTT)
PEO STRI
Project Manager Combined Arms Tactical Trainers
SFAE-STRI-PMCATT
12350 Research Parkway
Orlando, FL 32826-3276

Combat Service Support Communications (CSS Comms)
PEO EIS
PM Defense Communications and Army Transmission Systems
SFAE-PS-TS
9350 Hall Road
Building 1445
Ft. Belvoir, VA 22060

Command Post Systems and Integration (CPS&I) Standardized Integrated Command Post Systems (SICPS)
PEO C3T
Product Manager Command Post Systems & Integration (PdM CPS&I)
SFAE-C3T-WIN-CPSI
Redstone Arsenal, AL 35898

Common Hardware Systems (CHS)
PEO C3T
Product Director Common Hardware Systems (PD-CHS)
SFAE-C3T-WIN-CHS
Building 6007
Aberdeen Proving Ground, MD 21005

Common Remotely Operated Weapon Station (CROWS)
PEO Soldier
Project Manager Soldier Weapons
SFAE-SDR-SW
Building 151
Picatinny Arsenal, NJ 07806

Computer Hardware, Enterprise Software and Solutions (CHESS)
PEO EIS
PD Computer Hardware, Enterprise Software and Solutions
SFAE-PS-CH
9350 Hall Road
Building 1445
Ft. Belvoir, VA 22060

Counter Defilade Target Engagement (CDTE) - XM25
PEO Soldier
PEO Soldier Weapons
SFAE-SDR-SW
Picatinny Arsenal, NJ 07806

Counter-Rocket, Artillery, Mortar (C-RAM) / Indirect Fire Protection Capability (IFPC)
PEO C3T
C-RAM Program Directorate
SFAE-C3T-CR
Martin Road, Building 5250
Red Stone Arsenal, AL 35898

Countermine
PEO Ammunition
PM Close Combat Systems/PdM Countermine & EOD
SFAE-AMO-CCS
10205 Burbeck Road
Suite 100
Ft Belvoir, VA 22060-5811

Cryptographic Systems
PEO C3T
PD Cryptographic Systems
SFAE-CCC-COM
6007 Combat Drive
F5-148
APG, MD 21005

Defense Enterprise Wideband SATCOM System (DEWSS)
PEO EIS
PM Defense Communications and Army Transmission Systems
SFAE-PS-TS
9350 Hall Road
Building 1445
Ft. Belvoir, VA 22060

Distributed Common Ground System–Army (DCGS-A)
PEO IEW&S
PM Distributed Common Ground System–Army (DCGS-A)
SFAE-IEW-DCG
Building 6006
C5ISR Complex
Aberdeen Proving Ground, MD 21005-0001

Distributed Learning System (DLS)
PEO EIS
PD Distributed Learning System
SFAE-PS-DL
9350 Hall Road
Building 1445
Ft. Belvoir, VA 22060

Dry Support Bridge (DSB)
PEO CS&CSS
PM Bridging Systems
SFAE-CSS-FP-H
6501 East 11 Mile Rd.
Mail Stop 401
Warren, MI 43897-5000

Enhanced Medium Altitude Reconnaissance and Surveillance System (EMARSS)
PEO IEW&S
PdM Medium Altitude Reconnaissance and Surveillance Systems (MARSS)
SFAE-IEW-ARS
Building 6006, Room B1-125
Combat Drive
Aberdeen Proving Ground, MD 21005

Enterprise Email (EE)
PEO EIS
Product Director Enterprise Email
SFAE-PS-ES-EE
9350 Hall Road
Building 1445
Fort Belvoir, VA 22060

Excalibur (M982)
PEO Ammunition
PM Excalibur
SFAE-AMO-CAS-EX
Building 172
Buffington Road
Picatinny Arsenal, NJ 07806-5000

Family of Medium Tactical Vehicles (FMTV)
PEO CS&CSS
Product Manager–Medium Tactical Vehicles
SFAE-CSS
6501 E. 11 Mile Rd.
Warren, MI 43897-5000

Fixed Wing
PEO Aviation
DA Systems Coordinator–Fixed Wing
ASA (ALT) Aviation Directorate
SAAL-SAV
650 Discovery Dr
Huntsville, AL 35806

Force Protection Systems
Joint Program Executive Office (JPEO) for Chemical and Biological (CBD)
Joint Project Manager Guardian
SFAE-CBD-GN-F
E 2800 Bush River Road
Aberdeen Proving Ground , MD 21010-5424

Force Provider (FP)
PEO CS&CSS
PM Force Sustainment Systems
SFAE-CSS-FP-F
Kansas Street
Natick, MA 01760-5057

Force XXI Battle Command Brigade and Below (FBCB2)
PEO C3T
PM FBCB2
SFAE-C3T-FBC
6007 Combat Drive
4th Floor
Aberdeen Proving Ground, MD 21005-1846

Forward Area Air Defense Command and Control (FAAD C2)
PEO C3T
C-RAM Program Directorate
SFAE-C3T-CR
Building 5250
Redstone Arsenal, AL 35898-5000

General Fund Enterprise Business Systems (GFEBS)
PEO EIS
PM General Fund Enterprise Business System
SFAE-PS-GF
9350 Hall Road
Building 1445
Ft. Belvoir, VA 22060

Global Command and Control System–Army (GCCS–A)
PEO C3T
Product Manager, Strategic Mission Command
SFAE-C3T-MC-SMC
Building 6007
5th Floor
Aberdeen Proving Ground, MD 21005

Global Combat Support System–Army (GCSS–Army)
PEO EIS
PM GCSS-Army
SFAE-PS-GC
3811 Corporate Rd
Suite C
Petersburg, VA 23805

Ground Combat Vehicle (GCV)
PEO GCS
PM GCV
SFAE-GCS-GV
5500 Enterprise Dr.
Warren, MI 48043

Guardrail Common Sensor (GR/CS)
PEO IEW&S
PM Airborne Reconnaissance and Exploitation Systems (ARES)
SFAE-IEW-ARS
Building 6006
Combat Drive
Aberdeen Proving Ground, MD 21005

Guided Multiple Launch Rocket System (GMLRS) DPICM/Unitary/Alternative Warhead (Tactical Rockets)
PEO Missiles and Space
Precision Fires Rocket and Missile Systems
SFAE-MSLS-PF
Building 5250 Martin Road
Redstone Arsenal, AL 35898

Harbormaster Command and Control Center (HCCC)
PEO C3T
Product Manager Command Post Systems & Integration (PdM CPS&I)
SFAE-C3T-WIN-CPSI
Redstone Arsenal, AL 34898

Heavy Expanded Mobility Tactical Truck (HEMTT)/HEMTT Extended Service Program (ESP)
PEO CS&CSS
PM Heavy Tactical Vehicles
SFAE-CSS-TS-H
6501 E. 11 Mile Rd.
Mail Stop 429
Warren, MI 48397-5000

Heavy Loader
PEO CS&CSS
PM for Combat Engineer Materiel Handling Equipment
SFAE-CSS-FP-C
6501 E. 11 Mile Rd.
Warren, MI 48397-5000

HELLFIRE Family of Missiles
PEO Missiles and Space
JAMS Project Office
SFAE-MSLS-JAMS
5250 Martin Road
Redstone Arsenal, AL 35898

Helmet Mounted Night Vision Devices (HMNVD)
PEO Soldier
PM Soldier Sensors and Lasers
SFAE-SDR-SSL
10170 Beach Rd.
Building 325
Ft. Belvoir, VA 22060

High Mobility Artillery Rocket System (HIMARS) M142
Precision Fires Rocket and Missile Systems Project Office
SFAE-MSL-PF-FAL
Building 5250
Redstone Arsenal, AL 35898

High Mobility Engineer Excavator (HMEE) I and III
PEO CS&CSS
Product Manager Combat Engineer/MHE
SFAE-CSS-FP-C
6501 E. 11 Mile Rd
Warren, MI 48397-5000

High Mobility Multipurpose Wheeled Vehicle (HMMWV) Recapitalization (RECAP) Program
PEO CS&CSS
Product Director Light Tactical Vehicles
SFAE-CSS-JL-LT
6501 11 Mile Road MS 245
Warren, MI 43897

Improved Environmental Control Units (IECU)
PEO C3T
Project Manager Mobile Electric Power
SFAE-C3T-MEP-OPM
5850 Delafield Rd Bldg 324
Ft. Belvoir, VA 22060-5809

Improved Ribbon Bridge (IRB)
PEO CS&CSS
PM Bridging Systems
SFAE-CSS-FP-H
6501 E. 11 Mile Rd.
Mail Stop 401
Warren, MI 43897-5000

Improved Target Acquisition System (ITAS)
PEO Missiles and Space
PM Close Combat Weapon Systems
Project Office
SFAE-MSL-CWS-J
Redstone Arsenal, AL 35898

Improvised Explosive Device Defeat/ Protect Force (IEDD/PF)
PEO Ammunition
PM Close Combat Systems (CCS) /PdM
IED Defeat/Protect Force
SFAE-AMO-CCS
Bldg 183 Buffington Rd
Picatinny Arsenal, NJ 07806-5000

Installation Information Infrastructure Modernization Program (I3MP)
PEO EIS
PM Installation Information
Infrastructure Modernization Program
SFAE-PS-I3-MP
9350 Hall Road
Building 1445
Ft. Belvoir, VA 22060

Instrumentable–Multiple Integrated Laser Engagement System (I-MILES)
PEO STRI
Project Manager Training Devices
SFAE-STRI-PMTRADE
12350 Research Parkway
Orlando, FL 32826

Integrated Family of Test Equipment (IFTE)
PEO CS&CSS
Test, Measurement, and Diagnostic
Equipment Product Director
SFAE-CSS-JC-TM
Building 3651
Redstone Arsenal, AL 35898

Integrated Personnel and Pay System– Army (IPPS-A)
PEO EIS
PD Integrated Personnel and Pay
System - Army
SFAE-PS-IP
9350 Hall Road
Building 1445
Ft. Belvoir, VA 22060

Interceptor Body Armor (IBA)
PEO Soldier
Product Manager Soldier Protection and
Individual Equipment
SFAE-SDR-SPE
10170 Beach Rd.
Building 325
Ft. Belvoir, VA , VA 22060

Javelin
PEO Missiles and Space
PM Close Combat Weapon Systems
Project Office
SFAE-MSL-CWS-J
Redstone Arsenal, AL 35898

Joint Air-to-Ground Missile (JAGM)
PEO Missiles and Space
Joint Air to Ground Missile Product
Office
SFAE-MSL-JAMS-M
5250 Martin Rd.
Redstone Arsenal, AL 35898

Joint Battle Command–Platform (JBC-P)
PEO C3T
PM Force XXI Battle Command Brigade
and Below
SFAE-C3T-FBC
6007 Combat Drive
4th Floor
Aberdeen Proving Ground, MD 21005-1846

Joint Biological Point Detection System (JBPDS)
JPEO CBD
Joint Project Manager Biological
Defense
SFAE-CBD-BD-BDS
5183 Blackhawk Road
Aberdeen Proving Ground, MD 21010-5425

Joint Biological Tactical Detection System (JBTDS)
JPEO CBD
Joint Project Manager Biological
Defense
SFAE-CBD-BD-PD-FoS
5183 Blackhawk Rd (Bldg E3549)
Aberdeen Proving Grounds, MD 21010-5424

Joint Chem/Bio Coverall for Combat Vehicle Crewman (JC3)
JPEO CBD
Joint Project Manager Protection
JPM P
Suite 301
50 Tech Parkway
Stafford, VA 22556

Joint Chemical Agent Detector (JCAD) M4A1
JPEO CBD
Joint Project Manager NBC
Contamination Avoidance
SFAE-CBD-NBC
Building 2800
5183 Blackhawk Rd
Edgewood Area - Aberdeen Proving
Ground, MD 21010-5424

Joint Effects Model (JEM)
JPEO CBD
Joint Project Manager Information
System
JPM IS
4301 Pacific Highway
San Diego, CA 92110

Joint Effects Targeting System (JETS) Target Location Designation System (TLDS)
PEO Soldier
Project Manager Soldier Sensors and
Lasers
SFAE-SDR-SSL
10170 Beach Rd.
Building 325
Ft. Belvoir, VA 22060

Joint Land Attack Cruise Missile Defense Elevated Netted Sensor System (JLENS)
PEO Missiles and Space
SFAE-MSLS-CMDS-JLN
P.O. Box 1500
Huntsville, AL 35807

Joint Land Component Constructive Training Capability (JLCCTC)
PEO STRI
Project Manager Constructive Simulation
STRI-SFAE-PMCONSIM
12350 Research Parkway
Orlando, FL , FL 32826

Joint Light Tactical Vehicle (JLTV)
PEO CS&CSS
Joint Program Office Joint Light Tactical Vehicle (JLTV)
SFAE-CSS-JL
43087 Lake Street, NE
Building 301/2rd Floor/Mail Stop 640
Harrison Twp, MI 48045-4941

Joint Personnel Identification Version 2 (JPIv2)
PEO EIS
PM DoD Biometrics
SFAE-PS-BI
9350 Hall Road
Building 1445
Ft. Belvoir, VA 22060

Joint Precision Airdrop System (JPADS)
PEO CS&CSS
PM Force Sustainment Systems
SFAE-CSS-FP-F
Kansas St.
Natick, MA 01760-5057

Joint Service Aircrew Mask–Rotary Wing (JSAM RW) (MPU-5)
JPEO CBD
Joint Project Manager Protection
JPM P
50 Tech Parkway
Stafford, VA 22556

Joint Service General Purpose Mask (JSGPM) M-50/M-51
JPEO CBD
Joint Project Manager Protection
JPM P
Suite 301
50 Tech Parkway
Stafford, VA 22556

Joint Service Transportable Small Scale Decontaminating Apparatus (JSTSS DA) M26
JPEO CBD
Joint Project Manager Protection
JPM P
Suite 301
50 Tech Parkway
Stafford, VA 22556

Joint Tactical Ground Station (JTAGS)
PEO Missiles and Space
Lower Tier Project Office
SFAE-MSLS-LT
5250 Martin Road
Redstone Arsenal, AL 35898-8000

Joint Tactical Radio System Ground Mobile Radios (JTRS GMR)
PEO-C3T
Bldg 6007
Aberdeen Proving Ground, MD 21005

Joint Tactical Radio System Handheld, Manpack, Small Form Fit (JTRS HMS)
PEO-C3T
Bldg 6007
Aberdeen Proving Ground, MD 21005

Joint Tactical Radio System Network Enterprise Domain (JTRS NED)
TBD

Joint Warning and Reporting Network (JWARN)
JPEO CBD
Joint Project Manager Information System
JPM IS
4301 Pacific Hwy.
San Diego, CA 92110

Kiowa Warrior
PEO Aviation
COL Robert Grigsby
SFAE-AV-ASH
5681 Wood Rd.
Redstone Arsenal, AL , AL 35898

Korea Transformation, Yongsan Relocation Plan, Land Partnership Plan (KT/YRP/LPP)
PEO EIS
PD Korea Transformation/Yongsan Relocation Plan/Land Partnership Plan
SFAE-PS-I3-KT
9350 Hall Road
Building 1445
Ft. Belvoir, VA 22060

Lakota/UH-72A
PEO Aviation
PM Lakota
PEO AVN-UH-LUH
Lakota/UH-72A
Lakota PM
Huntsville, AL 35898-5000

Light Capability Rough Terrain Forklift (LCRTF)
PEO CS&CSS
Product Manager, Combat Engineer/ Material Handling Equipment
SFAE-CSS-FP-C
6501 E. 11 Mile Rd.
Mail Stop 401
Warren, MI 48397-5000

Lightweight 155mm Howitzer System (LW155)
PEO Ammunition
JPMO Towed Artillery Systems
SFAE-AMO-TAS
Building 151
Picatinny Arsenal, NJ 07806

Lightweight Counter Mortar Radar (LCMR)
PEO IEW&S
LTC Robert Thomas, PM RADARS
SFAE-IEW&S
6001 Combat Drive
Aberdeen Proving Ground, MD 21005

Lightweight Laser Designator Rangefinder (LLDR) AN/PED-1 & AN/ PED-1A
PM Soldier Sensors and Lasers
Soldier
10170 Beach Road
Building 325
Ft. Belvoir, VA 22060

Line Haul Tractor
PEO CS&CSS
PM Heavy Tactical Vehicles
SFAE-CSS-TV-H
Mail Stop 429
6501 E. 11 Mile Rd.
Warren, MI 48397-5000

Load Handling System Compatible Water Tank Rack (Hippo)
PEO CS&CSS
PM Petroleum and Water Systems
SFAE-CSS-FP-P
6501 E. 11 Mile Rd.
Mail Stop 111
Warren, MI 43897

Longbow Apache (AH-64D) (LBA)
PEO Aviation
PM Apache
SFAE-AV
Building 5681
Redstone Arsenal, AL 35898

M109 Family of Vehicles (FOV) (Paladin/FAASV, PIM SPH/CAT)
PEO GCS
HBCT
SFAE-GCS-HBCT
6501 E. 11 Mile Rd.
Warren, MI 48397

M1200 Armored Knight
PEO CS&CSS
Product Manager, Combat Engineer/
MHE
SFAE-CSS-FP-C
6501 E. 11 Mile Rd.
Mail Stop 401
Warren, MI 48397-5000

Medical Communications for Combat Casualty Care (MC4)
PM Medical Communications for Combat
Casualty Care (MC4)
SFAE-PS-MC
9350 Hall Road
Building 1445
Ft. Belvoir, VA 22060

Medical Simulation Training Center (MSTC)
PEO STRI
Project Manager Combined Arms
Tactical Trainers
SFAE-STRI-PMCATT
12350 Research Parkway
Orlando, FL 32826-3276

Medium Caliber Ammunition (MCA)
PEO Ammunition
PM Maneuver Ammunition Systems
SFAE-AMO-MAS
Picatinny Arsenal, NJ 07806

Medium Extended Air Defense System (MEADS)
PEO Missiles and Space
Project Manager, Lower Tier Project
Office
SFAE-MSLS-LT-MEADS
PEO Missiles and Space
5250 Martin Road
Redstone Arsenal, AL 35898-8000

Meteorological Measuring Set-Profiler (MMS-P)/Computer Meteorological Data-Profiler (CMD-P)
PEO IEW&S
Product Manager Meteorological
and Target Identification Capabilities
(MaTIC)
SFAE-IEWS-NCSP-MaTIC
Building 4504
Springfield Street
Aberdeen Proving Ground, MD 21005

Mine Protection Vehicle Family (MPVF), Area Mine Clearing System (AMCS), Interrogation Arm
PEO CS&CSS
Product Manager Assured Mobility
Systems
SFAE-CSS-MRA
6501 E. 11 Mile Rd.
Warren, MI 43897-5000

Mine Resistant Ambush Protected Vehicles (MRAP), Army
PEO CS&CSS
JPO MRAP
SFAE-CSS-MR
6501 E. 11 Mile Rd
Warren, MI 48397

Modular Fuel System (MFS)
PEO CS&CSS
PM Petroleum and Water Systems
SFAE-CSS-FP-P
6501 E. 11 Mile Rd.
Mail Stop 111
Warren, MI 48397

Mortar Systems
PEO Ammunition
PM Combat Ammunition Systems
SFAE-AMO-CAS-MS
Picatinny Arsenal, NJ 07806

Movement Tracking System (MTS)
PEO C3T
PM Force XXI Battle Command Brigade
and Below (FBCB2), PD MTS
SFAE-C3T-FBC
6007 Combat Drive
4th Floor
Aberdeen Proving Ground, MD 21005-
1846

MQ-1C Gray Eagle Unmanned Aircraft System (UAS)
PEO Aviation
Project Manager Unmanned Aircraft
Systems (UAS)
SFAE-AV-UAS
5300 Martin Road
Redstone Arsenal, AL 35898-5000

Multiple Launch Rocket System (MLRS) M270A1
PEO Missiles and Space
Precision Fires Rocket and Missile
Systems Project Office
SFAE-MSL-PF-FAL
Building 5250
Redstone Arsenal, AL 35898

NAVSTAR Global Positioning System (GPS)
PEO IEW&S
Product Director Positioning, Navigation and Timing (PD PNT)
SFAE-IEW&S-NS-PNT
6006 Combat Drive
B2101
Aberdeen Proving Ground, MD 21005

Nett Warrior (NW)
PEO Soldier
Project Manager Soldier Warrior
SFAE-SDR-SWAR
10125 Kingman Rd
Building 317
Ft. Belvoir, VA 22060

Night Vision Thermal Systems - Thermal Weapon Sight (TWS)
PM Soldier Sensors and Lasers
PEO Soldier
10170 Beach Road
Building 325
Ft. Belvoir, VA , VA 22060

Non-Intrusive Inspection Systems (NIIS)
JPEO CBD
Joint Project Manager Guardian
SFAE-CBD-GN
E 2800 Bush River Road
Aberdeen Proving Ground , MD 21010-5424

Nuclear Biological Chemical Reconnaissance Vehicle (NBCRV)– Stryker Sensor Suites
JPEO CBD
Joint Project Manager NBC Contamination Avoidance
SFAE-CBD-NBC-R
Building 2800
5183 Blackhawk Road
Aberdeen Proving Ground, MD 21010-5425

One Semi-Automated Force (OneSAF)
PEO STRI
Project Manager Constructive Simulation
SFAE-STRI-PMCONSIM
12350 Research Pkwy.
Orlando, FL , FL 32826

Palletized Load System (PLS) and PLS Extended Service Program (ESP)
PEO CS&CSS
PM Heavy Tactical Vehicles
SFAE-CSS-TS-H
6501 E. 11 Mile Rd.
Mail Stop 429
Warren, MI , MI 48397-5000

PATRIOT Advanced Capability–3 (PAC-3)
PEO Missiles and Space
Project Manager, Lower Tier Project Office
SFAE-MSLS-LT
Bldg. 5250, Martin Road
Redstone Arsenal, AL 35898-8000

Precision Guidance Kit (PGK)
PEO Ammunition
PM Combat Ammunition Systems
SFAE-AMO-CAS
Picatinny Arsenal, NJ 07806

Prophet
PEO IEW&S
PM PROPHET
SFAE-IEW EWP
Building 4504
Aberdeen Proving Ground, MD 21005

Rocket, Artillery, Mortar (RAM) Warn
PEO C3T
C-RAM Program Directorate
SFAE-C3T-CR
Martin Road, Building 5250
Redstone Arsenal, AL 35898

Rough Terrain Container Handler (RTCH)
PEO CS&CSS
Product Manager Combat Engineer/MHE
SFAE-CSS-FP-C
6501 E. 11 Mile Rd.
Mail Stop 401
Warren, MI 48397-5000

RQ-11B Raven Small Unmanned Aircraft System (SUAS)
PEO Aviation
Product Manager Small Unmanned Aircraft Systems (SUAS), Project Manager Unmanned Aircraft Systems (UAS)
SFAE-AV-UAS-SU
5300 Martin Road
Redstone Arsenal, AL 35898-5000

RQ-7B Shadow Tactical Unmanned Aircraft System (TUAS)
PEO Aviation
Project Manager Unmanned Aircraft Systems (UAS)
SFAE-AV-UAS
5300 Martin Road
Redstone Arsenal, AL 35898-5000

Secure Mobile Anti-Jam Reliable Tactical Terminal (SMART-T)
PEO C3T
PM WIN-T
SFAE-WIN-SAT
6010 Frankford Street
Aberdeen Proving Ground, MD 21005

Sentinel
PEO Missiles and Space
SFAE-MSLS
Redstone Arsenal, AL 35898

Single Channel Ground and Airborne Radio System (SINCGARS)
PEO C3T
Product Director NS
Aberdeen Proving Ground, MD 21005

Small Arms–Precision Weapons
PEO Soldier
Project Manager Soldier Weapons
SFAE-SDR-SW
Building 151
Picatinny Arsenal , NJ 07806

Small Arms–Crew Served Weapons
PEO Soldier
PM Soldier Weapons
SFAE-SDR-SW
PEO Soldier
Picatinny Arsenal, NJ 07806

Small Arms–Individual Weapons
PEO Soldier
PM Soldier Weapons
SFAE-SDR-SW
PEO Soldier
Picatinny Arsenal, NJ 07806

Small Caliber Ammunition
PEO Ammunition
PM Maneuver Ammunition Systems
SFAE-AMO-MAS
Picatinny Arsenal, NJ 07806

Stryker Family of Vehicles
PEO GCS
Project Manager–Stryker Brigade
Combat Team
SFAE-GCS-BCT MS 325
6501 E. 11 Mile Rd.
Warren, MI 48397

Sustainment System Mission Command (SSMC)
PM Sustainment System Mission
Command (SSMC)
SFAE-C3T-MC-SSMC
Hopper Hall, 6007 Combat Drive
5th Floor
Aberdeen Proving Ground, MD 21005

T-9 Medium Dozer
PEO CS&CSS
Product Manager, Combat Engineer/
Material Handling Equipment
SFAE-CSS-FP-C
6501 E. 11 Mile Rd.
Mail Stop 401
Warren, MI 48397-5000

Tactical Electric Power (TEP)
PEO C3T
Project Manager Mobile Electric Power
SFAE-C3T-MEP-OPM
5850 Delafield Road Bldg 324
Fort Belvoir, VA 22060-5809

**Tactical Mission Command (TMC)/
Maneuver Control System (MCS)**
Project Manager Mission Command
SFAE-C3T-MC-TMC
Bldg 6007, Floor 5
Aberdeen Proving Ground, MD 21005

Test Equipment Modernization (TEMOD)
PEO CS&CSS
Product Director, Test, Measurement,
and Diagnostic Equipment
SFAE-CSS-JC-TM
Building 3651
Redstone Arsenal, AL 35898

Tank Ammunition
PEO Ammunition
PM Maneuver ammunition Systems
SFAE-AMO-MAS
Picatinny Arsenal, NJ 07806

**Transportation Coordinators' –
Automated Information for Movements
System II (TC-AIMS II)**
PEO EIS
PD Transportation Information Systems
SFAE-PS-TI
9350 Hall Road
Building 1445
Ft. Belvoir, VA 22060

**Tube-Launched, Optically-Tracked,
Wire-Guided (TOW) Missiles**
PM Close Combat Weapon Systems
Project Office
SFAE-MSL-CWS-T
PM Close Combat Weapon Systems
Project Office
Redstone Arsenal , AL 35898

Unified Command Suite (UCS)
JPEO CBD
Joint Project Manager Guardian
SFAE-CBD-GN
Building E4465
5183 Blackhawk Road
Aberdeen Proving Ground, MD 21010

Unit Water Pod System (Camel II)
PEO CS&CSS
PM Petroleum and Water Systems
SFAE-CSS-FP-P
6501 E. 11 Mile Road
Mail Stop 111
Warren, MI 43897

**Warfighter Information Network–
Tactical (WIN-T) Increment 1**
PEO C3T
Building 6010 Frankford St
APG, MD 21005

**Warfighter Information Network–
Tactical (WIN-T) Increment 2**
PEO C3T
Building 6010 Frankford St
APG, MD 21005

**Warfighter Information Network–
Tactical (WIN-T) Increment 3**
PEO C3T
Building 6010 Frankford St
APG, MD 21005

**XM1216 & XM1216 E1 Small Unmanned
Ground System (SUGV)**
PEO GCS
RS JPO/PM UGV
SFAE-GCS-UGV
6501 E. 11 Mile Rd
Warren, MI 48397-5000

XM7 Spider
PEO Ammunition
PM Close Combat Systems
SFAE-AMO-CCS
Bldg 183 Buffington Rd
Picatinny Arsenal, NJ 07806